I0051644

Thermal Claddings for Engineering Applications

The text presents advances in the field of thermal claddings for protection against erosion, corrosion, and wear in hydraulic turbines, automobiles, agricultural equipment, power plant, chemical industries, and jet engines. It further discusses different cladding techniques such as electron beam, oxy-fuel, arc welding processes, and microwave hybrid heating. It explains the mechanism for failure of materials and cladding and emphasizes the protection mechanism.

This book:

- Discusses the design and simulation of thermal claddings and the use of algorithms to predict the process parameters and performance of developed clads using artificial intelligence and machine learning.
- Presents the tribological behaviour of novel wear-resistant thermal claddings for the components used in construction, mining, drilling, and hydropower plant.
- Showcases high-temperature oxidation, corrosion, and erosion-resistant thermal claddings for power plants, the automotive sector, and jet engines.
- Highlights the application of the thermal cladding process in remelting the existing surface to enhance the surface properties.
- Examines post-heat treatment procedures on thermal claddings for improving the microstructure and tribological properties.

The text is primarily written for senior undergraduate, graduate students, and academic researchers in the fields of mechanical engineering, manufacturing engineering, industrial engineering, and production engineering.

Thermal Claddings for Engineering Applications

Edited by Lalit Thakur, Jasbir Singh and Hitesh Vasudev

CRC Press
Taylor & Francis Group
Boca Raton London New York

CRC Press is an imprint of the
Taylor & Francis Group, an **informa** business

Front cover image: Lalit Thakur, Jasbir Singh and Hitesh Vasudev

First edition published 2024
by CRC Press
2385 NW Executive Center Drive, Suite 320, Boca Raton FL 33431

and by CRC Press
4 Park Square, Milton Park, Abingdon, Oxon, OX14 4RN

CRC Press is an imprint of Taylor & Francis Group, LLC

© 2024 selection and editorial matter, Lalit Thakur, Jasbir Singh and Hitesh Vasudev; individual chapters, the contributors

Reasonable efforts have been made to publish reliable data and information, but the author and publisher cannot assume responsibility for the validity of all materials or the consequences of their use. The authors and publishers have attempted to trace the copyright holders of all material reproduced in this publication and apologize to copyright holders if permission to publish in this form has not been obtained. If any copyright material has not been acknowledged please write and let us know so we may rectify in any future reprint.

Except as permitted under U.S. Copyright Law, no part of this book may be reprinted, reproduced, transmitted, or utilized in any form by any electronic, mechanical, or other means, now known or hereafter invented, including photocopying, microfilming, and recording, or in any information storage or retrieval system, without written permission from the publishers.

For permission to photocopy or use material electronically from this work, access www.copyright.com or contact the Copyright Clearance Center, Inc. (CCC), 222 Rosewood Drive, Danvers, MA 01923, 978-750-8400. For works that are not available on CCC please contact mpkbookspermissions@tandf.co.uk

Trademark notice: Product or corporate names may be trademarks or registered trademarks and are used only for identification and explanation without intent to infringe.

ISBN: 978-1-032-46054-3 (hbk)
ISBN: 978-1-032-74468-1 (pbk)
ISBN: 978-1-032-71383-0 (ebk)

DOI: 10.1201/9781032713830

Typeset in Sabon
by Newgen Publishing UK

Contents

3   Corrosion and microstructural behaviour of Inconel 625 microwave clad deposited on mild steel   51

GURBHEJ SINGH, AMIT BANSAL AND HITESH VASUDEV

4   Artificial intelligence revolutionizing the laser cladding industry   63

GAURAV PRASHAR AND HITESH VASUDEV

10 Application of thermal claddings for materials used as biomedical implants

HITESH VASUDEV AND AMRINDER MEHTA

11 Study on the thermal claddings used in biomedical implants

HARJIT SINGH, MUKHTIAR SINGH, MANINDER SINGH,
VINEET PUSHYA, HITESH VASUDEV AND AMRINDER MEHTA

14 Development of MoCoCrSi/fly ash composite cladding on stainless steel substrate through microwave irradiation   298

C. DURGA PRASAD, K. V. MANJUNATH, PREM KUMAR NAIK,
NAGABHUSHANA N. AND PRAKASH KUMAR

15 Advantages and applications of various surface engineering techniques   313

MUKHTIAR SINGH, MANINDER SINGH, HITESH VASUDEV AND
AMRINDER MEHTA

Preface

Thermal cladding technologies have been used for the development of alloys, ceramics, metallic and cermet-based claddings over different types of substrates owing to their simple, flexible, and cost-effective properties. These techniques are based on utilizing laser and electron beams, oxy-fuel/arc welding processes, and microwave hybrid heating. Thermal claddings are employed in the repair and re-fabrication/re-manufacturing of high-value components such as turbine blades and discs, rolling mills, tillage blades, propellers, boiler, and dockyard equipment. Thermal claddings also find extensive applications in a wide range of industrial equipment for wear and corrosion resistance such as the claddings for boilers, mining, drilling, hydroelectric power plant, wood processing, paper, and agricultural industries. Short corrosion-resistant claddings are used in thermal power plants and as a thermal barrier medium in automotive and gas turbine components.

Advances in the field of surface engineering has been complemented with the introduction of novel protective claddings for industrial applications. Technological developments and automation have also led to the precise fabrication of thermal claddings. The progress in thermal claddings has enabled providing resistance and shielding against wear and corrosion-related problems at room and elevated temperatures. Nowadays, super-alloys, high entropy alloys, and their nano-composites are employed by researchers for combating the wear, erosion, corrosion, and high-temperature oxidation of engineering components due to their structure-retaining property. These materials are costly and difficult to transform into useful products by conventional manufacturing processes. However, the fabrication of claddings with these advanced materials can be done effectively and economically to protect the components using thermal technologies.

The emerging trend in the field of surface engineering is the use of claddings to enhance the service life and performance of bio-implants inserted in the human body. Two well-known fields that employ biomedical implants are orthopaedics and dentistry. In clinical orthopaedics, implants and total joint replacements for hip, knee, shoulder, and ankle have received maximum attention in recent times. Furthermore, the application of various claddings enhances the wear and friction behaviour, biocompatibility, and hemocompatibility of metallic implants.

Therefore, this book intends to provide a comprehensive information on thermal claddings for various engineering applications. The book also provides detailed studies on the recent trends in thermal cladding technologies in terms of techniques, modelling, materials, characterization, and their control that can be useful for both academia and industries. This book focuses on the influence of materials and methods of thermal claddings on the microstructure, properties, corrosion, and wear performance during their service life. The book also covers the fundamental science and engineering aspects of thermal cladding technologies, historical developments, general applications, advantages, and challenges.

About the editors

Lalit Thakur is a faculty member of the Department of Mechanical Engineering at the National Institute of Technology, Kurukshetra, India. He obtained his M.Tech. and Ph.D. degrees from the Indian Institute of Technology (IIT) Roorkee, India. For the last 13 years, he has been continuously exploring new possibilities in the field of welding engineering and thermal spray technology. His current research areas are thermal spray coatings and tungsten inert gas (TIG) weld claddings for wear, corrosion, and bio-medical applications, as ferromagnetic lubricants, for advanced composite development by powder metallurgy, in stir casting, and in friction stir processing. He has authored more than 60 research articles and published them in various international journals (SCI/Scopus) and also presented some of them at conferences. He is also part of the editorial board of several journals such as *Applied Surface Science, Surface and Coatings Technology, Surface Engineering, Surface Topography, Tribology Transactions, Wear, Materials Today Communications, Engineering Failure Analysis, Materials Research Express,* and *Surface Review & Letters.* Moreover, he is a reviewer of most of the well-known journals related to materials science, surface engineering, and coatings technology. He has recently authored a book titled *Thermal Spray Coatings* (CRC Press, Taylor & Francis Group). He is also a life member of the Institution of Engineers India (IEI) and Indian Structural Integrity Society (INSIS). He has also been granted two patents in the field of thermal spray coatings and friction stir processing. He has guided many M.Tech. and Ph.D. scholars in the emerging areas of thermal spraying, advanced manufacturing, and welding technology.

Jasbir Singh is an assistant professor at Gurukula Kangri University, Haridwar, India. He has received his M.Tech and Ph.D. degrees from the National Institute of Technology, Kurukshetra, India. His area of research is thermal weld cladding, especially for the development of new materials and claddings to enhance the service life of engineering equipment that are exposed to extreme wear conditions. His research on weld claddings has been extensively published in highly regarded journals such as *Surface Coatings and Technology, International Journal of Refractory Metals and Hard Materials*, and *Surface Engineering*. He has more than 10 international publications in various international journals and conferences to his credit. He has been teaching for more than ten years.

Hitesh Vasudev is currently working as Associate Professor at Lovely Professional University, Phagwara, India. He has received Ph.D. in Mechanical Engineering from Guru Nanak Dev Engineering College (affiliated to I.K. Gujral Punjab Technical University (IKGPTU), Jallandhar), Ludhiana, India in 2018. His area of research is thermal spray coatings, especially for the development of new materials used for high-temperature erosion and oxidation resistance and microwave processing of materials. His research in thermal spray coatings has been extensively published in highly regarded journals such as *Surface Coatings and Technology, Materials Today Communications, Engineering Failure Analysis, Journal of Cleaner Production, Surface Topography: Metrology and Properties, Surface Reviews & Letters*, and *Journal of Failure Prevention and Control, International Journal of Surface Engineering* and interdisciplinary materials science under the flagship of various publication groups such as Elsevier, Taylor & Francis, Springer Nature, IGI Global, and InTech Open. Moreover, he is a dedicated reviewer of reputed journals such as *Surface Coatings and Technology, Ceramics International, Journal of Material Engineering Performance, Engineering Failure Analysis, Surface Topography: Metrology and Properties Material Research Express, Engineering Research Express* and IGI global journals. He has authored more than 30 papers that have been published in various international journals and conferences. He has also contributed 15 book chapters in various books related to surface engineering and manufacturing processes. He has also been granted a unique patent in the field of thermal spraying. He has been teaching for more than eight years. He received a "Research Excellence" award in 2019 at Lovely Professional University, Phagwara, India. He has organized a national conference and has been a part of many international conferences.

Contributors

Amit Bansal, I. K. Gujral Punjab Technical University, Kapurthala, India.

Ajeet Kumar Bara, Department of Production and Industrial Engineering, BIT Sindri, Jharkhand, India.

Harjot Singh Gill, University Center for Research and Development, Chandigarh University, Mohali, India.

Neeraj Kamboj, Department of Mechanical Engineering, National Institute of Technology, Kurukshetra, Haryana, India.

Kashif Hasan Kazmi, Department of Production and Industrial Engineering, BIT Sindri, Jharkhand, India.

Prakash Kumar, Department Mechanical Engineering, National Institute of Technology, Jamshedpur, India.

K. V. Manjunath, Department Mechanical Engineering, Siddaganga Institute of Technology, Tumakuru, Karnataka, India.

Amrinder Mehta, Research and Development Cell, Lovely Professional University, Phagwara, India.

Nagabhushana N., Department of Mechanical Engineering, New Horizon College of Engineering, Bengaluru, Karnataka, India.

Prem Kumar Naik, Department of Mechanical Engineering, AMC Engineering College, Bengaluru, Karnataka, India.

C. Durga Prasad, Department of Mechanical Engineering, RV Institute of Technology and Management, Bengaluru, Karnataka, India.

Gaurav Prashar, Rayat Bahra Institute of Engineering & Nano Technology, Hoshiarpur, VPO Bohan, Distt. Hoshiarpur, (Punjab) India.

Vineet Pushya, School of Mechanical Engineering, Lovely Professional University, Phagwara, Punjab, India.

Gyan Sagar, Department of Metallurgical Engineering, BIT Sindri, Jharkhand, India.

Sumit K. Sharma, Department of Metallurgical Engineering, BIT Sindri, Jharkhand, India.

Amarish Kumar Shukla, Indian Maritime University, Kolkata Campus, Taratala, Kolkata, West Bengal, India.

Gurbhej Singh, Amritsar Group of Colleges, Amritsar, India.

Harjit Singh, School of Mechanical Engineering, Lovely Professional University, Phagwara, Punjab, India.

Jasbir Singh, Department of Mechanical Engineering, Gurukul Kangri (Deemed to be University) Haridwar, Haridwar, Uttrakhand, India.

Jashanpreet Singh, University Centre for Research and Development, Chandigarh University, Mohali, Punjab, India.

Mandeep Singh, School of Mechanical Engineering, Lovely Professional University, Phagwara, Punjab, India.

Maninder Singh, School of Mechanical Engineering, Lovely Professional University, Phagwara, Punjab, India.

Mukhtiar Singh, School of Mechanical Engineering, Lovely Professional University, Phagwara, Punjab, India.

Lalit Thakur, Department of Mechanical Engineering, National Institute of Technology, Kurukshetra, Haryana, India.

Hitesh Vasudev, School of Mechanical Engineering, Lovely Professional University, Phagwara, Punjab, India.

Aim and scope

- **Introduction:** Historical developments, fundamentals of thermal cladding technologies, science and engineering aspects of thermal cladding technologies and their comparison with other surface engineering processes.
- **Recent progress in thermal claddings:**

  1. Tribological behaviour of novel wear-resistant thermal claddings for the components used in construction, mining, drilling, hydropower plant, wood processing, paper and agriculture industries.
  2. Advanced erosion and corrosion-resistant thermal claddings for sandblasting plant, chemical, and oil industries.
  3. High-temperature oxidation, corrosion, erosion-resistant thermal claddings for power plants, the automotive sector, boiler, and jet engines.
  4. Novel thermal barrier claddings for automobile and jet engines.
  5. Thermal claddings for repair/re-fabrication/re-manufacturing of worn-out high-value components.
  6. Engineered thermal claddings for biomedical implants.
  7. Nanostructured thermal claddings for improved life and performance of engineering components.
  8. Application of thermal cladding process in remelting the existing surface to enhance the surface properties.
  9. Post-heat treatment procedures on thermal claddings for improving the microstructure and tribological properties.

Chapter 1

Fundamentals and applications of thermal claddings

Jasbir Singh, Lalit Thakur and Hitesh Vasudev

1.1 INTRODUCTION

The life of engineering components is usually governed by the properties of the surface region as the failure is initiated from this region. Surface degradation can be defined as the loss of performance of an engineering component. There is a critical minimum level of performance for any equipment. For example, in case of an engine with worn-out cylinders, the wear can increase the clearance space between the cylinder and piston to a great extent and hence there is poor compression of combustible gases. Due to this, the engine is unable to pull the car, or the pick-up of the car reduces drastically. Mechanical degradation occurs at a rate that varies with service conditions and failure occurs if performance falls below the critical level. The degraded components, which cause the breakdown of the machine and loss of valuable working time, need to be replaced with new components. The aim of scientific studies focusing on the degradation of engineering materials is to predict the rate of decline in equipment performance. Figure 1.1 depicts the loss of performance of machine performance with respect to time [1].

The loss of equipment or machine performance is inevitable, therefore, some control measures need to be employed. Some of the prominent methods employed to prevent the failure from surface degradation are

(i) Fabricating the equipment by using a material having superior properties
(ii) Providing proper lubrication between the mating surfaces
(iii) Improving the surface and near-surface region properties of existing material (surface engineering)

Advanced materials having superior properties are very costly and require complex fabrication. The conventional techniques are unable to process these materials, therefore, sophisticated tools and equipment are required for their processing. Moreover, to understand the behaviour of these materials under actual service conditions, excessive testing and experimentation may

DOI: 10.1201/9781032713830-1

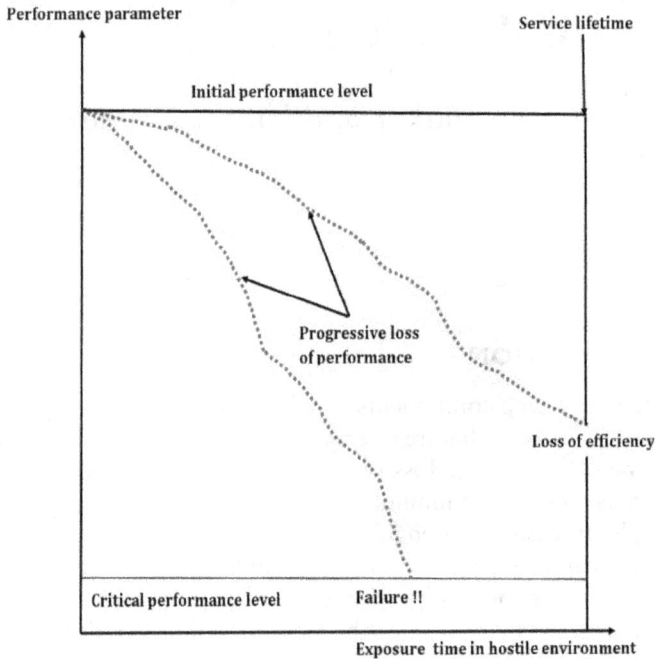

Figure 1.1 Performance vs service time.

be required for a prolonged period. The improvement in some surface properties may cause a reduction in other properties of materials. Many advanced
materials having high tensile strength compromise fatigue strength because
of enhancement in brittleness.

In the case of lubrication, sometimes it is very challenging to provide sufficient amount of lubrication and to maintain a uniform thickness between
the mating surfaces. Moreover, it is difficult to identify the optimum value
of volume needed for lubrication needed for a particular geometry. Many
times, the lubricant may react with the surfaces and reduce the equipment
or machine's performance [2].

Among the employed methods, surface engineering has been extensively
used because of its simplicity, flexibility, and economical approach [3,4]. It
allows the fabrication of protective claddings over the low-cost substrate,
which exhibits better surface properties compared to the substrate. Further,
it can be employed for the repair of worn-out materials. Surface engineering
can be defined as the interdisciplinary branch of science used to protect
machines and components from fracture, corrosion, and wear. Basically, it
is a process of developing the surface to have better tribological and mechanical properties as compared to the substrate to enhance the lifespan and

performance of equipment and machines. According to the *ASM Surface Engineering Handbook*, surface engineering is defined as "the treatment of surface and near-surface regions of a material to allow the surface to perform functions that are distinct from those functions demanded from the bulk of the material" [5].

The main objective of surface engineering is to ensure that the equipment's or machine's performance remains above the critical level during the lifespan of the engineering system. The claddings may be formed on the substrate by utilizing different processes such as physical, chemical, mechanical, thermal, thermo-mechanical, thermo-chemical, and electro-chemical. The specific type of cladding for a specific application and of required thickness can be obtained by employing different techniques or a combination of techniques. Figure 1.2 shows the classification of coating based on the manufacturing methods [6].

The most common methods which employ thermal energy to produce the cladding are thermal-spraying processes, microwave cladding, laser cladding, electron beam cladding, and welding techniques such as tungsten inert gas welding (TIG), metal inert gas welding (MIG) and plasma arc welding (PAW) [7,8]. Thermal cladding uses heat for the melting of cladding material and thereafter, a clad layer is formed over the substrate upon solidification of the molten material. The heat required for cladding can be obtained from microwave, laser, TIG, MIG, and PAW. The nature of cladding produced by each process instils different kinds of properties such as microstructure, internal stress, hardness, and toughness, which give result in mechanical and tribological behaviours. Using these techniques,

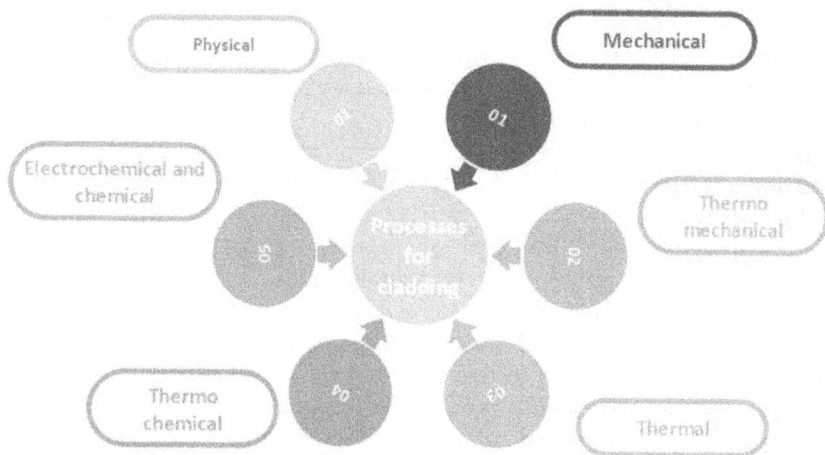

Figure 1.2 Classification of cladding based on manufacturing processes.

it is possible to obtain the enhanced properties of the surfaces. However, surface characteristics such as hardness, porosity, impurities, morphology, and composition depend on the specific method employed for cladding fabrication. The selection of the cladding technique primarily depends on the adaptability of the base material to the intended technique, functional requirements, availability, and cost of the equipment [1].

1.2 HISTORICAL DEVELOPMENT OF THERMAL CLADDINGS

Surface engineering can be traced back to ancient times. The primeval Greek and Chinese origin people were employing heat treatment processes such as tempering and case hardening to convert organic materials into a solid form [9]. The new advancement in surface engineering including the use of alloying and cladding were found in 18th and 19th century literature works and in patents. Some of the prominent developments in cladding techniques are listed in the following subsections.

1.2.1 Thermal-spraying processes

A Swiss scientist named Max Ulrich Schoop was granted a patent on thermal spraying in 1882. The patent provides detailed knowledge regarding the feeding of Pb and Sn wires through a redesigned oxyacetylene welding torch [10]. In 1908, Max Ulrich Schoop developed an electric arc spraying machine that possess better control and equipment, which was capable of spraying materials such as zinc, steel, and stainless steel.

In 1937, Quinlan and Grobel were awarded the first patent on thermal spraying for high-temperature corrosion protection in boilers. During the 1950s, significant development occurred on DC atmospheric plasma spraying that can employ argon for the generation of plasma stream. On further improvement in plasma spraying, it was extensively used in aircraft engines from 1970 to 1980.

In the early 1980s, Browning and Witfield uses rocket engine technology and develop high-velocity oxy-fuel technology. From 1994 to 2010, another technique, that is, high-velocity air-fuel spraying was evolved. This technique allows the use of a variety of fuels such as natural gas, hydrogen, propane, and propylene for different applications. In 1990s, another process called cold spraying was developed. This process utilizes kinetic energy for feedstock powder deposition and maintains the processing temperature below the melting point of feedstock powder. Cold spraying minimizes the residual stresses and heat-affected zone, which leads to better cladding properties [11]. Another thermal spraying process, namely warm spraying process, gained the attention of many researchers. Warm spraying coating is fabricated successive impacts of powder particles onto the substrate. The temperature produced during the coating process is below the melting of powder. For proper bond formation, a velocity greater than critical velocity is employed [12].

1.2.2 Thermal barrier coatings

The preliminary trials were carried out on thermal barrier coatings (TBCs) in the mid-1970s [13]. In the 1980s, TBCs were successfully tested on vanes and blades of aircraft gas turbines. In the 1990s, another variant of TBCs, that is physical vapour deposition, was successfully tested and deployed for commercial applications. From 2000 onwards, many advances lead to further enhancement in coating properties and their applications in diverse fields [14].

1.2.3 Laser claddings

In the 1960s, the invention by Maiman [15] of a working prototype of a laser was a breakthrough in science. During 1970–1975, the new advances in laser technology such as high-power CO_2 laser led to the use of laser for material processing, that is, cutting, hardening, and welding. In the 1970s, Gnanamuthu used laser cladding at Rockwell International Corporation to investigate the feasibility of ceramic cladding on a metallic substrate.

In 1981, the first commercial use of laser cladding was reported on hardfacing of Nimonic turbine blades of the jet engine, which was used in Rolls Royce. In 1986, an innovative process namely stereolithography was patented. This process employed the UV laser to cure the photopolymer material. Thereafter, in 1988, the first stereolithography machine was manufactured and sold for use in industries. In the 1990s, laser cladding was employed in repair and maintenance work. Laser cladding was used for coating and rebuilding of airfoil section or worn-out turbine vanes and in tip of turbine blade. In 19th and 20th centuries, laser cladding was getting attention in the field of additive manufacturing for layer formation and coating on substrates for their application in different fields [15–17].

1.2.4 Electron beam claddings

In 1879, W. Crookes presented a demonstration of an electron beam that could melt materials [18]. In 1897, J.J. Thompson used cathode rays, that is, a stream of high-velocity electrons for the melting of metals. In 1907, M. Pirani filed a patent on electron beam melting, which employs electron beams for melting purposes [19]. In 1968, the first electron beam welding machine was manufactured for industrial applications. From 1995 to 2001, electron beam cladding found extensive applications in various fields in which laser beam is replaced by electron beam for cladding purposes [20].

1.2.5 Microwave cladding

In 1832, Michael Faraday presented the electromagnetic phenomena that have wave-like motion. In 1846, Maxwell proposed mathematical models to demonstrate the electromagnetic phenomena, that is, the interaction of magnetic and electric and fields. In 1903, Hulzmeyer filed a patent for detecting the obstructions and navigation of ships by employing radio signals. In 1949, Percy Spencer came up with the phenomenon of microwave ovens, which utilized microwaves for heating, cooking, and food preparation. During 1950–1980, the microwave was used for food processing, wood curing, pathology, tumour detection, and ceramic processing. And 2000 onwards, microwaves were employed for newer applications such as sintering, cladding, joining, casting, and additive manufacturing [21].

1.2.6 Metal and tungsten gas arc cladding

In 1801, Humphry Davy discovered the short pulsed electric arc and thereafter, the continuous electric arc was developed by Vasily Petrov in 1802 [22]. In 1890, C. L. Coffin invented and was awarded a U.S. patent on arc welding processes in which metal electrodes are used for welding. This initial work of Coffin laid the foundation for metal and tungsten arc welding. In 1896, arc cladding on metals was proposed by I. W. Spenser. In 1941, Russell Meredith patented a machine for gas tungsten arc welding. From 1950 onwards, a lot of technological progress in welding technology lead to improvements in cladding techniques such as the use of automation, robotic arm, and additive manufacturing.

1.3 CLASSIFICATION OF CLADDINGS

These broad differences between the different cladding produced by different processes have been presented in Table 1.1 [23]. Some of the prominent cladding techniques have been discussed in detail.

1.3.1 Thermal spray processes

Thermal spraying is defined as the process in which material in a molten or semi-molten state or solid fine particles are deposited on the base material. A stream of these particles is employed for the spraying process and clad layer is deposited when these particles impinge on base material and deform plastically on impact. Figure 1.3 shows a schematic diagram of the thermal spray process [10]. Electrical energy or combustion flame can be utilized to melt or fuse a different kinds of materials. A very high-quality surface cladding can be produced by using a thermal spraying process. It has the potential to fabricate a large variety of shapes, sizes, and types of clad

Table 1.1 Comparison of thermal claddings

Weld Cladding process	Dilution (%)	Deposition rate (kg/h)	Thickness (mm)	Distortion	Precision	Integrity
Laser	1–5	0.2–7	0.2–2.0	Low	High	High
SMA	15–25	0.5–2.5	1.6–10	Medium	Low	High
MIG	15–20	2.3–11	1–6	Medium	Low	High
TIG	10–20	0.5–3.5	0.5–3.0	High	Medium	Medium
SA	10–50	5–25	2–10	High	Low	High
Plasma	5–30	0.5–7	0.1–0.2	Low	Medium	Low
Flame	1–10	0.45–2.7	0.8–2	High	Low	Medium
HVOF	Low (1–3)	1–5	0.3–1.5	Low	Low	Medium
Plasma Arc	Medium (3–10)	2.5–6.5	1–5	Medium	Medium	Medium

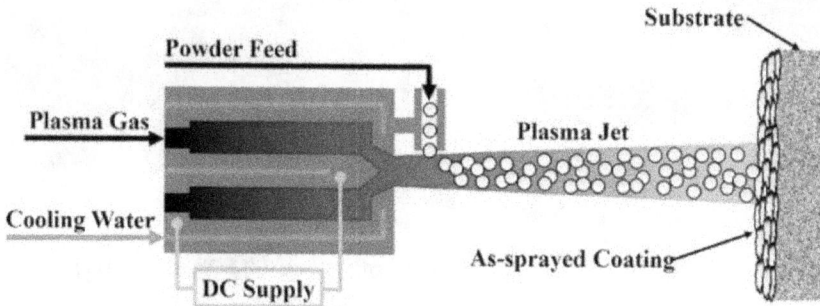

Figure 1.3 The schematic diagram of the thermal spray process [25].

Table 1.2 Classification on the basis of methods of generation [26]

Energy input	Variation	Spray technique
Combustion	Continuous Explosive	Flame spraying, high-velocity oxygen fuel spraying, detonation –gun
Electric Discharge	DC arc Pulsing arc	Arc spraying, atmospheric plasma spraying, vacuum plasma spraying
	HF (RF) Glow discharge	RF plasma spraying
Decompression of gas	--	Cold-gas spraying method

layers with variable thicknesses. Thermal spraying is used heavily in a wide variety of industrial fields for improving the performance and properties of substrate [24]. Table 1.2 shows the classification of the method of arc generation.

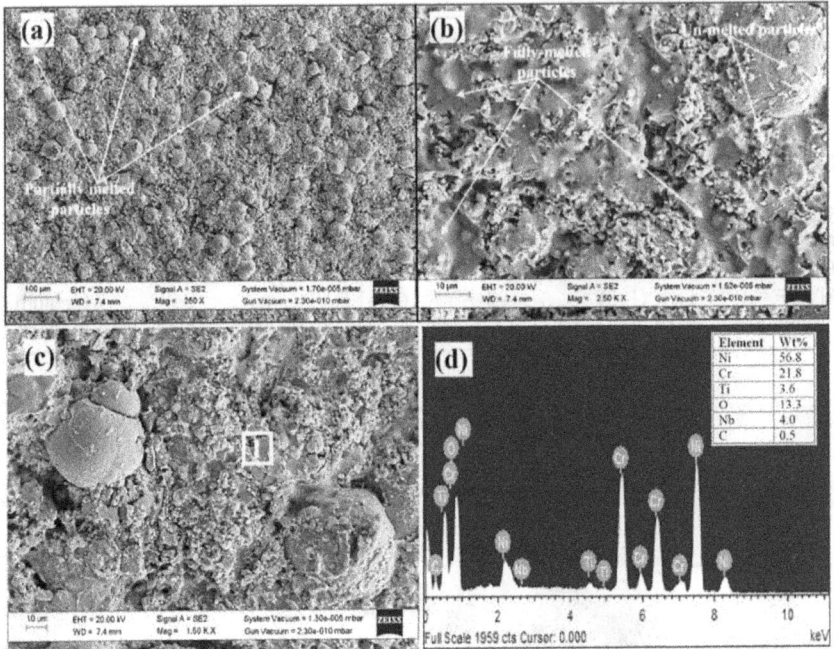

Element	Wt%
Ni	56.8
Cr	21.8
Ti	3.6
O	13.3
Nb	4.0
C	0.5

Figure 1.4 (a–c) SEM micrograph at different magnifications; (d) EDS analysis [27].

Vasudev et al. discussed the erosive performance of high-velocity oxygen fuel (HVOF) sprayed bi-layer alloy-718/NiCrAlY coating [27]. Microstructural examination of coating has been presented in Figure 1.4. HVOF coating exhibited the presence of both the partially melted and fully melted particles over the substrate. Micrographs showed the visible interface between partially and fully melted particles. Cracks were absent from the coating due to plastic deformation of particles. EDS analysis confirmed the presence of all the constituents of feedstock powder. However, microstructure and mechanical and tribological properties of thermal spayed coatings are largely dependent on the process parameters employed for the coating such as spray process, feedstock powder feed rate, spray distance and carrier gas flow rate [28,29]. Spray process and spray distance are significant parameters to influence the coating morphology and properties of coating [30].

1.3.1.1 Features of thermal spraying processes [10]

- A variety of materials can be utilized for thermal spraying processes depending upon the requirement and application of surface modification. A wide range of materials ranging from metals, plastics, ceramics etc. are used.

- Thicknesses varying from micron to millimetres for small, large, and complex parts can be easily modified.
- The microstructure and tribological properties of the substrate remain unchanged due to low dilution because the temperature for thermal spraying is quite low.
- This process can be employed in low vacuum, inert gas, and atmospheric conditions depending upon the intended application and requirement.

1.3.1.2 Advantages [28]

Thermal-spraying cladding processes have many superior features as compared to other surface engineering processes.

- Thermal spray claddings are hard and durable and can sustain extreme levels of wear and tear especially in bearings and gears.
- These claddings protect the base material from chemical attack, corrosion, and various environmental degradations.
- These claddings act as a thermal barrier and prevent heat interaction with surroundings which makes them ideal for energy efficiency.
- The cladding process is extremely fast and can be used for a wide variety of substrate materials.
- On-site repair and maintenance can be carried out and the required thickness of clad layer can be achieved.
- Thermal spray cladding is employed as thermal barrier coating because of its high hardness, excellent chemical and electrical properties, and high thermal stability.

1.3.1.3 Disadvantages [28]

Although there are numerous advantages of the thermal spray process, there are some disadvantages as well.

- The thermal spray process produces high temperatures in the cladding process, which makes it difficult to bond properly with the substrate in a sealed environment.
- These processes are not used for certain types of materials due to their tendency to react undesirably in high temperature and pressure conditions.
- A loud noise, fume, and dust are formed during the spray process, which affects human health.
- The high running and initial cost of thermal spray equipment make these processes expensive and often limited to application in an expensive equipment.
- In order to operate thermal spray equipment, highly skilled labour is required.

1.3.1.4 Applications

The thermal spray process finds extensive application in repairing, restoring original dimensions, rebuilding, retrofitting, and applying to the materials to prevent them from wear and corrosion. A list of applications areas where these coating has been used are as follows [31]:

- Thermal spray coatings have been employed to prevent wear in grinding hammers (tobacco industry), sucker rod couplings (oil industry), aircraft flat tracks (air-frame component), and core mandrels (dry cell batteries).
- Aluminium spraying has been utilized to prevent corrosions in off-shore oil rigs for flare stacks, wellhead assemblies, and other steel components.
- The thermal spray process is used to develop coatings for ortho-paedic appliances as functional bioactive agents which support the healing process.
- The use of thermal spray process is used in electronics and sensing applications. The clad layer itself acts as a functional component and is used in the fabrication of conventional electronic items.
- Thermal spray claddings are used in thermal gas turbines to reduce the clearance between the stationary shrouds and blades employed in turbine and compressor sections.
- The thermal spray process (mainly HVOF) is utilized in oil and gas industry equipment to prevent corrosion and wear, which are the prime factors to cause leakage and failure.
- In the area of renewable energy, thermal-sprayed zinc and zinc aluminium coatings are used for onshore and offshore installations for corrosion protection.

1.3.1.5 Key challenges [10]

1. The major challenge in the automobile industry is to develop a suitable material and cost-effective coating process for cylindrical bores in order to build a fully integrated coating process.
2. To develop a system that can repair or rebuild the high-thickness structures employed in gas turbines which are used in aerospace and thermal power plants.
3. To identify the relationship between the operating parameters, microstructure, and behaviour of corrosion and wear-resistant coating.
4. Modelling the thermal spray coating to minimize porosity and dilution.
5. To design and develop a fully simulated model to optimize the process parameters which gives a cost-effective and efficient method.

6. Cost-effective equipment, machines, and testing facilities are required to predict the microstructure and behaviour of deposited thermal-sprayed coatings.
7. To redesign the traditional and standard methods to assess the mechanical and tribological properties in a better way.
8. Difficulties in coating complex shapes and shadowed areas, which are out of line of sight [10].

1.3.1.6 Future scope [32]

1. Newer improved materials such as composites, high-strength materials, and nanomaterials for aerospace and biomedical applications.
2. Spray forming for high-temperature ceramics used in superconductors.
3. Utilization of thermal spraying processes for additive manufacturing to fabricate complex-shaped geometries with shorter lead times.
4. Continuous improvement in fabrication technique, testing, and real-time sensors for better control.
5. To produce smart coatings with self-healing properties to work under severe working conditions.

1.3.2 Laser cladding process

The use of lasers for the fabrication of cladding has grown drastically in recent years. Laser cladding aims to fabricate a completely new layer on a substrate with minimum dilution and strong interfacial bond [33]. Laser treatment enhances the surface properties of the substrate material. It allows the use of cheap substrate material with poor surface properties. Laser cladding is defined as the fabrication of an extraneous layer on the substrate with minimal surface melting, resulting in the formation of a strong metallurgical bond between the substrate and with an entirely different clad layer. This cladding process works on the principle of using a coherent beam light source of high intensity that impinges upon the substrate material, which results in surface layers getting modified. This process can be carried out in a vacuum or protective gas atmosphere. Figure 1.5 depicts the laser cladding process [33]. All variations of laser cladding are characterized by fast heating and cooling rates, which results in a rapidly solidified clad layer, where both the microstructure and the distribution of the different elements of cladding material could be controlled by the operating parameters. Cladding material can be supplied to the laser beam by using wire feeding or preplaced powder or blown powder technique as shown in the figure. The use of coaxial injection of cladding powder for laser cladding has been reported in several studies [33]. It has been observed that there is certain minimal power required for substrate melting.

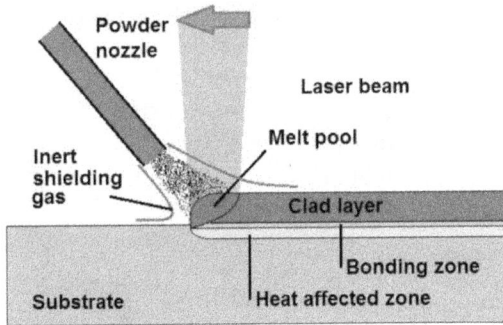

Figure 1.5 Laser cladding process [34].

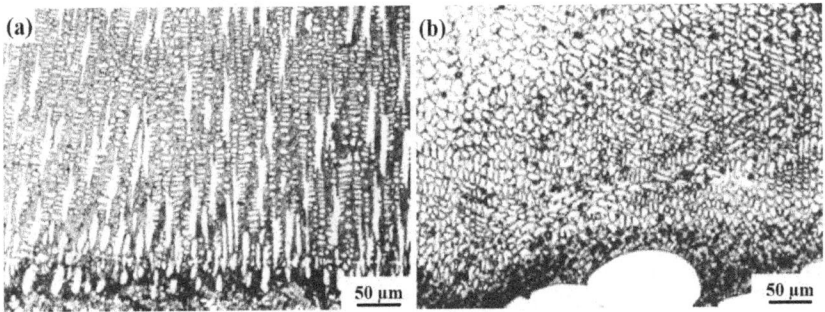

Figure 1.6 Microstructural examination (a) without nano-Y_2O_3 and (b) with 0.5 % nano-Y_2O_3 [35].

Mingxi and Li et al. [35] examined the microstructure of Co-based cladding with the addition of nano-Y_2O_3. Cladding morphology showed the presence of dendrites and equiaxed crystallization through out the thickness due to the solidification (Figure 1.6). However, the fine grain structure prevented the growth dendrites due to the addition of nano-Y_2O_3 particles. The effect of process parameters such as laser power, scan speed, and powder feed rate on microstructure and mechanical properties was examined by Zhu et al. The optimal use of laser power reduces the formation of voids, cracks, and produces good quality clad layer. Similarly scan speed influences the dilution ratio of cladding [36].

1.3.2.1 Advantages [15]

Although considered to be an expensive process, laser cladding offers several advantages over other conventional thermal cladding processes.

- The clad layer consists of novel microstructure and superior surface properties as compared to other processes because of the formation of metastable and non-equilibrium phases, which are formed as a result of rapid heating and quenching.
- Laser cladding is able to provide highly directional and precisely controlled beams of required shape, that is, spot, line, and area.
- It is possible to clad the intrinsic shapes that are difficult to approach because of the directionality of the laser beam.
- Laser beam is able to provide large amounts of energy in a very small region for a very short duration.
- Laser cladding is extremely environment-friendly because it is chemical-free and has no requirement for post-processing treatment.
- It can be easily automated.
- Laser beam is highly localized, which results in a smaller heat-affected zone and hence minimum dilution and distortion of the substrate.

1.3.2.2 Disadvantages [15]

Laser cladding offers numerous advantages over conventional cladding processes, but the process has some drawbacks as well.

- High investment cost.
- The high sensitivity of the laser cladding process results in poor reproducibility even with small changes in working conditions.
- Low efficiency of laser sources.
- Poor clad quality due to random or periodic disturbances between processing cycles with similar operating parameters.
- The fabrication speed of the laser cladding process is relatively low compared to other techniques such as thermal and plasma spraying.

1.3.2.3 Applications

The significance of the laser cladding process has increased in recent years because of the diversified potential to carry out material processing, such as low-volume manufacturing, repair, prototyping, and metallic cladding [15].

- Laser cladding is used extensively for cladding of industrial parts, which are subjected to abrasion, adhesion, corrosion, and oxidation. For example, (i) engine valve seats, (ii) moulds, (iii) tool hardfacing, and (iv) shaft used in drilling.
- Laser cladding employed for the repair and restoration of high-valve equipment, that is, military equipment, turbine blades, and tools.
- Laser cladding is used in additive manufacturing for the customization of products and rapid fabrication of mass complex tools and

components. Laser Engineering Net Shaping™ is one of the popular metal forming processes utilizing laser cladding to develop near-net shaped metal components.

- Laser cladding can be employed to repair the turbine blade without any need to remove the blades from the rotor.
- Laser cladding has promising applications in the development of space-related equipment, that is, robotic arm components used in spaceships.

1.3.2.4 Key challenges

- Optimization of operating parameters of laser cladding to obtain the homogenous distribution of material composition. Therefore, it is imperative to have a precise control of the powder feed rate and mixing of different powders.
- The development of new materials such as multi-performance composites, nanomaterials, high wear, and corrosion-resistant materials for mining, petrochemicals, aerospace, and biomedical applications [36].
- The identification and control of defects such as dilution, void, and cracks are essential for sound cladding for engineering applications.
- The simulation and development of computer technology for in-depth analysis of flow field, stress, and temperature field of cladding.
- Small, portable, and in-situ laser cladding machines are required for the repair and maintenance of small precision parts.
- The post-processing required for laser cladding because of the problems of dimensional accuracy, surface quality, and residual stress to meet the industrial needs.

1.3.2.5 Future scope [37]

- The joint simulation of flow, residual stress, and temperature fields of claddings need to be explored as laser cladding is a multi-field inter-action process.
- The growth of the microstructure depends upon the temperature produced during the cladding process. Monitoring and controlling of temperature field is crucial for optimizing the microstructure, mech-anical, and tribological properties of the clad layer, which further improves the powder utilization.
- The effort should be focused on increasing the processing speed of laser cladding, which is slower as compared to other techniques such as thermal and plasma spraying.
- The performance of laser cladding can be explored in combination with other newer technologies such as mega data and artificial intelligence.
- Hybrid laser processing, that is, laser cladding in conjunction with ultrasonic-assisted cutting and additive manufacturing is gaining

attention. Hybrid technologies enable the fabrication of thin-walled and precisely designed clad layers with density near to 100% [33].

1.3.3 Electron beam cladding process

In electron beam cladding, a high-energy electron beam is used to fabricate the cladding on the substrate. An electron gun is used to generate an electron beam. A heated cathode and an anode are commonly used in the cannon. Electrons are emitted and propelled towards the anode by a high-voltage potential difference when the cathode is heated. Electromagnetic lenses are used to concentrate an electron beam into a small region. These lenses use magnetic fields to control the electron flight, allowing for accurate beam focusing on the target area. Cleaning and, if necessary, preheating the substrate material are performed. This enables optimal cladding layer adhesion and helps to reduce residual strains. Metal powder, frequently in the form of fine particles, is transported to the desired location. This can be accomplished with the help of a powder delivery system, which regulates the flow of powder and assures a steady supply to the cladding zone. The concentrated electron beam scans across the surface of the substrate in a precise pattern. Scanning can be accomplished by shifting the electron beam source or the substrate. The shape and dimensions of the cladding layer are determined by the scanning process. Figure 1.7 shows the clad layer deposited by using electron beam cladding on a titanium substrate [38].

The electron beam transfers energy to the metal powder as it interacts with it, rapidly heating and melting it. The molten particles are then driven onto the surface of the substrate, where they fuse and solidify to form a cladding layer. The electron beam's strong localized heat ensures quick solidification and minimum heat transmission to the adjacent areas. Once deposited, the molten metal rapidly cools and solidifies to form a solid cladding layer.

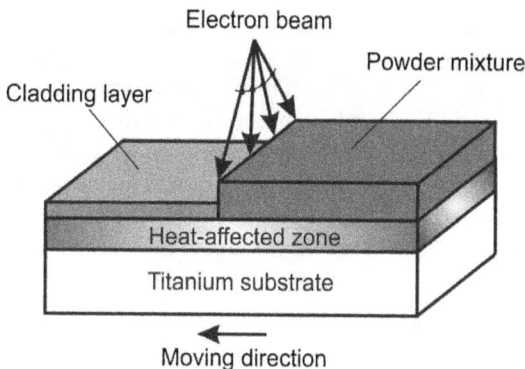

Figure 1.7 Schematic diagram of electron beam cladding [38].

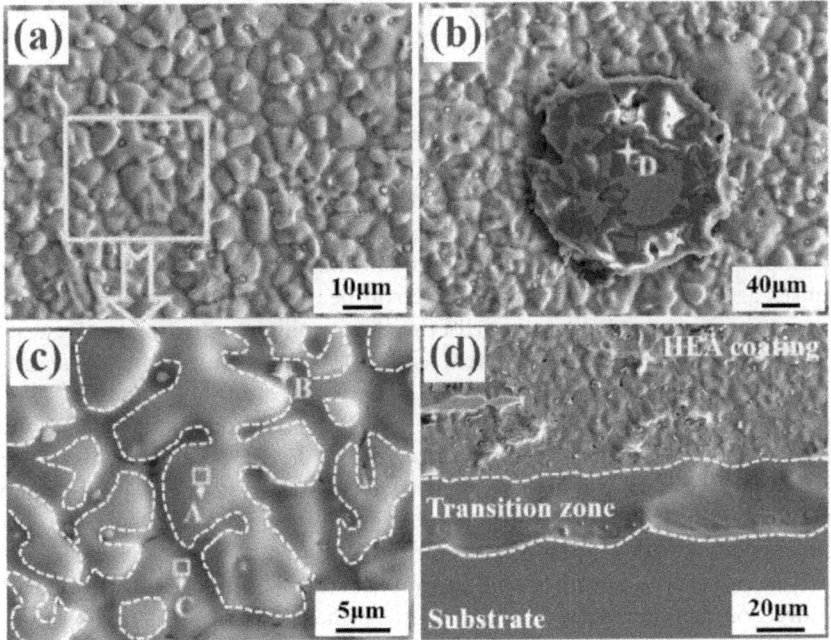

Figure 1.8 SEM images at different magnifications [39].

Cooling speeds can be varied to alter the clad layer's microstructure and material qualities. To optimize the intended material qualities, additional cooling methods such as water quenching or controlled heating may be used.

The microstructural analysis of AlCrTiNbMo electron beam cladding exhibited the coarse dendrite within the cladding having uniform distribution of feedstock powder constituents (Figure 1.8) [39]. EDS analysis confirmed that the dendritic layers were enriched with Mo, Nb, and Al elements, while Ti, Al, and Cr were predominant in the interdendritic regions. However, the variation in dilution, elemental composition, microstructure, and mechanical and tribological properties can be obtained by varying the process parameters such as current, voltage, welding time, and number of passes [20]. Current and welding time being the most influential parameters can increase or decrease the dilution, which in turn affects the wear behaviour of cladding. High dilution degrades the wear resistance of cladding and vice versa.

1.3.3.1 Advantages [18]

- High precision and control in the cladding process.
- Precise and controlled deposition of high-quality coatings with minimal distortion.

- Electron beam cladding provides a high-energy density heat source, allowing the cladding material to melt and solidify quickly.

1.3.3.2 Limitations [18]

- Limited build size
- High capital cost
- Material constraints
- Surface roughness

1.3.3.3 Applications [40]

- Aerospace component repair
- Oil and gas equipment refurbishment
- Power generation component protection
- Tool and die manufacturing
- Automotive part repair
- Nuclear industry component cladding
- Medical implant manufacturing

1.3.3.4 Key challenges

- To increase the material deposition rate in order to develop the high-thickness cladding on the substrate in extreme wear conditions.
- To increase the automation so that robots can be used to deposit the claddings on complex parts.
- Development of new cladding materials, which can withstand high temperatures during cladding processes.
- To ensure successful integration of analytical instrumentation to obtain reproducibility, stability, and dimensional tolerances.

1.3.3.5 Future scope

- Development of low-cost, efficient, and small-sized cladding machine to carry out the repair and maintenance on-site [41].
- Development of nanostructured and biomedical claddings.
- Development of advanced online monitoring systems to check the accuracy, precision, and quality of cladding.
- Fully simulated model of electron beam operation to design the stable and efficient process.

1.3.4 Thermal barrier cladding (TBC) process

Thermal barrier claddings operate on the premise of insulating and protecting high-temperature components. The coatings are made up of a ceramic

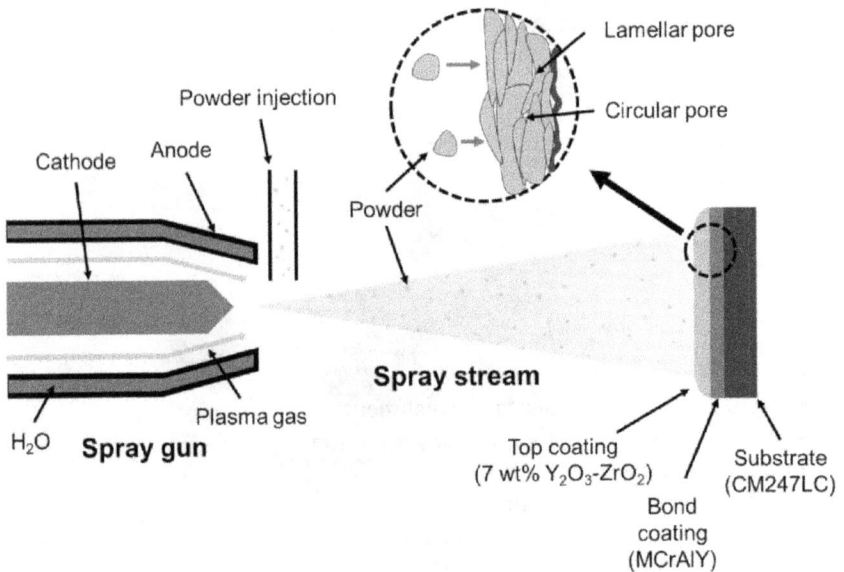

Figure 1.9 Thermal barrier coating process [42].

topcoat, which is commonly constructed of yttria-stabilized zirconia (YSZ). Because of its low heat conductivity, the ceramic topcoat works as an insulating layer. When heated, the topcoat creates a temperature gradient across the coating, which reduces heat conduction to the underlying substrate. The coefficients of thermal expansion of the ceramic topcoat and metallic bond coat are different. This mismatch causes compressive stresses in the ceramic topcoat, which helps to prevent crack development and increases coating endurance. The metallic bond coat protects the underlying substrate from oxidation, preventing degradation and preserving the coating system's integrity. Figure 1.9 shows the development of TBCs using the plasma spray process.

Optimization of parameters in TBCs is essential for the development of good quality coating. The main factors influencing the performance of coating are current, voltage, feed rate, scan speed, spray distance, and number of layers. Kumar and Pandey [43] discussed the effect of each parameter on coating properties. Current and spray distance were the most influencing factors to decide the microstructure and coating properties. Coating properties were affected by substrate temperature and interaction between the top and bond coat layer [44]. Beele et al. [14] discussed the microstructural performance of TBCs by using different processes. Electron beam physical-vapour-deposited TBCs exhibited better performance due to development of columnar microstructure (Figure 1.10). These columnar structures have higher stress compliance and prevent the stress build-up.

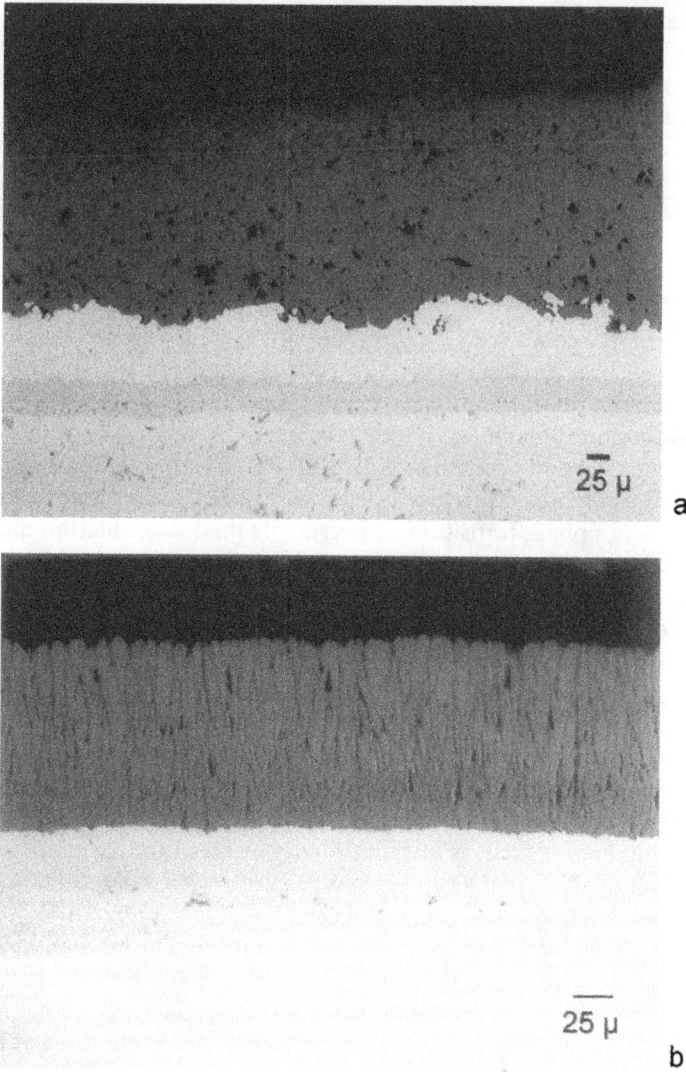

Figure 1.10 Micrographs of TBCs exhibiting (a) laminar and (b) columnar structure [14].

1.3.4.1 Advantages [13]

- TBCs provide greater thermal stability, allowing components to operate at higher temperatures and improve performance and efficiency.
- TBCs can offer thermal shock resistance, shielding components from fast temperature fluctuations and accompanying mechanical loads.

- TBCs can act as a protective barrier against corrosive conditions, extending component life.
- TBCs can improve the efficiency of energy conversion systems like gas turbines and jet engines by limiting heat transmission to the substrate.
- TBCs provide excellent thermal insulation properties, limiting heat transfer and protecting high-temperature components from thermal stress and fatigue.

1.3.4.2 Limitations [13]

- Thermal conductivity differences in thermal barrier coatings can cause heat gradients inside the coating and impair performance in specific applications.
- Adhesion difficulties between the ceramic topcoat and the metallic bond coat can occur, resulting in coating delamination or spallation.
- TBCs have a maximum temperature at which the ceramic topcoat can undergo phase transitions, lowering its thermal insulation qualities.
- TBCs can degrade over time owing to reasons such as heat cycling, erosion, and environmental exposure.
- Coating integrity and durability may need monitoring and maintenance.

1.3.4.3 Applications [45]

- TBCs can be used to enhance thermal efficiency and minimize heat loss in industrial furnace components.
- TBCs are widely employed in gas turbine engines to protect components such as turbine blades, vanes, and combustor liners from the high temperatures experienced during operation.
- Thermal barrier coatings are used to enhance efficiency, minimize emissions, and extend component life in power production systems such as steam and gas turbines.
- TBCs are used in automobile engines to protect exhaust manifolds and other high-temperature components, which reduces heat dissipation and improves engine performance.

1.3.4.4 Key challenges [14]

- TBCs evolved from insulating layers to complex designs for temperature reduction. However, scatter data is obtained during laboratory and engine testing due to complex design [45].
- Improvement in laboratory and component testing, and simulation of engine working conditions helps the prediction of failure and life expectancy.

- The requirement of further advancement in traditional processes of thermal spraying can bring uniformity in microstructure, process efficiency, flexibility, and cost reduction.
- One of the main concerns is the availability of rare earth elements for the eternity and also at affordable prices.
- The development of new materials can further increase the maximum bond coat temperature that can be employed.

1.3.4.5 Future scope [44]

- Life prediction, validation, and simulation of engine performance continued to be explored.
- There are a few techniques available that can determine the residual stress in TBCs; therefore, a simple, convenient, and cost-effective method needs to be developed [44].
- There is a need to update the physics and mechanism behind the failure of next-generation TBCs before their implementation in the real environment.
- To explore the TBCs in a hot corrosion environment and the effect of whisker reinforcement on TBCs.

1.3.5 Microwave cladding process

Microwave cladding is a method that uses microwave radiation to deposit or fuse a layer of cladding material onto a substrate. Microwave cladding is a low-cost method for improving the surface characteristics of metallic components. In recent years, it has evolved as a method for treating a wide range of materials. This method has gained popularity in recent years because it aids inhomogeneous heating throughout the material at the molecular level.

The material to be deposited, such as metal powder or wire, is prepared in a cladding-ready state. Microwave radiation is created and concentrated on the surface of the substrate or substance to be deposited. The microwave radiation quickly warms the substance, melting or fusing it to the substrate. A regulated procedure, such as powder feeding, wire feeding, or other approaches, is utilized to deposit the melted or fused material onto the substrate surface. The deposited material cools and hardens on the substrate, generating a surface-adhering cladding layer. The microwave cladding process has been presented in Figure 1.11 [46].

Gupta and Sharma [47] carried the microstructural examination of Ni+ WC microwave clad on mild steel specimen. The results indicated the uniform distribution of tungsten carbide particles in a nickel-based matrix. Moreover, columnar dendritic structure was obtained from clad layer due to growth of the finer grains in a direction opposite to the heat flow [48].

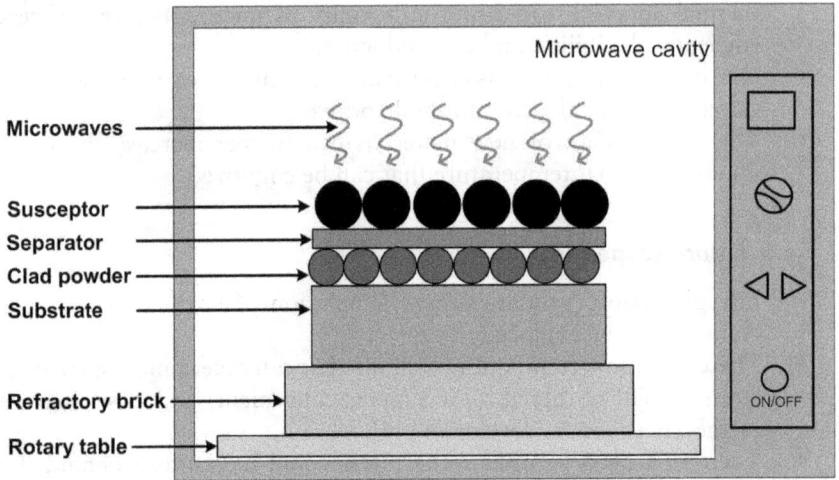

Figure 1.11 Microwave cladding process [46].

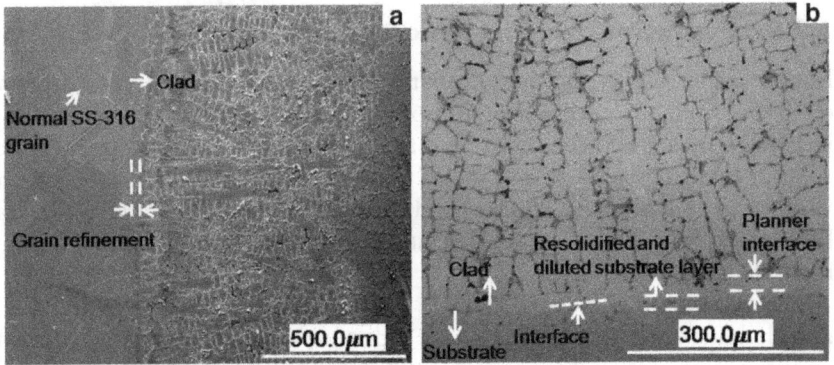

Figure 1.12 (a) Microstructure of clad layer; (b) back-scattered-electron image exhibiting cellular structure [47].

An equiaxed grain structure having 200 µm of grain size can be seen in Figure 1.12. The porosity in microwave cladding was minimal as compared to other processes due to the uniform heating in the microwave. The microstructure and properties of clad layer depends on the process parameters such as material used, exposure time, and microwave power employed during the cladding process [49].

1.3.5.1 Advantages [21]

- Microwave radiation can offer fast and volumetric heating.
- Microwave cladding can provide localized heating, which reduces the heat-affected zone and the possibility of deformation or other thermal-related difficulties in the substrate.
- Microwave cladding may be used on a variety of materials such as metals, alloys, and certain ceramics.
- Enabling rapid material melting or fusing and shorter processing times than traditional cladding techniques.
- Microwave cladding provides precision control over the heating and deposition processes, allowing for perfect material placement and layer thickness control.

1.3.5.2 Limitations [46]

- It might be difficult to have perfect control over the layer thickness in microwave cladding.
- To obtain uniform and constant layer thickness, the deposition process may need to be optimized. Microwave cladding is best suited for materials that can be heated successfully using microwave radiation.
- Some materials may react poorly to microwave heating or may be incompatible with the procedure. Microwave cladding requires specialized equipment, such as microwave generators and applicators, which can be costly, rendering the procedure inaccessible to small-scale enterprises.

1.3.5.3 Applications [21]

- Microwave cladding can be used to repair and refurbish damaged or worn-out components, therefore prolonging their service life and lowering replacement costs.
- It is used in surface protection and restoration.
- Functional coatings, such as wear-resistant or heat-resistant layers, can be applied by microwave cladding.
- Microwave cladding can be used in additive manufacturing procedures to add extra material layers to existing structures or to create complicated shapes.

1.3.5.4 Key challenges

Some of the challenges associated with the processing of microwaves are listed below [50]:

- The absorption rate of materials varies with an increase in temperature, which causes certain complications with the recording and analysis of different parameters during real-time measurement.
- The modelling and simulation of electron beam cladding models are very challenging due to the scarcity of experimental data.
- Few irregularities in the properties of composites have been noticed due to the cumulative effect of temperature rise, hence material properties change significantly because of non-uniform heating.
- Oxidation of matrix segments in metal–matrix composites due to high temperature produced by highly localized electron beam.
- The moisture absorption phenomenon in polymer composites and densification in metal–matrix composites needs to be addressed.
- Microwave exposure is hazardous to human life; therefore appropriate measures should be taken to prevent exposure and leakages.

1.3.5.5 Future scope

Microwave cladding still provides huge potential for the development of improved coating for different applications [48].

- Optimization of working parameters of microwave cladding.
- The development of newer materials, nanomaterials, and composites leads to improved surface properties.
- Microwave cladding in conjunction with additive manufacturing will provide significant enhancement in the properties of claddings.
- The macrostructure of microwave-processed cladding can produce crack-free layer with minimum porosity and high density.

1.3.6 Metal inert gas (MIG) cladding process

MIG cladding is a welding procedure that deposits a protective layer of material onto a substrate. It is also known as metal inert gas cladding or gas metal arc cladding (GMA cladding). The welding flame is fed with a continuous wire electrode. To protect the molten metal from ambient contamination, a shielding gas, often a combination of argon and carbon dioxide, is fed to the welding area. An electric arc forms between the wire electrode and the workpiece, resulting in extreme heat. The arc's heat melts the wire electrode, and the molten metal is deposited on the workpiece to produce the cladding layer. Figure 1.13 depicts the cladding process using the MIG welding technique.

Tribological and mechanical properties of MIG cladding mainly dependent of current, voltage, interaction between current and voltage, and scan speed [51]. Current is the dominant factor to influence the properties of cladding, while the effect of other factors was minimal. Microstructural analysis of

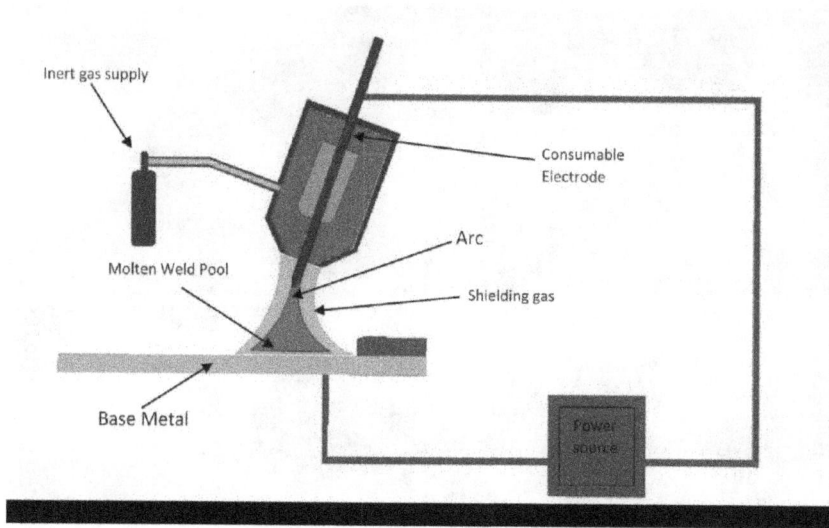

Figure 1.13 MIG cladding process.

boron-rich cladding on steel substrate revealed the higher concentration of Fe_2B on the top surface [52], even though some of the FeB phase was formed because of boron-rich atmosphere. SEM images in the backscattered mode (Figure 1.14(a,b)) exhibited the two regions, that is, light and dark areas. Light area represents a phase mixture of α-Fe and boride, whereas the dark area showed the primary borides. Optical micrograph exhibited the visible interface between cladding and substrate and the cladding region (Figure 1.14(c–e)).

1.3.6.1 Advantages [53]

- MIG cladding has high deposition rates, enabling speedy and efficient covering of huge surfaces.
- The technique has strong control over the deposition, allowing for precise coating thickness and cladding parameter control.
- Through the application of the cladding material, MIG cladding may improve the mechanical qualities of the substrate, such as hardness, wear resistance, and corrosion resistance.
- It is suited for a wide range of materials, including different metals and alloys, allowing for material flexibility in cladding applications.
- Because of its fast deposition rates and efficient material consumption, MIG cladding is a comparatively low-cost method.

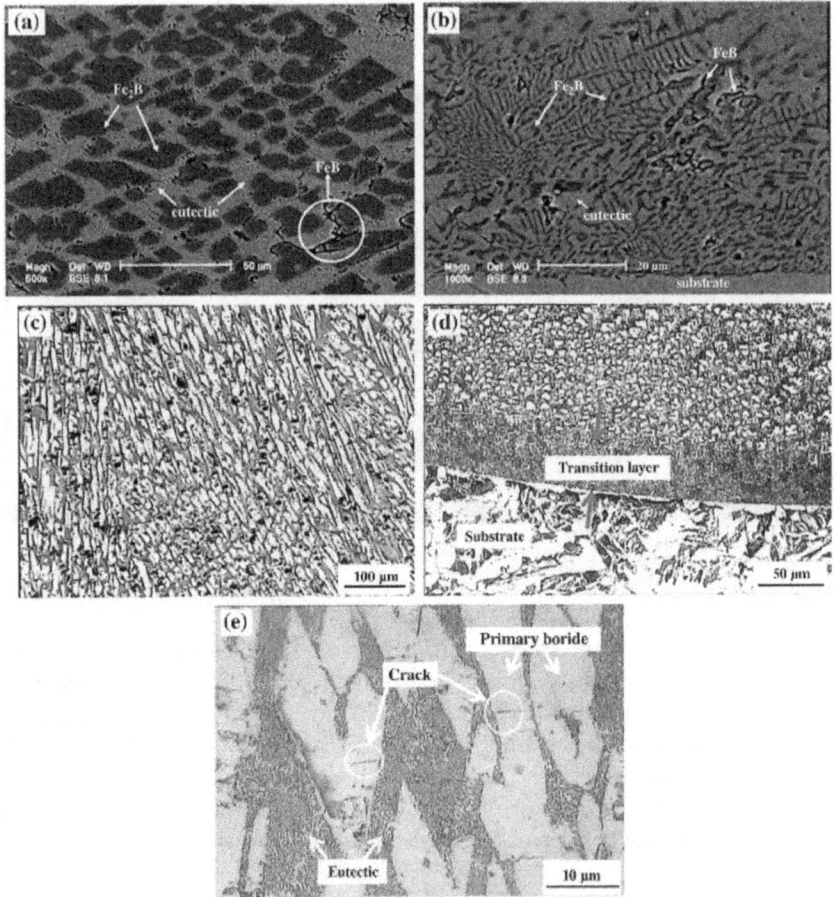

Figure 1.14 (a–b) SEM images; (c–e) optical micrograph of MIG cladding [52].

1.3.6.2 *Limitations [4]*

- To regulate the process parameters and provide consistent and high-quality cladding results, MIG cladding requires competent operators.
- If not correctly controlled, MIG cladding can cause a heat-affected zone in the substrate, which can result in changes in material characteristics and possible deformation.
- MIG cladding is primarily appropriate for thin cladding layers, rendering it unsuitable for applications requiring thick coatings. The size and style of the welding torch used in MIG cladding may restrict access to limited or complicated geometries, making cladding application in such places difficult.

1.3.6.3 Applications [54]

- MIG cladding is used to repair worn-out or damaged surfaces of components, therefore prolonging their service life and avoiding the need for complete replacement.
- It can be used to rebuild dimensions on damaged or undersized items, allowing for further machining or finishing.
- MIG cladding is used to provide heat and oxidation protection to high-temperature components such as furnaces, exhaust systems, and boilers.
- MIG cladding is frequently used to apply corrosion-resistant coatings to components exposed to harsh environments, such as pipelines, and chemical equipment.
- MIG cladding can be used to improve the wear resistance of engine valves, turbine blades, and tools.

1.3.6.4 Key challenges

- The selection of proper coating material for the development of corrosive and wear-resistant cladding becomes necessary as there is a scarcity of proven materials for MIG cladding [55].
- Optimization of process parameters to obtain the proper melting of material and uniform distribution of coating material throughout the surface of the substrate [55].
- The development of advanced material which exhibits uniform properties. The non-uniformity leads to non-uniform heating and cooling, which causes the generation of internal stresses and thermal degradation.
- The mathematical modelling and simulation of the cladding process involving complex and advanced material.

1.3.6.5 Future scope

- The optimal selection of process parameters for advanced materials such as nanomaterials and composites to minimize the dilution in claddings.
- The addition of high-entropy and rare-earth elements in the conventional cladding material leads to the fabrication of cladding with improved mechanical and tribological properties.
- The fabrication of compact and movable cladding set-up opens the possibility of on-site repair and maintenance of machines and equipment.
- To enhance the ability of MIG cladding to deposit a wide range of materials on flat and non-flat surfaces.
- The development of real-time testing and measurement test rig to obtain accurate and reliable information regarding the behaviour of cladding.

1.3.7 Tungsten inert gas (TIG) cladding process

The TIG weld cladding process uses a non-consumable tungsten electrode to generate the arc between the electrode and workpiece [56,57]. Shielding gases, such as argon or helium, are used to provide an inert gas atmosphere, which shields the tungsten electrode and molten weld pool. The welding torch moves along the entire surface of the substrate and the arc progressively melts the preplaced layer (cladding material), which gets deposited over the substrate. The cladding layer starts to form as soon as the weld pool solidifies and a metallurgical bond is formed between the cladding material and substrate. The deposition of clad layer over the substrate improves its properties. Figure 1.15 depicts the arc scanning over the preplaced powder layer.

The performance of WC-CoCr cladding processed using TIG cladding process is largely governed by several process parameters such as current, scan speed, standoff distance, and argon flow rate [56,57]. Current and scan speed are the dominant factors to influence the microstructure and mechanical and tribological properties. With increase in current and decrease in scan speed, heat input to cladding increases, which leads to the increase in dilution level. The microstructure of claddings processed at low current and high scan speed showed the presence of partially melted WC grains in CoCr matrix [58]. While the cladding processed at high current and low scan speed exhibited the dendritic microstructure (Figure 1.16), EDS analysis also showed a variation with a change in processing conditions.

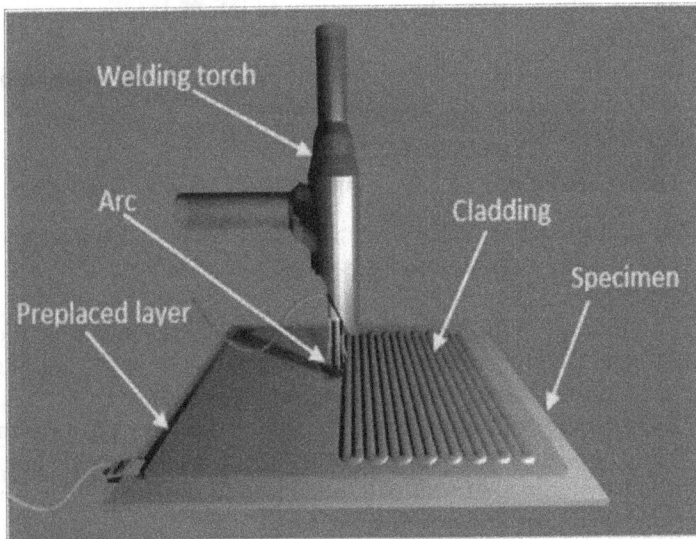

Figure 1.15 Schematic diagram of the TIG weld cladding process.

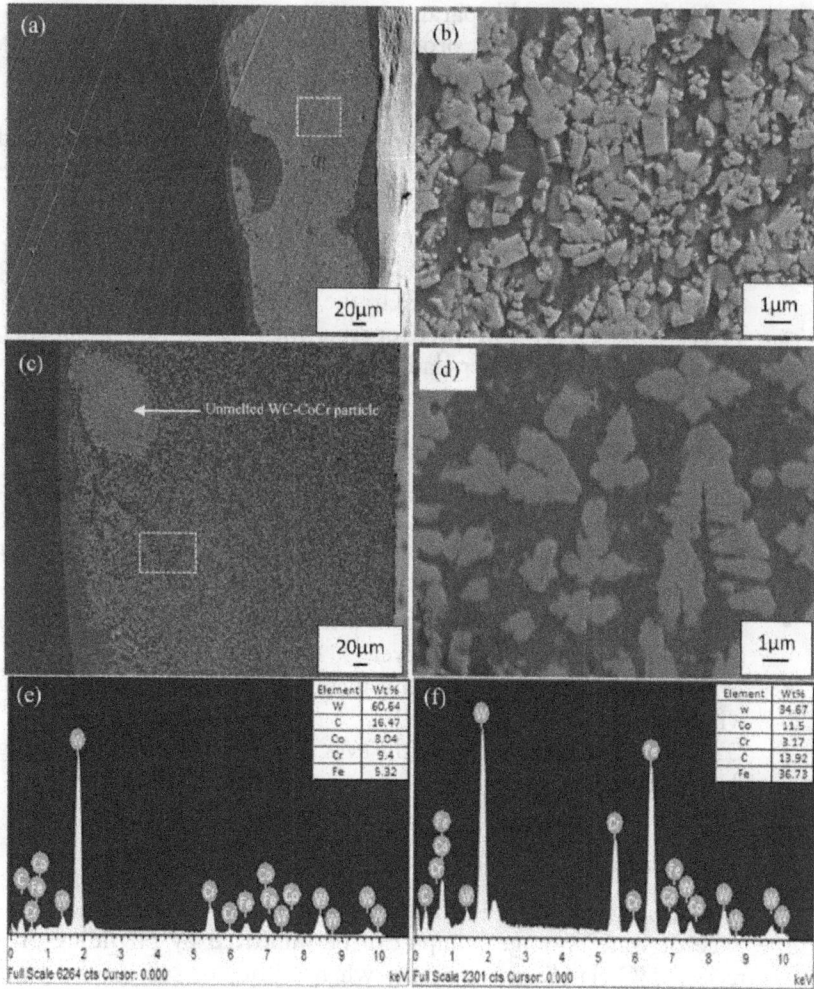

Figure 1.16 SEM images of cladding at different magnication at (a,b) low current and high scan speed and (c,d) high current and low scan speed and (e,f) EDS analysis at different processing conditions [56].

1.3.7.1 Advantages

- Many favourable features are also associated with TIG weld cladding, which includes high energy density, smaller standoff distance, low welding current, and protective inert gas atmosphere to protect the weld pool from contaminations.

- The deposition rate and thickness of the clad layer can be easily controlled by using suitable parameters [57].
- TIG cladding has a short processing time, is economical, and offers flexibility in operation as compared to plasma, laser, and HVOF process [56].
- A strong metallurgical bond is obtained between the clad layer and substrate, which results in excellent adhesion and ability to fabricate a composite hard and wear resistance clad layer.
- Other cladding techniques employ complicated set-ups along with restricted mobility of the equipment and low deposition rate making them less desirable as compared to the TIG cladding [59].
- TIG cladding can be easily used for repairing engineering components, which is difficult for other cladding techniques [57].

1.3.7.2 Disadvantages

- High heat input during the cladding process may damage the substrate and contribute to high residual stress.
- Dilution in TIG cladding is higher as compared to laser and HVOF processes [10].
- TIG cladding causes non-uniform distribution of arc, which results in inhomogeneity of the clad layer.
- Arc stability reduces in the low inert gas atmosphere and during windy days.
- Care should be taken to shield the eyes and skin from arc rays during the cladding process.

1.3.7.3 Applications

- In hydroelectric plant and mining industries, the life and performance of water storage pumps can be improved by providing protective TIG claddings, especially when a brine or saline solution is employed, which increases the severity of abrasive wear [60].
- In Pelton turbines, TIG cladding can be employed instead of hard chrome weld overlay for the protection of runners, nozzles, and needle valves from corrosion, cavitation, and wear.
- The exterior surface of the drill bit employed in oil drilling and earth-boring tools is subjected to excessive abrasive wear [7].
- In mining, ball mills, rock drills, and rock crushers are subjected to severe abrasion, which can be prevented by using TIG-added equipment [61].
- The wear and tear of tools and equipment employed in agriculture tools, mining, and sugar industry can be prevented and their life can be increased with the help of TIG claddings.

1.3.7.4 Key challenges

- The optimization of process parameters has not been explored for advanced micro- and nanostructured materials [56].
- The ability to reproduce the coatings consistently with minimum variation in cladding properties.
- Minimal or no post-processing requirement for claddings.
- To develop high-thickness cladding for the repair of machines and equipment subjected to wear, corrosion, and other surface degradation.
- To enhance the deposition efficiency to reduce the wastage of advanced materials.
- To automate the cladding process with minimal or no human intervention.

1.3.7.5 Future scope

- A powder injection technique can be employed to obtain the uniform thickness of TIG cladding, which in turn may produce cladding with low dilution, and uniform mechanical and tribological properties.
- The effect of variation in the percentage of rare-earth oxides on the microstructure, hardness, fracture toughness, and abrasive wear resistance of claddings can be explored.
- A study related to the heat treatment of developed claddings can be conducted to study its effect on their microstructure, properties, and wear behaviour.
- Mathematical models and simulations can be developed for a better understanding of the physics and microstructure of cladding.
- The process parameters can be optimized by using artificial intelligence-based modelling and optimization techniques.

1.4 CONCLUSION

Thermal claddings are flexible and economical and help develop claddings for improvement in the surface properties of engineering surfaces, machines, and equipment. Thermal claddings find applications in diverse fields such as automotive, power generation, railways, chemical processes, aerospace, and biomedical industries. With use of thermal claddings, it is possible to develop cheaper material with superior cladding having superior properties such as high hardness, toughness, corrosion, and wear resistance. It results in considerable cost savings and higher life expectancy. Thermal cladding processes have high competency and capability to influence manufacturing, research, and development sectors. The major technological advances such as automation, industrial robotics, additive manufacturing, and artificial intelligence lead to the development of high-quality claddings. With new improved technologies using advanced nanomaterials and composites, future applications of thermal cladding will keep growing.

CONFLICTS OF INTEREST

The authors declare that there is no conflicts of interest.

ACKNOWLEDGEMENT

The authors like to acknowledge the editors for providing the opportunity to write this chapter.

REFERENCES

[1] Batciielor A W, Lam L N and Chandrasekaran M *Materials Degradation and Its Control by Surface Engineering*. Imperial College Press, London.

[2] Hutchings I and Shipway P 2017 *Tribology*. Butterworth-Heinemann, Elsevier, Cambridge, United States.

[3] Dwivedi D K 2018 *Surface Engineering*. Springer (India) Pvt. Ltd., New Delhi, India.

[4] Singh R, Kumar D, Mishra S K and Tiwari S K 2014 Laser cladding of Stellite 6 on stainless steel to enhance solid particle erosion and cavitation resistance *Surf. Coatings Technol*. 251 87–97.

[5] Cotell C M, Sprague J A and Smidt F A 1994 *Surface Engineering*. ASM International, United States.

[6] Burakowski T and Wierzchon T 1999 *Surface engineering of metals*. CRC Press, United States. .

[7] Mellor B G 2006 *Surface coatings for protection against wear*. Woodhead Publishing Limited, Cambridge, England and CRC Press LLC, New York, United States.

[8] Paulo Davim J, Oliveira C and Cardoso A 2006 Laser cladding: An experimental study of geometric form and hardness of coating using statistical analysis *Proc. Inst. Mech. Eng. Part B J. Eng. Manuf*. 220 1549–54.

[9] Walia R S, Murtaza Q, Pandey S H and Tyagi A 2023 *Surface Engineering*. CRC Press, Taylor & Francis Group, United States.

[10] Thakur L and Vasudev H 2022 *Thermal Spray Coatings*. CRC Press, Taylor & Francis Group, United States.

[11] Srikanth A, Mohammed Thalib Basha G and Venkateshwarlu B 2020 A Brief Review on Cold Spray Coating Process *Mater. Today Proc*. 22 1390–1397.

[12] Kuroda S, Kawakita J, Watanabe M and Katanoda H 2008 Warm spraying – A novel coating process based on high-velocity impact of solid particles *Sci. Technol. Adv. Mater*. 9 033002.

[13] Miller R A 1997 Thermal barrier coatings for aircraft engines: History and directions. *J. Therm. Spray Technol*. 6 (1) 35–42.

[14]   Beetle W, Marijnissen G and Van Lieshout A 1999 The evolution of thermal barrier coatings – Status and upcoming solutions for today's key issues *Surf. Coatings Technol.* **120–121** 61-7.

[15]   Toyserkani E, Khajepour A and Corbin S 2005 *Laser Cladding*. CRC Press, United States.

[16]   Dubourg L and Archambeault J 2008 Technological and scientific landscape of laser cladding process in 2007 *Surf. Coatings Technol.* **202** (24) 5863-9.

[17]   Santo L 2008 Laser cladding of metals: a review *Int. J. Surf. Sci. Eng.* **2** (5) 327-36.

[18]   Węglowski M S, Błacha S and Phillips A 2016 Electron beam welding – Techniques and trends – Review. *Vacuum* **130** 72-92.

[19]   Węglowski M S, Błacha S and Phillips A 2012 – *Welding and Joining of Aerospace Materials* . Woodhead Publishing Ltd., Elsevier Ltd., United States.

[20]   Shellabear M and Nyrhilä O 2004 DMLS — Development History and State of the Art *LANE 2004 Conference, Germany*. Sept. 21–24, 2004,1–12.

[21]   Bansal A and Vasudev H 2023 *Advances in Microwave Processing for Engineering Materials*. CRC Press, Taylor & Francis Group, United States.

[22]   Anders A 2003 Tracking down the origin of arc plasma science I. Early pulsed and oscillating discharges *IEEE Trans. On Plasma Sci.* **31** (5) 1052-9.

[23]   Ion J C 2005 *Laser Processing of Engineering Materials*. Elsevier Butterworth-Heinemann, Oxford, England.

[24]   Pruncu C I, Aherwar A and Gorb S 2022 *Tribology and Surface Engineering for Industrial Applications*. CRC Press, Taylor & Francis Group, United States.

[25]   Fard A K, McKay G, Buekenhoudt A, Al Sulaiti H, Motmans F, Khraisheh M and Atieh M 2018 Inorganic membranes: Preparation and application for water treatment and desalination *Materials (Basel)*. **11**(1) 74.

[26]   Pawlowski L 2008 *The Science and Engineering of Thermal Spray Coatings*. John Wiley & Sons Ltd, England.

[27]   Vasudev H, Thakur L, Singh H and Bansal A 2020 An investigation on oxidation behaviour of high velocity oxy-fuel sprayed Inconel718-Al2O3 composite coatings *Surf. Coatings Technol.* **393** 125770.

[28]   Davis J R 2004 *Thermal Spray Technology*. ASM International, United States.

[29]   Thakur L and Arora N 2013 Sliding and abrasive wear behavior of WC-CoCr coatings with different carbide sizes *J. Mater. Eng. Perform.* **22** 574-83.

[30]   Pierlot C, Pawlowski L, Bigan M and Chagnon P 2008 Design of experiments in thermal spraying: A review *Surf. Coatings Technol.* **202** 4483-90.

[31]   Vardelle A, Moreau C, Akedo J, Ashrafizadeh H, Berndt C C, Berghaus J O, Boulos M, Brogan J, Bourtsalas A C, Dolatabadi A, Dorfman M, Eden T J, Fauchais P, Fisher G, Gaertner F, Gindrat M, Henne R, Hyland M, Irissou E, Jordan E H, Khor K A, Killinger A, Lau Y C, Li C J, Li L, Longtin J, Markocsan N, Masset P J, Matejicek J, Mauer G, McDonald

A, Mostaghimi J, Sampath S, Schiller G, Shinoda K, Smith M F, Syed A A, Themelis N J, Toma F L, Trelles J P, Vassen R and Vuoristo P 2016 The 2016 Thermal spray roadmap. *J. Therm. Spray Technol.* **25** (8) 1376–440.

[32]   Heberlein J V R *Thermal Spray Fundamentals*. Springer Science & Business Media, London.

[33]   Roy M 2013 *Surface Engineering for Enhanced Performance against Wear*. Springer, Wien Heidelberg.

[34]   Fry A T and Gorman D M 2016 Comparison of laser clad coatings under simulated 100 % biomass firing conditions, Proceedings of Baltica X Conference p 1–16.

[35]   Mingxi L, Yizhu H and Xiaomin Y 2006 Effect of nano-Y2O3 on microstructure of laser cladding cobalt-based alloy coatings *Appl. Surf. Sci.* **252** 2882–7.

[36]   Zhu L, Xue P, Lan Q, Meng G, Ren Y, Yang Z, Xu P and Liu Z 2021 Recent research and development status of laser cladding: A review *Opt. Laser Technol.* **138** 106915.

[37]   Vasudev H and Prakash C 2023 *Handbook of Laser-based Sustainable Surface Modification and Manufacturing Techniques*. CRC Press, Taylor & Francis Group, United States.

[38]   Lenivtseva O G, Bataev I A, Golkovskii M G, Bataev A A, Samoilenko V V. and Plotnikova N V. 2015 Structure and properties of titanium surface layers after electron beam alloying with powder mixtures containing carbon *Appl. Surf. Sci.* **355** 320–6.

[39]   Yu T, Wang H, Han K and Zhang B 2022 Microstructure and wear behavior of AlCrTiNbMo high-entropy alloy coating prepared by electron beam cladding on Ti600 substrate *Vacuum* **199** 110928.

[40]   Wang W, Zhang S, Xiao S, Sato Y S, Wang D, Liu Y, Liu D and Hu Q 2022 Microstructure and properties of multilayer WC-40Co coating on Ti-6Al-4V by electron beam cladding *Mater. Charact.* **183** 111585.

[41]   Kathuria Y P 2000 Some aspects of laser surface cladding in the turbine industry *Surf. Coatings Technol.* **132** 262–9.

[42]   Kim K and Kim W 2021 Effect of heat treatment on microstructure and thermal conductivity of thermal barrier coating *Materials (Basel)*. **14** 7801.

[43]   Kumar D and Pandey K N 2017 Optimization of the process parameters in generic thermal barrier coatings using the Taguchi method and grey relational analysis *Proc. Inst. Mech. Eng. Part L J. Mater. Des. Appl.* **231** (7) 600–10.

[44]   Thakare J G, Pandey C, Mahapatra M M and Mulik R S 2021 Thermal Barrier Coatings—A State of the Art Review *Met. Mater. Int.* **27** 1947–68.

[45]   Darolia R 2013 *Thermal barrier coatings technology: Critical review, progress update, remaining challenges and prospects* **58** (6) 315–48.

[46]   Singh B and Zafar S 2021 Microwave cladding for slurry erosion resistance applications: A review *Mater. Today Proc.* **46** 2686–90.

[47]   Gupta D and Sharma A K 2011 Development and microstructural characterization of microwave cladding on austenitic stainless steel. *Surf. Coatings Technol.* **205** 5147–55.

[48] Vasudev H, Singh G, Bansal A, Vardhan S and Thakur L 2019 Microwave heating and its applications in surface engineering: A review. *Mater. Res. Express* **6** 102001.

[49] Bansal A and Vasudev H 2023 *Advances in Microwave Processing for Engineering Materials*. CRC Press, Taylor & Francis Group, United States.

[50] Chandra Srivastava S, Murtaza Q and Kumar P 2020 Microwave cladding on metallic surfaces: A review *Mater. Today Proc.* **21** 1533–6.

[51] Lin C M, Su T L and Wu K Y 2015 Effects of parameter optimization on microstructure and properties of GTAW clad welding on AISI 304L stainless steel using Inconel 52M *Int. J. Adv. Manuf. Technol.* **79** 2057–66.

[52] Amushahi M H, Ashrafizadeh F and Shamanian M 2010 Characterization of boride-rich hardfacing on carbon steel by arc spray and GMAW processes *Surf. Coatings Technol.* **204** 2723–8.

[53] Prabhu R and Alwarsamy T 2017 Effect of process parameters on ferrite number in cladding of 317L stainless steel by pulsed MIG welding *J. Mech. Sci. Technol.* **31** (3) 1341–7.

[54] Weman K 2003 *Welding Processes Handbook*. Woodhead Publishing Limited, Cambridge, England and CRC Press LLC, New York, United States.

[55] Kannan T and Yoganandh J 2010 Effect of process parameters on clad bead geometry and its shape relationships of stainless steel claddings deposited by GMAW *Int. J. Adv. Manuf. Technol.* **47** 1083–95.

[56] Singh J, Thakur L and Angra S 2020 An investigation on the parameter optimization and abrasive wear behaviour of nanostructured WC-10Co-4Cr TIG weld cladding *Surf. Coatings Technol.* **386** 125474.

[57] Singh J, Thakur L and Angra S 2020 A study of tribological behaviour and optimization of WC-10Co-4Cr Cladding *Surf. Eng.* **37** 70–9.

[58] Singh J, Thakur L and Angra S 2020 Abrasive wear behavior of WC-10Co-4Cr cladding deposited by TIG welding process *Int. J. Refract. Metals Hard Mater.* **88** 105198.

[59] Peng D 2012 The effects of welding parameters on wear performance of clad layer with TiC ceramic *Ind. Lubr. Tribol.* **64** (5) 303–11.

[60] Singh J, Thakur L and Angra S 2022 Comparative analysis of micrometric and nano-metric WC-10Co-4Cr GTA cladding *Eng. Res. Express.* **4** 025041.

[61] Eroğlu M and Önalp S 2002 Tungsten inert gas surface modification of SAE 4140 steel *Mater. Sci. Technol.* **18** 1544–50.

Chapter 2

Improved room and high-temperature wear performance with Inconel 625 TIG weld cladding

Neeraj Kamboj and Lalit Thakur

2.1 INTRODUCTION

AISI-304 stainless steel (SS) is ideal for the chemical, oil, and gas sectors due to its superior corrosion and oxidation resistance properties. Low wear resistance prevents it from being used in many industries, including mining, drilling, and agriculture. The material gets degraded in these industries mainly due to abrasive, erosive, and surface fatigue wear [1,2]. Abrasive and adhesive wear are the primary types of degradation that significantly lead to various components' financial losses and maintenance costs [3,4]. Abrasive wear often occurs in two forms: (a) two-body abrasive wear — it happens when hard particles remove material from the opposite surface and (b) three-body abrasive wear happens when particles are unrestrained to roll and slide on a surface. Figure 2.1 illustrates the mechanism of two-body and three-body abrasive wear. Adhesive wear occurs when two surfaces come into contact with one another while being loaded, and both surfaces experience plastic deformation and eventually weld together. These bonds break as the sliding progresses, leaving cavities on one side and depressions on the other, which adds to wear [5]. Figure 2.2 illustrates the mechanism of adhesive wear.

Due to cyclic loading, surface fatigue wear weakens a material's surface [6,7]. According to the literature, wear on a component primarily happens on its surfaces. Therefore, the life and reliability of components can be enhanced by protecting their surfaces. Surface modification with the coating is the best method to combat these types of wear. It has been observed that various surface engineering techniques like thermal spray, high-velocity oxygen fuel (HVOF), and laser cladding are used to deposit the coatings/claddings of WC-CoCr, Inconel-600, Inconel-625, Inconel-718 materials and sometimes containing hard ceramics particles of TiC, SiC, and WC, in different compositions for improving the surface properties and wear resistance of other substrates [8–11]. Among these techniques, TIG cladding is the most economical, simple, and high deposition rate process for fabricating

DOI: 10.1201/9781032713830-2

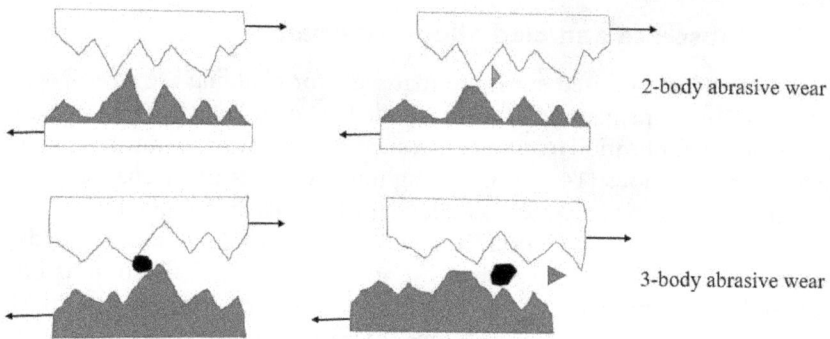

2-body abrasive wear

3-body abrasive wear

Figure 2.1 Two-body and three-body abrasive wear.

A

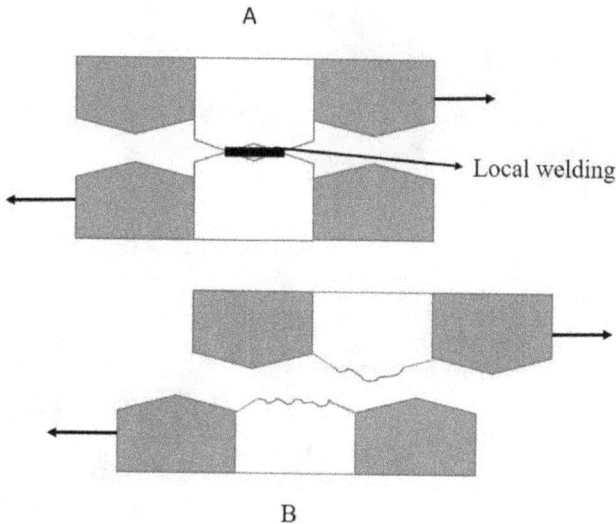

Local welding

B

Figure 2.2 Adhesive wear mechanism.

the clad layer on the different grades of stainless steel substrate [12,13]. The Inconel-625 alloys are a potential coating material owing to a unique mix of mechanical and tribological characteristics such as excellent wear, corrosion resistance, and thermal stability. Inconel-625 alloy is significantly employed in marine equipment, petrochemical components, heat exchangers, and power generation industries.

2.2 MATERIALS AND METHODS

2.2.1 Substrate and clad alloy materials

The most widely utilised substrate materials for cladding are steel. Nearly any wear-resistant material can be bonded to steel. Substrates are chosen based on the application requirements, such as extreme temperatures or abrasive wear conditions [14,15]. Cladding improves its surface characteristics without affecting a component's bulk characteristics. It is applied to surfaces prone to corrosion, oxidation, and wear. Inconel 625 is mostly used in cladding since it can be deposited on any substrate and function at both low and high temperatures [16]. The scanning electron microscopy (SEM), energy-dispersive X-ray spectroscopy (EDS) of stainless steel (304L) and Inconel 625 powder are shown in Figure 2.3, which depicts the spherical morphology of particles and particle size that varies between 20 and 70 μm.

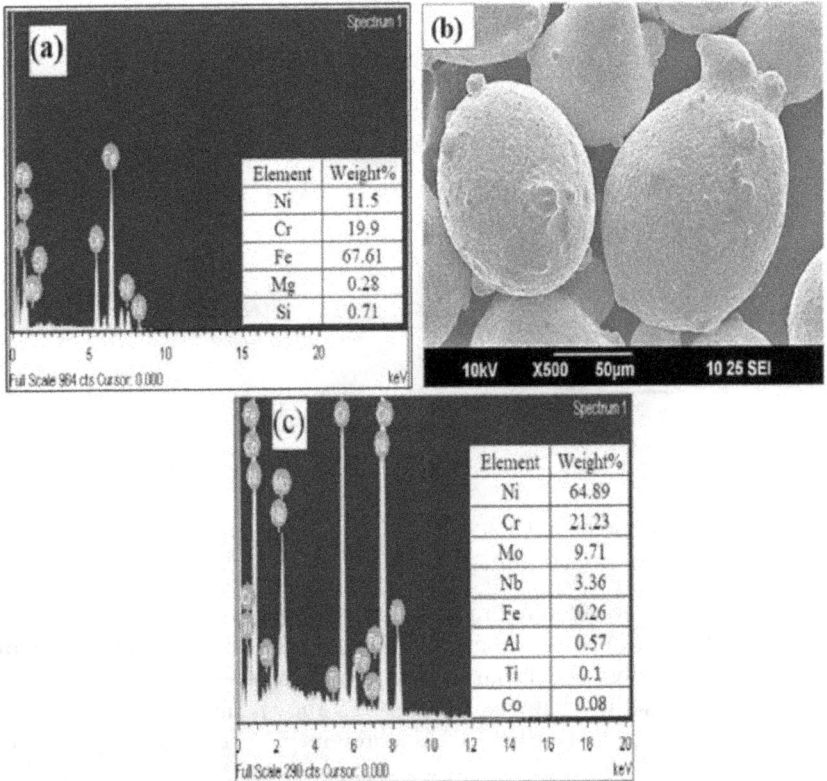

Figure 2.3 (a) EDS of stainless steel substrate, (b) SEM and (c) EDS of IN625 powder.

2.2.2 Cladding procedure

Figure 2.4 shows a flow diagram for TIG weld cladding. AISI 304 (SS) material was chosen as a substrate for the deposition of claddings. Firstly, the substrate material was rubbed with abrasive paper and then washed with acetone to increase the surface roughness and remove any type of impurities. Additionally, it aids in forming a strong bond between the substrate and the clad layer. Inconel 625 alloy was used as the clad layer powder feedstock material. Before the cladding deposition, a pre-placed layer of feedstock powder was deposited on the SS substrate. A semi-solid paste was made by adding the 5 wt. % polyvinyl alcohol (PVA) + distilled water binder slurry into the Inconel 625 powder. Thereafter this paste was evenly spread across the substrate material's surface, as shown in Figure 2.5, and then dried for 4 h in an oven maintained at 130°C.

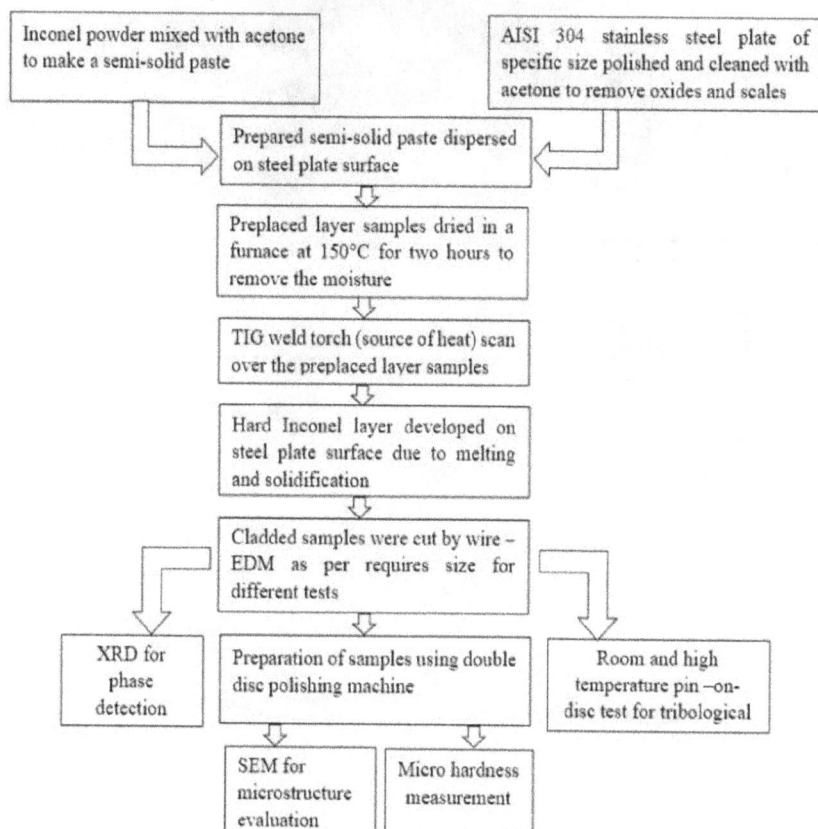

Figure 2.4 Flow diagram for TIG weld cladding.

Figure 2.5 Inconel 625 powder pre-placed sample [13].

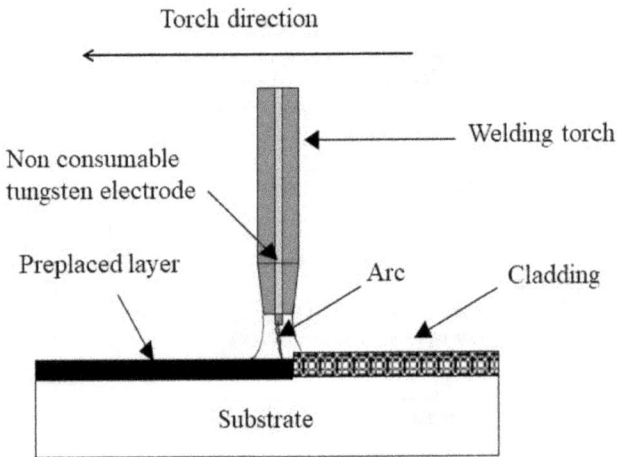

Figure 2.6 Schematic diagram of cladding by pre-placing method using TIG weld.

Table 2.1 Operating parameters

Voltage	20 V
Polarity	DCEN
Protective gas	Argon
Standoff distance	2.5 mm

Table 2.2 Detailed experimental conditions for TIG cladding process

Exp. no.	Current (A)	Torch scan speed (mm min-1)	Heat input (kJ/m)
1	60	200	172.97
2	60	250	138.46
3	60	300	115.2
4	70	200	201.80
5	70	250	161.53
6	70	300	134.4
7	80	200	230.63
8	80	250	184.61
9	80	300	153.6

After pre-placing the Inconel 625 layer of 0.9 mm thickness on the substrate material, the cladding was deposited over the substrate using an automatic TIG welding manipulator. Figure 2.6 illustrates the schematic diagram of the TIG weld cladding. In TIG weld cladding process, a tungsten electrode (2.5 mm diameter) of non-consumable type was used to produce the arc. For proper arc generation, the gap between the workpiece and electrode was maintained at 2.5 mm. The flow rate of shielding gas (argon) was set to 9 l/min. Table 2.1 provides the TIG weld cladding process parameters employed in this study. The welding current and scanning speeds were varied as illustrated in Table 2.2.

Using the relation $H = \eta(V.I/S.1000)$, we calculated the heat input where η is the heat absorption efficiency and its value is 48 [17], welding current (A), whereas S is the scan speed (mm/min).

2.3 CHARACTERISATION

Figure 2.7(a–c) displays the cladding samples produced by various TIG welding parameter settings. Thereafter, the specimens were cut from the centre of the clad layer for microstructural examination and hardness testing. The cut specimens were polished on a disc-polishing machine using various numbers of emery papers (220–1200 grit size). The cross-sectional microstructure of the different cladding samples and their worn-out morphology was examined with the help of scanning electron microscopy (SEM) imaging

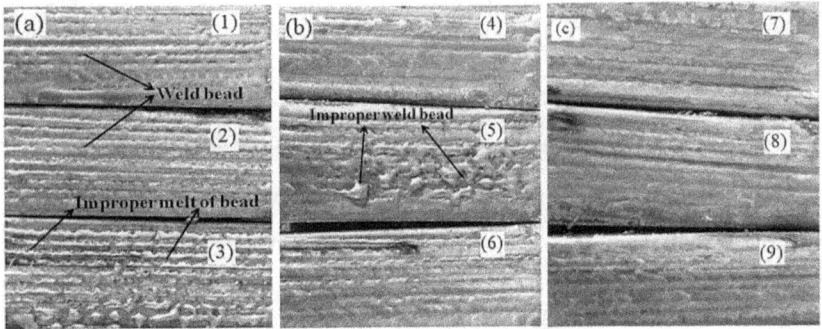

Figure 2.7 Images of samples developed at (a) 60 A current, (b) 70 A current, and (c) 80A current [13].

technique. The SEM images were captured using the JEOL JSM-6390LV machine, which also has the capability for recording energy-dispersive spectroscopy (EDS). The evolved phases in the developed claddings were identified with the help of XRD analysis. The XRD was carried out with the help of XRD equipment (D8 Advance, Bruker, Mumbai, India) using the CuKα radiation with a scan rate of 2° per minute and 2θ range of 20°–100°.

2.4 TESTS AND RESULTS

2.4.1 Dilution

A digital tool maker microscope was used to take pictures of the weld bead profile of specimens S2 and S8, which are displayed in Figure 2.8(a,b) (Sipcon Instrument Industries, India).

The low heat input situation is depicted in figure and it shows how a good weld bead profile and reduced bead width and depth were achieved. On the other hand, at high heat input conditions, an inappropriate weld bead profile with a higher breadth and depth was seen. The dilution rate increases as the heat input rises (from 138.46 to 184.6 kJ/m), resulting in deeper penetration and weld beads with a wider profile. This is due to the fact that when the heat input rises, more dissolution among the clad powder and substrate takes place, which causes the rate of dilution to rise correspondingly. The following equation has been used to compute the dilution brought on by the welding current [13].

$$D(\%) = \frac{Area\,X}{Area\,X + Area\,Y} \times 100$$

where D = Dilution (%)

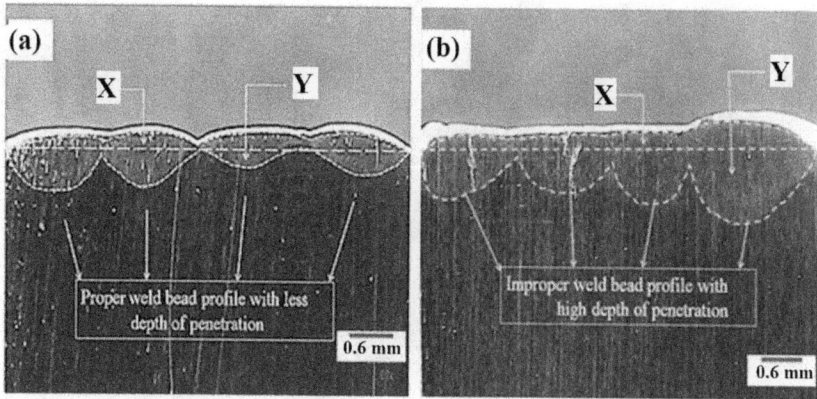

Figure 2.8 Optical microscope images (a) of maximum hardness sample and (b) of minimum hardness sample [13].

Cross-sectional areas have been estimated using ImageJ software. The dilution in Sample 2 is discovered to be 40%, whereas Sample 8 shows a 70% dilution.

2.4.2 Microhardness test

A Vickers' micro indentation hardness tester (XHVT-1000Z, Jinan Testing Equip. Corp., China) was used to determine the hardness of the samples at a normal load of 300 g and a dwell period of 15 s. The microhardness of Inconel 625 claddings was measured at the cross-section, and the mean value of microhardness was calculated by considering ten numbers of indents on the cross-section of claddings. The microhardness value of each sample has been reported in Figure 2.9. From the microhardness test, it was observed that the maximum hardness value was observed in the cladding developed at the current 60 A and 250 mm/min scanning speed (138.46 kJ/m of heat input). However, the minimum hardness was seen at a current of 80 A and 200 mm/min scanning speed (230.63 kJ/m of heat input) because the high heat input results in greater dilution and porosity in the cladding. From Table 2.2, it is clear that the hardness value was undetected at extremely low heat input values (60 A, 300 mm/min, 115.2 kJ/m). The clad layer was not formed due to the improper melting of powder particles, and there was a lack of inter-particle bonding between the powder particles and the adhesion of the clad layer with the substrate. The clad layer was not formed properly and exhibited visible porosity and improper bead formation. The improved hardness of the cladding may be attributed to the lower levels of dilution in the cladding.

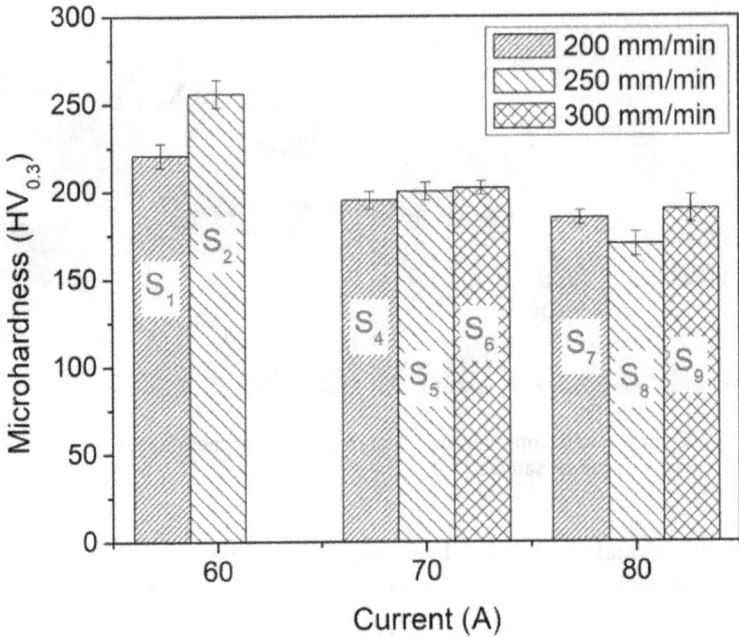

Figure 2.9 Microhardness of IN625 cladding at various values of welding current and scanning speed.

2.4.3 At room temperature and 650°C sliding wear test

Figure 2.9(a,b) illustrates the results of a sliding wear test utilising a pin-on-disc tribometer at ambient temperature and 650°C. The weight loss at room and 650°C of each sample has been illustrated in Figures 2.10 and 2.11.

Sample 2 was developed at a welding current of 60 A and a scanning speed of 250 mm/min, and it showed the least weight loss in both the ambient and 650°C conditions. Sample 8, on the other hand, was developed at a welding current of 80 A, and a scanning speed of 250 mm/min, and it showed the most weight loss. It demonstrates that, in contrast to scanning speed, the welding current significantly impacts the wear resistance of Inconel 625 cladding. Additionally, the weight loss at 650°C was significantly lower than at room temperature as shown in Figure 2.12.

Figure 2.13(a, b) and Figure 2.12(a, b) show the SEM images of worn-out surfaces of sample 2 and sample 8 at room temperature and at 650°C, respectively which are used to analyse the wear tracks produces during wear test.

In Figure 2.14(a), sample 2 depicts a consistent, smooth wear track throughout the clad surface. In contrast, sample 8 (Figure 2.9b) displays

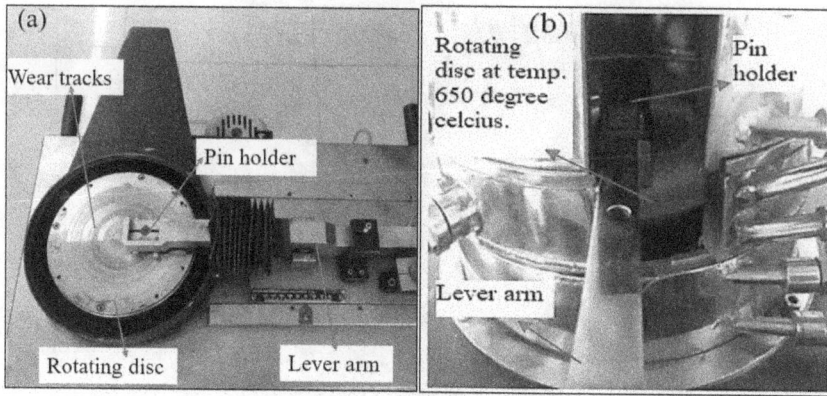

Figure 2.10 Pin-on-disc sliding set up for (a) ambient temperature and (b) at 650°C [13].

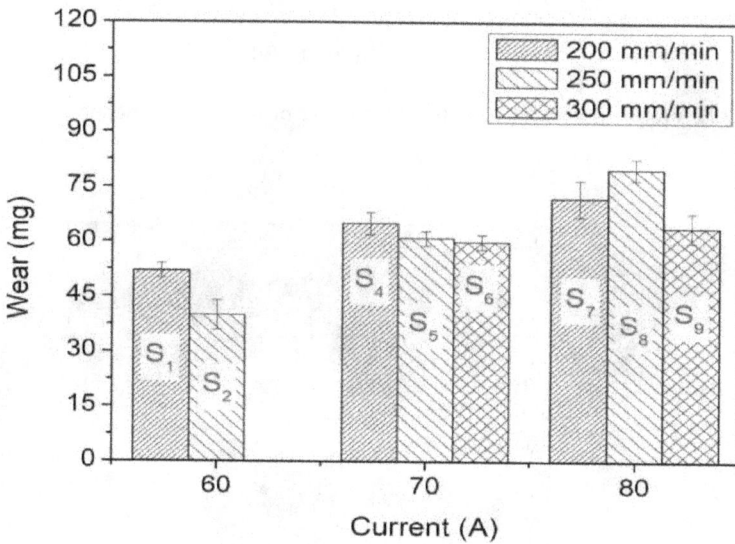

Figure 2.11 Weight loss at ambient temperature with different values of welding current and scan speed.

an uneven wear track with deep grooves that led to material ploughing and abrasion. At 650°C, when compared to other claddings generated under different parametric settings, sample 2 cladding once more showed the highest wear resistance, whereas sample 8 cladding demonstrated the lowest

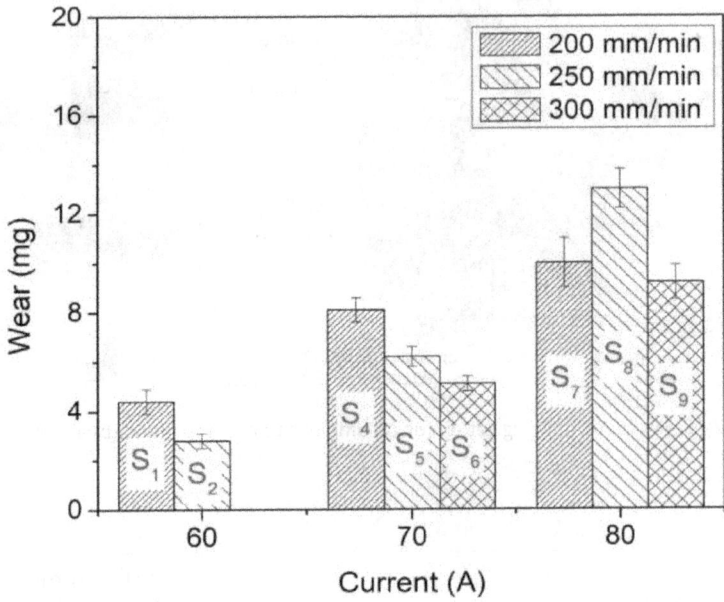

Figure 2.12 Weight loss at 650°C with various values of welding current and scan speed.

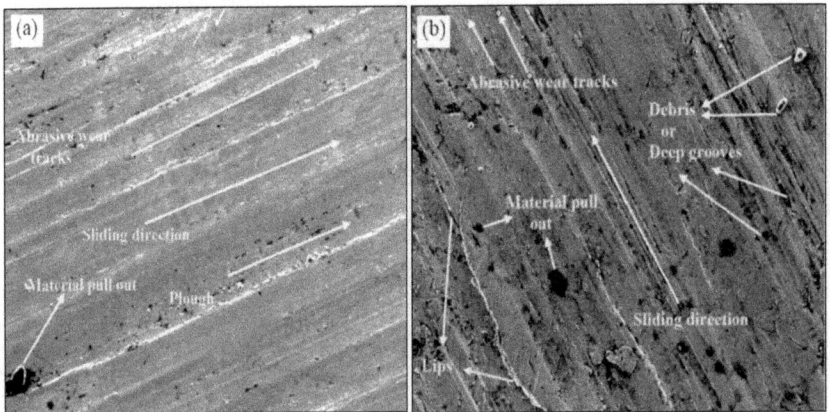

Figure 2.13 SEM images of the worn-out surface of (a) sample 2 and (b) sample 8 at ambient temperature.

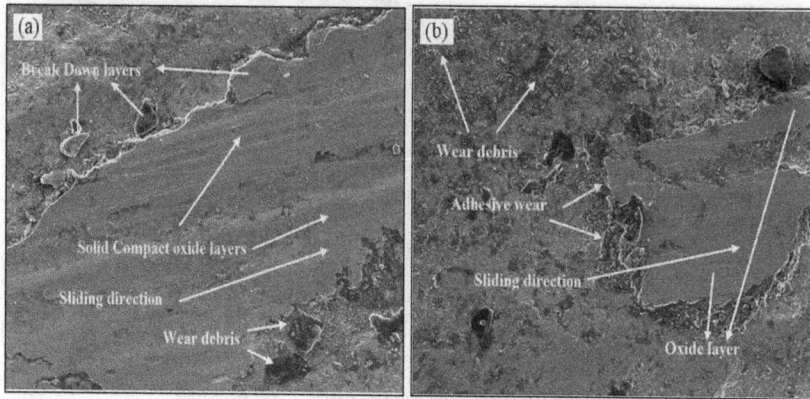

Figure 2.14 SEM images of worn-out surface: (a) sample 2 and (b) sample 8 at 650°C.

wear resistance. According to the SEM image, sample 2 developed a solid compact oxide layer and became firmly attached to the cladding surface throughout the sliding wear test. The cladding surface generated solid compact oxide layers that served as smooth load-bearing zones increasing wear resistance at high temperatures. More wear and debris formation on the surface resulted from sample 8 because a weak solid compact oxide layer was developed and was only retained in a short region compared to sample 2.

Figures 2.15(a–c) show the EDS report of the substrate for samples 2 (least wear) and 8 (maximum wear), respectively. The EDS spectra of both samples showed that, as the welding current increased, Ni and Cr elements decreased while the Fe concentration increased. The transfer of Fe from the substrate beneath to the Inconel 625 cladding is responsible for the rise in Fe content. Due to the prolonged solidification process by the high welding current, more substrate matrix and Inconel 625 cladding elements are mixed together. Therefore, longer solidification times result in more diluted iron in the cladding. At a lower welding current, a rich clad composition (EDS report) with less Fe diffusion is achieved. Sandeep et al. [8] reported the wear performance of Inconel 625 cladding using the electrode weaving technique. The researchers concluded that a lower base metal dilution is accountable for better wear resistance. Z. Liu et al. [18] also investigated the wear performance of HVOF-sprayed Inconel 625 coatings. It was reported that dendritic structures obtained in the Inconel 625 coatings are responsible for high wear performance. H. Kashani et al. [19] also developed the coatings of Inconel 625 on H11 steel substrate in order to improve its room and high-temperature wear behaviours. The results revealed that a very low wear rate at high temperatures has been observed in the cladding due to the formation of smooth compacted oxide layers during the wear test.

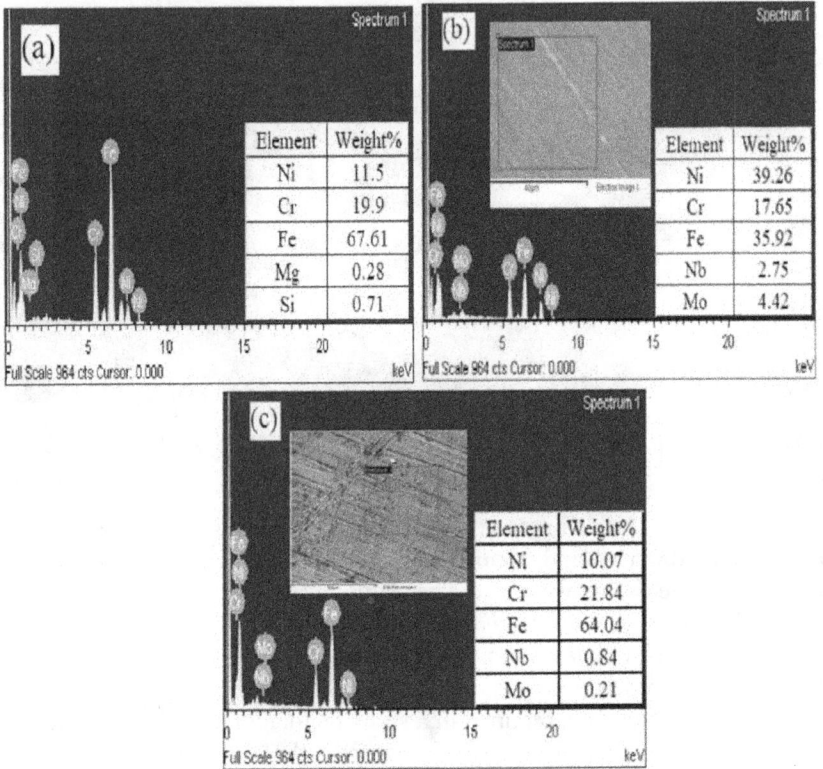

Element	Weight%
Ni	11.5
Cr	19.9
Fe	67.61
Mg	0.28
Si	0.71

Element	Weight%
Ni	39.26
Cr	17.65
Fe	35.92
Nb	2.75
Mo	4.42

Element	Weight%
Ni	10.07
Cr	21.84
Fe	64.04
Nb	0.84
Mo	0.21

Figure 2.15 EDS report of (a) stainless steel substrate, (b) sample 2, and (c) sample 8.

A. Evangeline et al. [20] deposited a cladding of Inconel 625 on 316L substrate to investigate the microstructure of the cladding. High wear resistance and microhardness in the cladding results from the grain refinement and existence of a rich clad composition.

To produce higher wear-resistant cladding, the cladding layer's dilution should be under control. This dilution is mostly controlled by the welding process parameters, with the welding current being the critical factor. Additionally, it was discovered that Inconel 625 alloy cladding exhibits higher levels of wear resistance under conditions of sliding wear at high temperatures compared to ambient temperature.

2.5 CONCLUSION

The following findings from the current research can be drawn:

- The TIG cladding is used as a practical, affordable way to increase the life of numerous industrial components.
- In TIG weld cladding, heat input is the main factor governing the claddings' mechanical and tribological properties.
- The key factor affecting the dilution and porosity of the claddings is welding current.
- Inconel 625 alloy cladding is successfully deposited on stainless steel (304L), improving its microhardness and wear resistance properties approximately twice.

REFERENCES

[1] N. Axén and K. H. Zum Gahr, "Abrasive wear of TiC-steel composite clad layers on tool steel," *Wear*, vol. 157, no. 1, pp. 189–201, 1992, doi: 10.1016/0043-1648(92)90197-G

[2] V. Jankauskas, M. Antonov, V. Varnauskas, R. Skirkus, and D. Goljandin, "Effect of WC grain size and content on low stress abrasive wear of manual arc welded hardfacings with low-carbon or stainless steel matrix," *Wear*, vol. 328–329, pp. 378–390, 2015, doi: 10.1016/j.wear.2015.02.063

[3] K. Feng *et al.*, "Improved high-temperature hardness and wear resistance of Inconel 625 coatings fabricated by laser cladding," *J. Mater. Process. Technol.*, vol. 243, pp. 82–91, 2017, doi: 10.1016/j.jmatprotec.2016.12.001

[4] Q. B. Nguyen *et al.*, "Effect of impact angle and testing time on erosion of stainless steel at higher velocities," *Wear*, vol. 321, pp. 87–93, 2014, doi: 10.1016/j.wear.2014.10.010

[5] M. M. Stack, "Some frictional features associated with the sliding wear of the nickel-base alloy N80A at temperatures to 250 "C," vol. 176, pp. 185–194, 1994.

[6] N. Kamboj and L. Thakur, "A study of the processing and characterization of RSM optimized YSZ-Inconel625 wear-resistant TIG weld cladding," *Surf. Topogr. Metrol. Prop.*, vol. 10, no. 4, 2022, doi: 10.1088/2051-672X/aca345

[7] H. Vasudev, L. Thakur, H. Singh, and A. Bansal, "Erosion behaviour of HVOF sprayed Alloy718-nano Al_2O_3 composite coatings on grey cast iron at elevated temperature conditions," *Surf. Topogr. Metrol. Prop.*, vol. 9, no. 3, 2021, doi: 10.1088/2051-672X/ac1c80

[8] S. S. Sandhu and A. S. Shahi, "Metallurgical, wear and fatigue performance of Inconel 625 weld claddings," *J. Mater. Process. Technol.*, vol. 233, pp. 1–8, 2016, doi: 10.1016/j.jmatprotec.2016.02.010

[9] M. Masanta, S. M. Shariff, and A. Roy Choudhury, "A comparative study of the tribological performances of laser clad TiB2-TiC-Al_2O_3 composite coatings on AISI 1020 and AISI 304 substrates," *Wear*, vol. 271, no. 7–8, pp. 1124–1133, 2011, doi: 10.1016/j.wear.2011.05.009

[10] G. Ramírez, E. Jiménez-Piqué, A. Mestra, M. Vilaseca, D. Casellas, and L. Llanes, "A comparative study of the contact fatigue behavior and associated damage micromechanisms of TiN- and WC:H-coated cold-work

tool steel," *Tribol. Int.*, vol. 88, pp. 263–270, 2015, doi: 10.1016/j.triboint.2015.03.036

[11]   J. Singh, L. Thakur, and S. Angra, "Effect of argon flow rate and standoff distance on the microstructure and wear behaviour of WC-CoCr TIG cladding," *J. Phys. Conf. Ser.*, vol. 1240, no. 1, 2019, doi: 10.1088/1742-6596/1240/1/012162

[12]   G. Rasool and M. M. Stack, "Wear maps for TiC composite based coatings deposited on 303 stainless steel," *Tribol. Int.*, vol. 74, pp. 93–102, 2014, doi: 10.1016/j.triboint.2014.02.002

[13]   N. Kamboj and L. Thakur, "A study of processing and high-temperature sliding wear behaviour of Inconel-625 alloy TIG weld cladding," *Int. J. Mater. Eng. Innov.*, vol. 13, no. 1, p. 1, 2022, doi: 10.1504/ijmatei.2022.10045603

[14]   A. Kumar Das, "Recent developments in TIG torch assisted coating on austenitic stainless steel: A critical review," *Mater. Today Proc.*, vol. 57, pp. 1846–1851, 2022, doi: 10.1016/j.matpr.2022.01.077

[15]   R. Ranjan and A. Kumar Das, "Protection from corrosion and wear by different weld cladding techniques: A review," *Mater. Today Proc.*, vol. 57, pp. 1687–1693, 2022, doi: 10.1016/j.matpr.2021.12.329

[16]   A. Evangeline and P. Sathiya, "Structure–property relationships of Inconel 625 cladding on AISI 316L substrate produced by hot wire (HW) TIG metal deposition technique," *Mater. Res. Express*, vol. 6, no. 10, 2019, doi: 10.1088/2053-1591/ab350f

[17]   J. Singh, L. Thakur, and S. Angra, "A study of tribological behaviour and optimization of WC-10Co-4Cr cladding," *Surf. Eng.*, vol. 37, no. 1, pp. 70–79, 2021, doi: 10.1080/02670844.2020.1745367

[18]   Z. Liu, J. Cabrero, S. Niang, and Z. Y. Al-Taha, "Improving corrosion and wear performance of HVOF-sprayed Inconel 625 and WC-Inconel 625 coatings by high power diode laser treatments," *Surf. Coatings Technol.*, vol. 201, no. 16–17, pp. 7149–7158, 2007, doi: 10.1016/j.surfcoat.2007.01.032

[19]   H. Kashani, A. Amadeh, and H. M. Ghasemi, "Room and high temperature wear behaviors of nickel and cobalt base weld overlay coatings on hot forging dies," *Wear*, vol. 262, no. 7–8, pp. 800–806, 2007, doi: 10.1016/j.wear.2006.08.028

[20]   A. Evangeline and P. Sathiya. "Cold metal arc transfer (CMT) metal deposition of Inconel 625 superalloy on 316L austenitic stainless steel: Microstructural evaluation, corrosion and wear resistance properties. *Materials Research Express*, vol. 6, no.6, 066516, 2019, doi: 10.1088/2053-1591/ab0a10

Chapter 3

Corrosion and microstructural behaviour of Inconel 625 microwave clad deposited on mild steel

Gurbhej Singh, Amit Bansal and Hitesh Vasudev

3.1 INTRODUCTION

One of the widely utilised techniques for the improvement of surface properties as well as material function without modifying the majority of the properties is known as surface engineering. The protection against corrosion is one of the research's critical aims for the ferrous alloy (stainless steel; mild steel), which is employed in the gas as well as oil industries. Various engineering applications use stainless steel as one of the common materials. Various methods can be used for ferrous alloys' surface modification, including thermal spraying, chemical vapour deposition (CVD), physical vapour deposition (PVD), heat treatment, laser cladding, microwave cladding, etc. [1,2]. Because of factors like ease of operation as well as being able to build a variety of materials on the substrate, one of the most common techniques that is widely used nowadays is thermal spraying, although limitations associated with the thermal spray technique include porosity and splats' poor mechanical bonding [3], Furthermore, when high accuracy is required for performing the cladding on small components, laser cladding is used, which entails a high cost of operation. Furthermore, in the laser cladding process, during the rapid melt pool solidification, there is a tendency for crack formation [4,5]. Further, the corrosion is induced in stainless steel, which is one of the common materials. Various methods can be used for ferrous alloy's surface modification that are thermal spraying, CVD, PVD, heat treatment, laser cladding and microwave cladding [1, 2]. Because of factors like ease of operation as well as being able to build a variety of materials on substrate, one of the most common techniques that is used widely nowadays is thermal spraying although limitations associated with thermal spray technique includes porosity and splats' poor mechanical bonding [3]. Furthermore, when high accuracy is required for performing the cladding on the small components, then laser cladding is used, which entails a high cost of operation. Furthermore, in the laser cladding process, during the rapid melt pool solidification, there exists a tendency for crack formation [4,5]. Further, corrosion in the pitting as well as crevice form was

DOI: 10.1201/9781032713830-3

easily initiated at the crack formation sites associated with laser cladding. Presently, there exists one more popular technique, namely, microwave processing of materials, as it has various advantages like volumetric heating and uniform heating, as well as improved mechanical properties because of its enhanced microstructural characteristics. Furthermore, in the materials' microwave processing, at the atomic level, heat is produced, which further results in enhanced productivity as well as lower consumption of energy. Also, during this process, the material, due to the atomic-level interaction, generated volumetric heat inside it in comparison to conventional surface heating techniques where conductive modes of heat transfer are used. Thus, the material resulted in a decreased thermal gradient inside it, which results in enhanced functional properties along with decreased residual stresses inside the material [6,7]. First, Sharma et al. [8] gave the microwave heating application as microwave cladding for enhancement of metallic material's functional properties in patent form. The microwave hybrid heating (MHH) technique's principle was used by the authors for developing the microwave clad. It was reported by the authors that enhanced mechanical as well as tribological properties are exhibited by the microwave-induced clads. Furthermore, it has been noticed that absolute diffusion bonding has been displayed by the developed clads with the substrate regardless of any interfacial cracking [8]. Subsequently, there exists various research studies that emphasise the enhancement of the metallic material's surface properties by utilising some microwave cladding that is cost-effective and has various advantages like low power consumption, low material wastage, and the ability to produce quality cladding that has improved tribological performance [9]. Inconel-625, a nickel-based superalloy, is a widely used material with an exceptional combination of mechanical qualities and corrosion resistance [10–14]. By developing an IN microwave-cladded layer, the corrosion resistance of base mild steel can be increased. The main purpose of this study is to develop microwave cladding of IN625 on mild steel by using domestic microwaves of 900 watts at a frequency of 2.45 GHz. Scanning electron microscopy, X-ray diffraction, and Vicker's microhardness were all used to characterise the developed IN625 clads. The corrosion resistance of IN62-clad samples was further investigated using the potentiodynamic polarisation method in a 3.5 wt% NaCl solution.

3.1.1 Material detail

As a cladding powder, Inconel-625, a commercially available Ni-based superalloy with exceptional hardness and corrosion resistance, was used. Figure 3.1 depicts a typical SEM micrograph of Inconel-625 powder morphology. The powder particles are mostly spherical, with a diameter distribution of 405 m. As a base material, mild steel, one of the most extensively used engineering materials, was chosen. The chemical compositions (weight

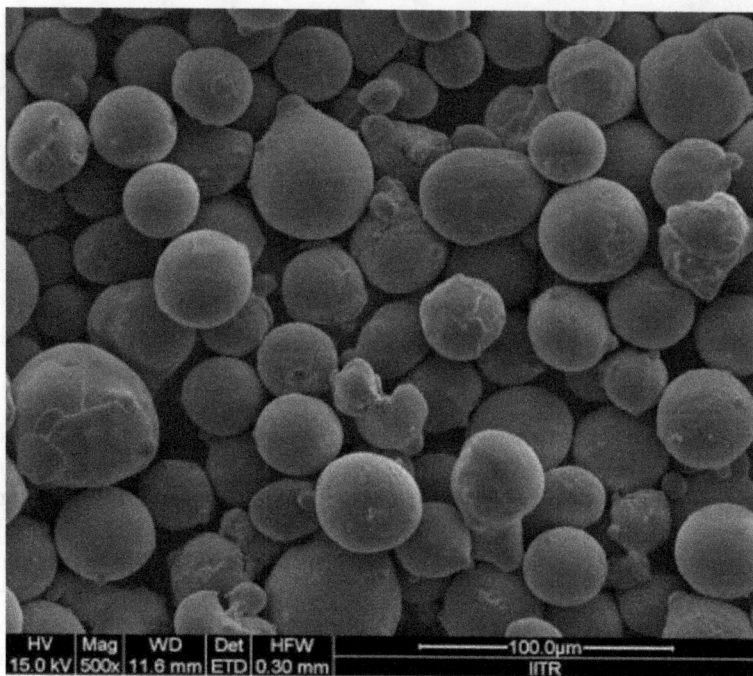

Figure 3.1 The shape of Inconel-625 powder may be seen in a typical secondary electron picture.

Table 3.1 The elemental composition (wt%) of the materials used for development of cladding [36]

Composition	Cr	Mo	Co	Nb+Ta	Fe	Si	C	Mn	Ni
Substrate (SS316)	17.0	2.0	......	.....	Bal	0.75	0.08	2.0	8.0
Inconel-625 powder	25	9	1.5	5	5.5	0.5	0.1+	0.5	Bal

percentages) of the substrate and Inconel-625 powder as determined by the optical emission spectrometer (OEM) are listed in Table 3.1.

3.2 EXPERIMENTAL PROCEDURE

Powder made of Inconel-625, a nickel-based alloy, was baked onto a substrate made of mild steel in a domestic microwave oven with power settings of 900 W and frequency settings of 2.45 GHz. There is more information available that provides a more in-depth explanation of the microwave cladding technique [15]. On top of the substrate, an even layer of Inconel-625 powder with an average thickness of 1 mm was applied. After that, an alumina plate with a thickness of 0.5 mm was put on top of the powder that had been previously deposited. After that, the susceptor powder, which was charcoal, was placed on top of the alumina plate. The whole assembly was maintained inside of the microwave chamber and subjected to electromagnetic microwave radiation. Several iterations of trial runs were carried out before settling on the optimal process parameters (power: 900 W; time = 10 min). In order to properly deposit the cladding, the fundamentals of microwave hybrid heating (MHH) were utilised. The susceptor (charcoal), which has a high loss tangent value and quickly couples with microwaves when they are at room temperature, transmits heat to the clad powder via the alumina plate using the conduction method of heat transmission. When the temperature goes over the threshold, the powder particles begin to absorb microwave radiation, and this causes them to melt at higher temperatures. Due to the limited skin depth of the base metal, however, the melting of the metal could only occur at a shallow depth. Following the conclusion of the experiment, the molten pool will be used to create the clad layer that will cover the mild steel base. Figure 3.2 displays a schematic diagram that illustrates the experimental setup that was utilised for the creation of cladding using a microwave approach that was efficient and cost-effective. Moreover, a diagrammatic representation of the microwave-induced clad specimens may be shown in Figure 3.3, as depicted. In addition, the power absorbed during microwave exposure due to the internal electric field

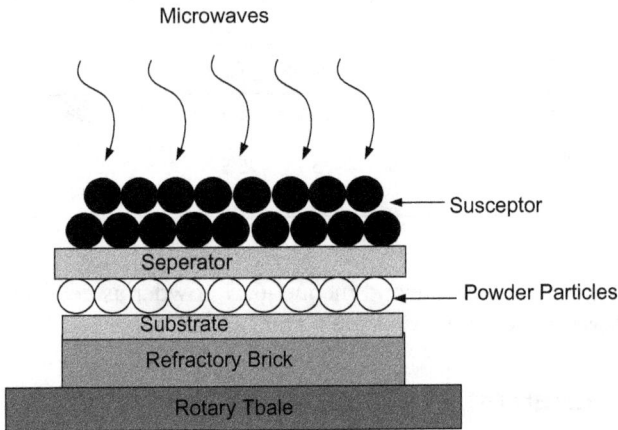

Figure 3.2 Schematic view of the arrangement adopted to modify the surface layer of MS with IN625.

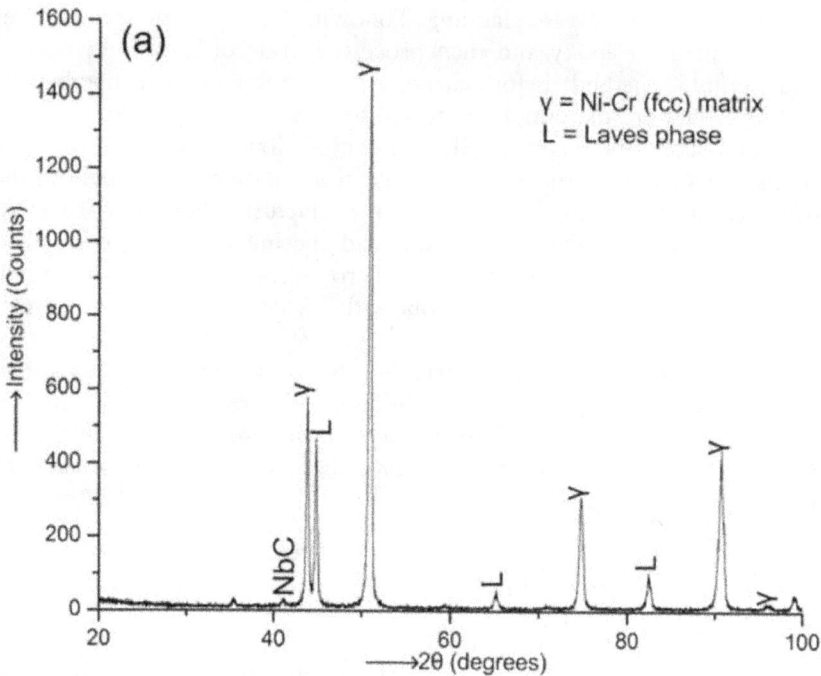

Figure 3.3 XRD pattern of microwave-induced cladding of the IN625 [15].

component may be computed by applying the following formula. This is possible since the power is inversely proportional to the square of the electric field.

$$P = 2\pi f \, \varepsilon_0 \varepsilon''_{eff} |E|^2 \tag{3.1}$$

where $\varepsilon_0 = 8.86\times10^{-12}$ F/m, permittivity of free space
ε''_{eff} = relative effective dielectric loss vector
f = frequency of microwaves (Hz)
E = electric field (V/m)

3.3 MICROSTRUCTURAL AND MECHANICAL CHARACTERISATION

By using Cu-k radiation in a linked -two arrangement and utilising XRD, it was possible to investigate the different phases that had developed in the clad sample. In order to conduct microstructural investigation, the clad samples were first cut using a slow-speed diamond cutter along the direction of the transverse cladding. Following this step, the cut samples were mounted on epoxy and then processed in accordance with normal metallographic method. Before carrying out a cross-sectional microstructural investigation, the samples that had been polished were ultrasonically treated in alcohol. We used the SEM in conjunction in order to investigate the microstructural features (phases) that were present all throughout the cross-section of the samples that were manufactured. For the purpose of determining the microhardness of the clad specimens along the depth in the transverse cross-sectional view, the micro indentation method was used. The micro-indentation was carried out with a Vickers indenter at a normal load of 50 g for 10 s as a function of the distance from the top surface towards the cladding. The microstructural features in other processes such as thermal spray are different from cladding and here the less porosity can also be obtained [16–25]. Moreover, the microhardness is even higher side and the control over the parameters is bit difficult in cladding as compare to other coating techniques [26–38].

3.4 RESULTS AND DISCUSSION

3.4.1 XRD study

Figure 3.3 depicts an XRD spectrum of a microwave-induced cladding of the IN625 on mild steel. From Figure 3.3, it has been deduced that there is a presence of secondary phases such as Laves and carbide in addition to the main fcc solid solution matrix. The presence of these secondary phases was attributed to the various metallurgical changes that occurred in the modifying clad layer at elevated temperatures. During hybrid MH, there is a complete melting of IN625 powder along with melting of the top surface layer of SS-304 was occurred simultaneously. The highly reactive elements like niobium (Nb) in the molten pool has strong affinity towards carbon (C) at elevated temperatures, which led to the creation of a various carbide phases (NbC) in the modify clad layer.

3.4.2 Microstructural analysis

The Inconel-625 clad was effectively deposited on a mild steel (MS) substrate using a unique MHH process, as shown in the current study. Discussion of the findings follows. The following sections will explain the findings that were obtained after completing a variety of characterisation tests on the Inconel 625 clad that had been deposited. A typical SEM image (Figure 3.4a) shows how electromagnetic microwave radiation caused the clad layer to form on the mild steel base. On the MS base, the covering with a thickness of about 1 mm was seen to be deposited. Figure 3.4a shows that the perfect

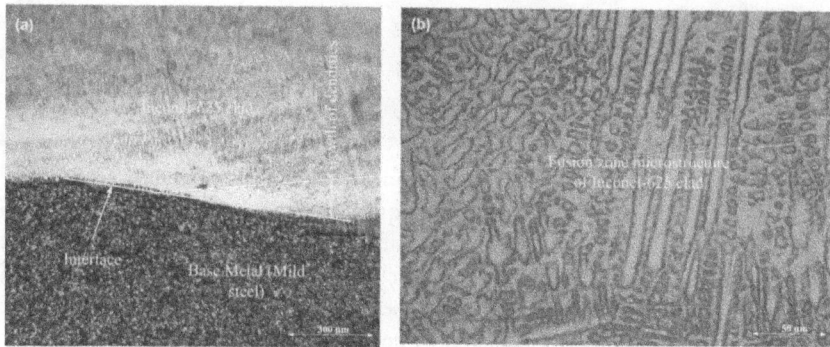

Figure 3.4 The formed Inconel-625 clad on the MS (a) and the fusion zone (b) at higher magnifications on a SEM image [15].

diffusion bonding between the clad powder and the base was made possible by heating the powder particles all the way down. The metallic connection between the substrate and the deposited clads occurred when elements moved back and forth across the surface between the deposited clad and the substrate. It was seen that the cladding was well attached and had no flaws or cracks at the edges. Because microwave hybrid heating causes uniform cooling, the microstructure that formed in the clad areas (shown in Figure 3.4b) was uniform and thick, with very few holes and pores. Inconel-625 is a highly alloyed material and it usually forms in a branching way. As shown in Figure 3.3a, the columnar dendritic grain grew epitaxially from the base in the opposite direction of the heat flow. The base works as a heat sink in microwave coating. So, when clad powder solidifies, most of the melting pool cools down through the base. Thus the dendrites grain growth perpendicular to the substrate was observed. Because of "uniform cooling," which is part of the microwave hybrid heating (MHH) process, the dendrites seem to be spread out evenly in the covered parts of the substructure.

3.4.3 Porosity

In the modify clad layer, the porosity evaluation is crucial since the presence of porosity directly affects the wear performance of the clad specimen. The porosity in the modify layer was determined by using Dewinter material plus software (version 5.1). The ten images were taken in the modify layer at different regions at 500× using optical microscope for the calculation of porosity. The measured value is the average porosity reported for ten images in the current work. The measured average porosity in the modify layer was found to be less than 2%. The volumetric and uniform heating attributed related to hybrid MH are responsible for considerable low level of porosity in the modify clad layer.

3.4.4 Vickers microhardness

The procedure explained in Section 3.2 was followed to evaluate the microhardness across the cross-section of the modify clad layer. The hardness measurements were taken at equal distance (100 μm) from the top of the modified layer towards the base metal (Figure 3.5). The average value microhardness of the modify clad layer was measured to be 410±45 HV, which is significantly higher than the base metal hardness (240±45 HV). The increased hardness of the clad layer is beneficial in terms of increased wear resistance ability of the modify clad layer because the increased hardness of the material generally leads to a proportionate increase in its wear resistance ability although the effect of hardness is not straightforward.

3.5 CORROSION PERFORMANCE

Corrosion is evaluated by the Tafel extrapolation technique, including the corrosion current (Icorr). Table 3.2 shows how Tafel plots (Figure 3.6) were used to figure out the different corrosion factors (E corrosion potential and i corrosion current density). The fact that the E value of the clad specimen is

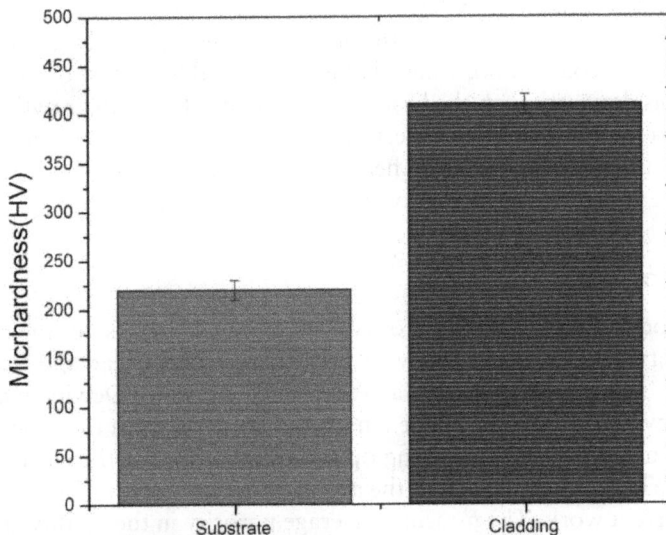

Figure 3.5 Microhardness value of Substrate and IN 625 Cladded samples.

Table 3.2 Corrosion kinetic parameters that were really observed, as estimated potential (Ecorr) corrosion rate (CR), anodic Tafel slope (a), and cathodic Tafel slope (c)

Samples	Icorr (μA)	Ecorr	CR (mpy)	βa (V/decade)	βc (V/decade)
Substrate (MS)	2.4	-730	2.3	276.0e-3	279.3e-3
Inconel-625 cladding	1.5	-330	1.50	475.7e-3	190.6e-3

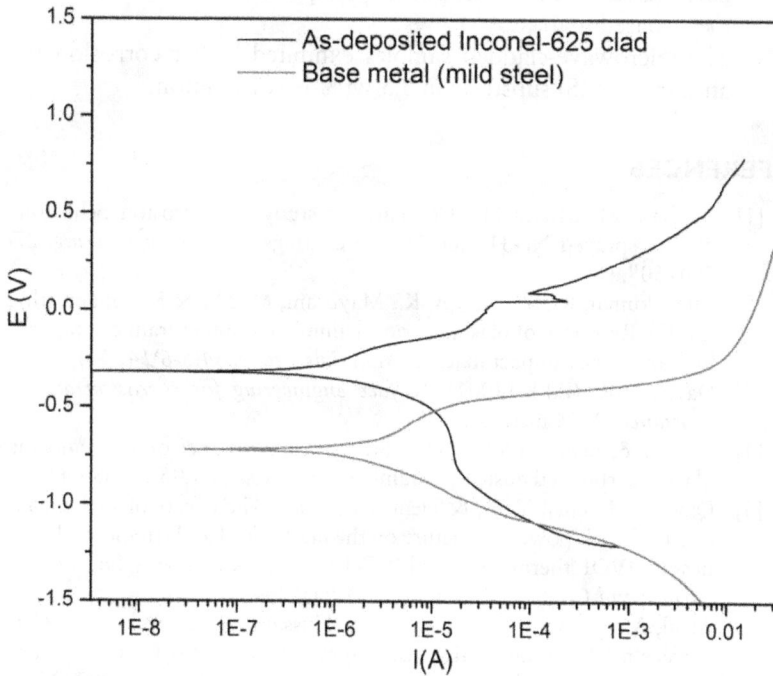

Figure 3.6 Inconel-625 cladding and substrate (MS) Tafel plots [15].

much higher than that of the base metal (BM) shows that the clad specimen is more resistant to rust. Also, the corrosion current density i for BM is very high, which shows that BM corrodes a lot in the 3.5wt% NaCl solution.

3.6 CONCLUSIONS

1. The IN625 cladding on the metallic substrates (SS-304) was carried out in a multimode domestic microwave at 2.45 GHz at 900 W by using the hybrid MH approach.

2. The fabricated clad layer exhibited perfect diffusion bonding with the mild steel substrate through mutual diffusion of elements. The clad microstructure displayed reinforcements of hard carbide and Laves phases in a tougher metal-based matrix.

3. The fabricated clad layer was free from any interfacial gaps and cracking. The fabricated clad layer exhibited significantly less porosity, which is less than 2%.

4. The fabricated clad layer had considerably higher hardness than the substrate hardness. The improved microhardness was due to various hard (Laves and carbide) phases precipitated at the GB region of the metal-based matrix.

5. The microwave-cladded samples exhibited higher corrosion resistance from (MS) substrate in 3.5 wt% NaCl solution.

REFERENCES

[1] Verdian, M. M. (2011). Comparative study on corrosion behaviour of plasma sprayed Ni–3Ti and Ni–3Al coatings. *Surface engineering*, 27(7), 504–508.

[2] Vijayakumar, K., Sharma, A. K., Mayuram, M. M., & Krishnamurthy, R. (2002). Response of plasma-sprayed alumina–titania ceramic composite to high-frequency impact loading. *Materials letters*, 54(5-6), 403–413.

[3] Davis, J. R. (Ed.). (2001). *Surface engineering for corrosion and wear resistance*. ASM international.

[4] Sun, Y., & Bell, T. (2002). Dry sliding wear resistance of low temperature plasma carburised austenitic stainless steel. *Wear*, 253(5-6), 689–693.

[5] Qiao, Y., Fischer, T. E., & Dent, A. (2003). The effects of fuel chemistry and feedstock powder structure on the mechanical and tribological properties of HVOF thermal-sprayed WC–Co coatings with very fine structures. *Surface and Coatings Technology*, 172(1), 24–41.

[6] Afzal, M., Ajmal, M., Khan, A. N., Hussain, A., & Akhter, R. (2014). Surface modification of air plasma spraying WC–12% Co cermet coating by laser melting technique. *Optics & Laser Technology*, 56, 202–206.

[7] Zhou, S., Huang, Y., Zeng, X., & Hu, Q. (2008). Microstructure characteristics of Ni-based WC composite coatings by laser induction hybrid rapid cladding. *Materials Science and Engineering: A*, 480(1–2), 564–572.

[8] Zhou, S., Zeng, X., Hu, Q., & Huang, Y. (2008). Analysis of crack behavior for Ni-based WC composite coatings by laser cladding and crack-free realization. *Applied Surface Science*, 255(5), 1646–1653.

[9] Vasudev, H., Singh, G., Bansal, A., Vardhan, S., & Thakur, L. (2019). Microwave heating and its applications in surface engineering: a review. *Materials Research Express*, 6(10), 102001.

[10] Arora, H. S., Grewal, H. S., Veligatla, M., Singh, H., & Mukherjee, S. (2014). Microwave assisted in situ composite coating using metallic glass precursor. *Surface engineering*, 30(11), 779–783.

[11] A. K. Sharma and D. Gupta: 'A method of cladding/coating of metallic and nonmetallic powders on metallic substrates by microwave irradiation', Indian Patent application No. 527/Del/2010.

[12] Bansal, A., Zafar, S., & Sharma, A. K. (2015). Microstructure and abrasive wear performance of Ni-WC composite microwave clad. *Journal of Materials Engineering and Performance*, 24, 3708–3716.

[13] Hebbale, A. M. (2015). Microstructural characterization of Ni based cladding on SS-304 developed through microwave energy. *Materials Today: Proceedings*, 2(4–5), 1414–1420.

[14] Singh, G., Vasudev, H., & Arora, H. (2020). A short note on the processing of materials through microwave route. In Advances in Materials Processing: Select Proceedings of ICFMMP 2019 (pp. 101-111). Springer Singapore.

[15] Singh, G., Vasudev, H., Bansal, A., & Vardhan, S. (2020). Microwave cladding of Inconel-625 on mild steel substrate for corrosion protection. *Materials Research Express*, 7(2), 026512.

[16] Vasudev, H., Thakur, L., Singh, H., & Bansal, A. (2022). Effect of addition of Al_2O_3 on the high-temperature solid particle erosion behaviour of HVOF sprayed Inconel-718 coatings. *Materials Today Communications*, 30, 103017.

[17] Mehta, A., Vasudev, H., & Singh, S. (2020). Recent developments in the designing of deposition of thermal barrier coatings–A review. *Materials Today: Proceedings*, 26, 1336–1342.

[18] Prashar, G., Vasudev, H., & Technology, C. (2022). Structure-property correlation and high-temperature erosion performance of Inconel625-Al_2O_3 plasma-sprayed bimodal composite coatings. Surface and Coatings Technology, 439, 128450.

[19] Prashar, G., Vasudev, H., & Thakur, L. (2021). Influence of heat treatment on surface properties of HVOF deposited WC and Ni-based powder coatings: a review. *Surface Topography: Metrology and Properties* 9(4), 043002.

[20] Singh, P., Bansal, A., Vasudev, H., & Singh, P.. (2021). In situ surface modification of stainless steel with hydroxyapatite using microwave heating. 9(3), 035053.

[21] Vasudev, H., Prashar, G., Thakur, L., & Bansal, A. (2022). electrochemical corrosion behavior and microstructural characterization of HVOF sprayed inconel718-Al_2O_3 composite coatings. *Surface Review and Letters 29(2)*, 2250017.

[22] Prashar, G., & Vasudev, H. (2021). High temperature erosion behavior of plasma sprayed Al_2O_3 coating on AISI-304 stainless steel.

[23] Vasudev, H., Singh, P., Thakur, L., & Bansal, A. (2020). Mechanical and microstructural characterization of microwave post processed Alloy-718 coating. *Materials Research Express 6(12), 1265f1265*.

[24] Vasudev, H., Thakur, L., Singh, H., & Bansal, A. (2018). Mechanical and microstructural behaviour of wear resistant coatings on cast iron lathe machine beds and slides. Kovove Materialy 56(1), 55–63.

[25] Singh, G., Vasudev, H., Bansal, A., & Vardhan, S. (2021). Influence of heat treatment on the microstructure and corrosion properties of the Inconel-625

clad deposited by microwave heating. *Surface Topography: Metrology and Properties*, 9(2), 025019.

[26] Bansal, A., Vasudev, H., Sharma, A. K., & Kumar, P (2019). Investigation on the effect of post weld heat treatment on microwave joining of the Alloy-718 weldment. *Materials Research Express* 6(8), 086554.

[27] Singh, M., Vasudev, H., & Kumar, R (2020). Microstructural characterization of BN thin films using RF magnetron sputtering method. *Materials Today: Proceedings*, 26, 2277–2282.

[28] Vasudev, H., Prashar, G., Thakur, L., Bansal, A., & Prevention. (2021). Electrochemical corrosion behavior and microstructural characterization of HVOF sprayed Inconel-718 coating on gray cast iron. *Journal of Failure Analysis and Prevention* 21, 250–260.

[29] Singh, J., Vasudev, H., & Singh, S. (2020). Performance of different coating materials against high temperature oxidation in boiler tubes–A review. *Materials Today: Proceedings* 26, 972–978.

[30] Prashar, G., Vasudev, H., & Thakur, L. (2022). High-temperature oxidation and erosion resistance of Ni-based thermally-sprayed coatings used in power generation machinery: *A review. Surface Review and Letters*, 29(3), 2230003.

[31] Singh, M., Vasudev, H., & Kumar, R. (2021). Corrosion and tribological behaviour of bn thin films deposited using magnetron sputtering. *International Journal of Surface Engineering and Interdisciplinary Materials Science (IJSEIMS)*, 9(2), 24–39.

[32] Prashar, G., & Vasudev, H. (2022). Surface topology analysis of plasma sprayed Inconel625-Al$_2$O$_3$ composite coating. *Materials Today: Proceedings*, 50, 607–611.

[33] Singh, M., Vasudev, H., & Singh, M. (2022). Surface protection of SS-316L with boron nitride based thin films using radio frequency magnetron sputtering technique. *Journal of Electrochemical Science and Engineering*, 12(5), 851–863.

[34] Prashar, G., Vasudev, H., & Bhuddhi, D. (2022). Additive manufacturing: expanding 3D printing horizon in industry 4.0. 1–15.

[35] Prashar, G., & Vasudev, H. (2022). Structure–property correlation of plasma-sprayed Inconel625-Al$_2$O$_3$ bimodal composite coatings for high-temperature oxidation protection. *Material Research Express* 31(8), 2385–2408.

[36] Singh, G., Vasudev, H., Bansal, A., & Vardhan, S. (2020). Microwave cladding of Inconel-625 on mild steel substrate for corrosion protection. *Materials Research Express*, 7(2), 026512.

[37] Singh, S., Sajwan, M., Singh, G., Dixit, A. K., & Mehta, A. (2023). Efficient surface detection for assisting Collaborative Robots. *Robotics and Autonomous Systems*, 161, 104339.

[38] Singh, G., Singh, H., Sharma, Y., Vasudev, H., & Prakash, C. (2023). Analysis and optimization of various process parameters and effect on the hardness of SS-304 stainless steel welded joints. *International Journal on Interactive Design and Manufacturing (IJIDeM)*, 1–8.

Chapter 4

Artificial intelligence revolutionizing the laser cladding industry

Gaurav Prashar and Hitesh Vasudev

4.1 INTRODUCTION

Laser cladding (LC) is a powerful additive manufacturing technique that uses a laser beam to deposit material onto a substrate to create a new surface layer. This process has many benefits, including its precision and efficiency, making it a popular choice in industries like aerospace, automotives, and medicine. The LC process involves melting a powdered material, such as metal or ceramic, using a laser beam. The molten material is then deposited onto a substrate, creating a new surface layer [1–4]. The laser beam can be precisely controlled, allowing for the creation of complex geometries and the use of multiple materials. Figure 4.1 shows the commonly used types of laser cladding techniques (coaxial and preplaced powder). One of the key benefits of LC is its efficiency [5]. The process can be carried out quickly and with minimal waste, making it a cost-effective solution for many manufacturing applications. Additionally, the process can be carried out with a high degree of precision, ensuring that the resulting surface layer meets exact specifications. Another advantage of LC is its ability to improve the properties of the underlying substrate. By depositing a new surface layer with different material properties, laser cladding can improve wear resistance, corrosion resistance, and other key characteristics. This can extend the life of components and reduce maintenance costs. Laser cladding can also be used for repair and remanufacturing applications. By depositing a new surface layer onto worn or damaged components, laser cladding can restore them to their original dimensions and properties. This can be a cost-effective alternative to replacing the entire component.

There are several different types of laser cladding techniques, including coaxial, lateral, and directed energy deposition as discussed in the introduction of this chapter. Coaxial laser cladding involves the laser beam and material feed tube being coaxially aligned, while lateral laser cladding involves the laser beam and material feed being directed at the substrate from different angles. Directed energy deposition involves depositing material onto a moving substrate to create a 3D structure. Overall, laser

DOI: 10.1201/9781032713830-4

63

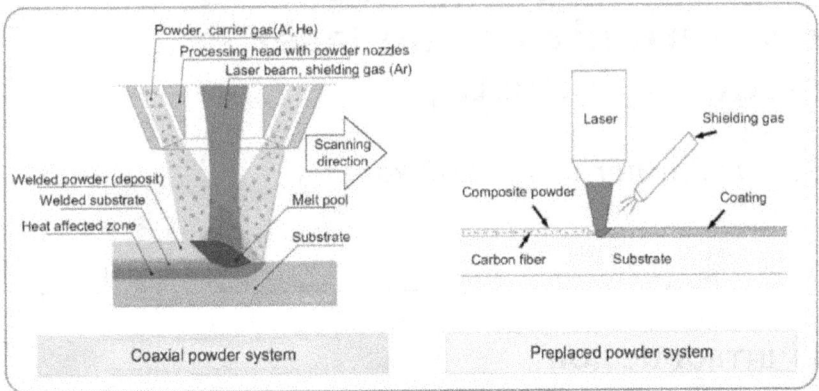

Figure 4.1 Two commonly used techniques used in laser cladding process [5].

cladding is a versatile and efficient additive manufacturing technique that has many benefits for a wide range of industries. Its precision, efficiency, and ability to improve material properties make it an attractive choice for applications where high-quality surfaces are required [5].

But in Industry 4.0, artificial intelligence (AI) is quickly becoming an integral part of the manufacturing industry, and laser cladding is no exception. AI can be used to automate and optimize this process. Firstly, AI can be used to automate the cladding process by controlling the laser parameters. By implementing AI-based feedback control systems, the laser parameters can be adjusted to ensure that the process is operating at a safe and efficient level. AI can also be used to detect welding defects in real time and prevent them from occurring. This can help to reduce the risk of weld failure and optimize the process. Secondly, AI can be used to optimize the LC process. AI-based algorithms can be used to accurately predict the best laser parameters for a given application and determine the most optimal cladding patterns [5]. Additionally, AI can be used to identify the most suitable materials for a given application and to tailor the laser parameters to the specific characteristics of the materials. AI-based algorithms can be used to recognize and classify defects in real time and to identify potential causes.

4.2 ARTIFICIAL INTELLIGENCE

AI is the science of getting machines to act and think like humans. It is a broad field that encompasses many sub-fields, including machine learning, natural language processing, robotics, and more. The commonalities between different branches of AI and their scope are shown schematically in Figure 4.2. At its core, AI is about enabling computers to make decisions that

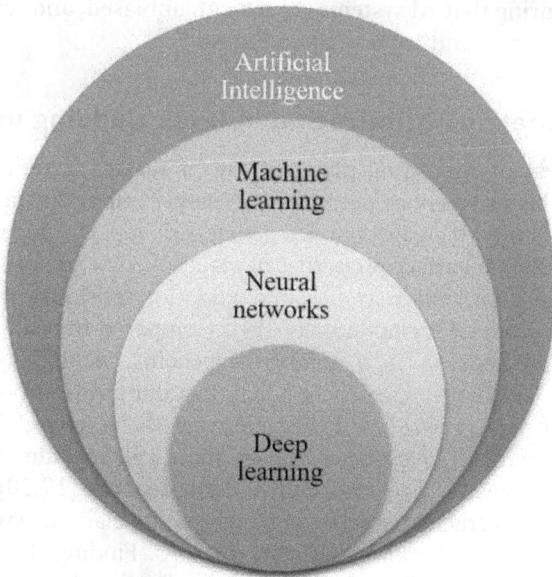

Figure 4.2 Illustration demonstrating the general scope of AI.

are similar to those made by humans. To do this, computers must be able to learn from data, recognize patterns, and make decisions that are based on their understanding of the data they've been given. This is done through the use of algorithms, which are sets of instructions that tell the computer how to process and act on the data [6,7]. The most common types of AI such as supervised learning involves training the AI to recognize patterns in data by providing it with labeled examples. Unsupervised learning uses unlabeled data to discover patterns and uncover hidden relationships. The goal of AI is to create machines that can think and behave like humans [8]. This requires machines to be able to reason and make decisions, much like humans do. AI has the ability to revolutionize the LC industry.

In addition to machine learning, AI systems can also be designed to use a technique called deep learning (DL). It is a subset of ML that uses artificial neural networks (ANN), which are modeled after the structure of the human brain. DL is used in applications like recognition of speech and image, image, and also natural language processing [9,10].

One of the challenges of AI is ensuring that the algorithms used are ethical and unbiased. AI systems can be influenced by the data they are trained on, which can lead to biases and discrimination. To mitigate this, AI developers must ensure that their systems are designed to be fair and impartial. In conclusion, AI is based on the principle of machine learning, which enables machines to learn from data and make decisions based on that data [11]. However, ensuring that AI systems are ethical, unbiased, and secure is essential for their success and widespread adoption.

4.2.1 Parametric optimization in laser cladding using AI

LC is a process of depositing material onto a surface using a high-energy laser beam. This process has become increasingly important in Industry 4.0 because of its ability to create complex geometries, repair damaged parts, and enhance the different surface properties of materials. The substrate's hardness, wear resistance, corrosion resistance, and oxidation resistance are intended to be improved by the laser cladding [12–14]. Compared to traditional surface modification methods, it has a number of benefits, including smaller heat-affected zones, less dilution, better bonding (metallurgical) with the substrate, and fine grain sizes [15–18]. But the quality of the resulting cladding layer depends on several LC parameters (more than 19), including laser power, scanning speed, powder flow rate, and standoff distance [19,20]. Optimizing these parameters is crucial for achieving high-quality cladding layers and minimizing defects like holes and cracks in coatings. Finding the ideal process parameters can be accomplished in part by investigating the impact of process parameters on the characteristics of the cladding layer using the response surface technique, regression methodology, and Taguchi method [21]. However, manual optimization of these parameters is a time-consuming and expensive process. AI has emerged as a promising tool for automating the optimization of LC parameters. Parametric optimization in LC using AI involves using ML algorithms to identify the optimal laser cladding parameters for a given material and application [22,23]. As shown in Figure 4.3, the technique specifically involves the sequence for the AI cladding [6–8].

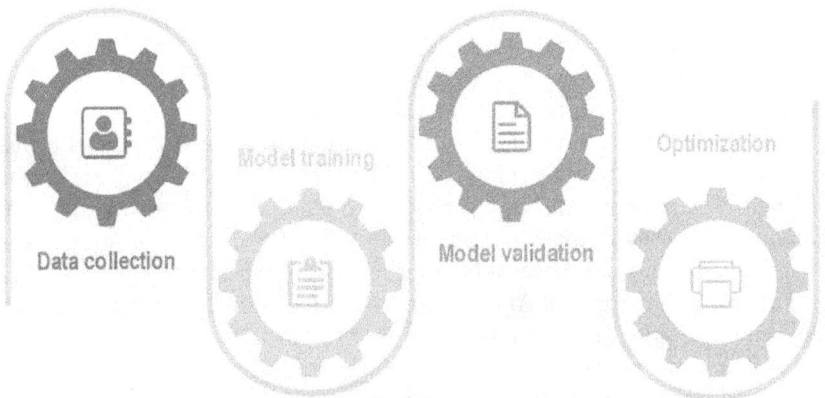

Figure 4.3 The steps followed for the AI for laser cladding [6–8].

- Data collection: The first step in AI-based optimization is to collect data on the performance of different laser cladding parameters. This data can be obtained through experimental trials or simulations. The data should include information on the quality of the cladding layer, such as porosity, hardness, and adhesion strength, as well as the process parameters used to achieve the layer.
- Model training: Once the data is collected, an ML model can be trained to predict the performance of different LC parameters. There are several ML algorithms that can be used for this purpose, including neural networks, decision trees, and support vector machines. The model is trained using the collected data, with the LC parameters as the input and the quality of the cladding layer as the output.
- Model validation: After the ML model is trained, it is important to validate its performance using additional data that was not used in the training process. This step ensures that the model can accurately predict the performance of the LC parameters for new data.
- Optimization: Once the ML model is validated, it can be used to optimize the LC parameters for a given material and application. This involves using the model to predict the performance of different LC parameters and selecting the parameters that result in the best quality cladding layer.

There are several benefits to using AI-based optimization for laser cladding parameters. First, it can significantly reduce the time and cost associated with manual optimization. Second, it can help identify optimal parameters that may not be easily identifiable through trial and error. Finally, it can improve the overall quality and consistency of the resulting cladding layers.

However, there are also some limitations to using AI-based optimization for LC parameters. One limitation is that the model is only as good as the data used to train it. If the data is limited or biased, the model may not be able to accurately predict the performance of the LC parameters. Another limitation is that the model may not be able to capture all of the complex interactions that occur during the LC process. Therefore, it is important to use AI-based optimization as a tool for guiding the optimization process rather than relying on it as the sole decision-making tool.

In conclusion, parametric optimization in LC using AI has the potential to significantly improve the quality and efficiency of the LC process. By automating the optimization process, manufacturers can achieve high-quality cladding layers while reducing the time and cost associated with manual optimization [5]. However, it is important to use AI-based optimization as a tool for guiding the optimization process rather than relying on it as the sole decision-making tool.

4.2.2 Artificial neural network model in laser cladding

ANN models have been used in various fields of engineering, including laser cladding. These models can be used to predict the outcome of LC processes, optimizing different LC parameters discussed previously. As illustrated in Figure 4.4, ANNs can be used in a variety of ways to improve the performance of laser cladding operations.

The ANN model is trained on a dataset of input and output pairs where the input is the set of process parameters and the output is the resulting characteristics of the coating, such as microstructure, hardness, and wear resistance. The model learns to recognize patterns in the input–output data and can then make predictions on new data that it has not seen before.

The accuracy of an ANN model depends on the quality and quantity of the training data, as well as the architecture and parameters of the network. The input data should be carefully selected to capture the relevant features of the process, while the output data should be accurately measured and representative of the desired coating properties. ANN models can be used to optimize the LC process, reducing the number of experimental trials required to achieve the desired coating properties. They can also be used for process monitoring and control, detecting and correcting deviations from

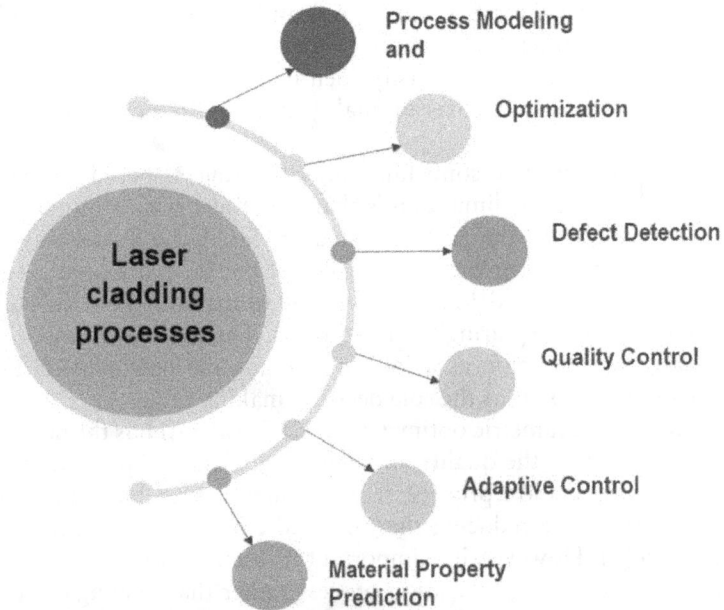

Figure 4.4 Various ways to improve the performance of laser cladding processes.

the desired process parameters in real time. From an experimental comparison, Song et al. [24] discovered that the ANN model is superior than the response surface model for explaining the nonlinear relationship among LC process parameters and geometry of cladding. Li et al. [25] discovered that under the ideal LC process parameters, the coating final microstructure comprised of a basic BCC solid solution phase, and no significant segregation of internal elements took place. In order to investigate the correlation between the LC process parameters and the cladding layer size, Caiazzo et al. [26] used an ANN. The findings demonstrate that this technique is capable of obtaining the necessary process parameters for particular cladding layer sizes. Overall, ANN models offer a powerful tool for improving the efficiency and quality of laser cladding processes, but they take little longer for learning and training.

4.2.3 Support vector machines (SVMs)

SVMs are a type of ML algorithm that can be used in LC to predict and control the quality of the final product. SVMs are particularly useful in LC because they can handle complex datasets and nonlinear relationships between the input variables and the output variables. In laser cladding, SVMs can be used to predict the optimal parameters for the process based on the desired quality of the final product. For example, SVMs can be used to predict the optimal laser power, laser speed, and material composition for a given substrate. SVMs work by identifying a hyperplane that separates the data into different classes or categories. The hyperplane is chosen so that the margin between the classes is maximized. SVM modeling was utilized by Chen et al. [27] to investigate how process parameters affected coating characteristics. According to the findings, the three most important process variables are the laser spot diameter, preset powder thickness, and power. The results presented by Zhang et al. [28] demonstrate that the optimal process parameters determined by this method can result in coatings with greater performance when compared to other methodologies, such as the response surface methodology. Overall, SVMs can be a powerful tool for controlling and optimizing the LC process, leading to improved product quality and reduced manufacturing costs. However, they are only effective when working with little amounts of data, and when dealing with several data sets, they take a long time to address the issue and produce subpar results.

4.3 NOVEL NON-DOMINATED SORTING GENETIC ALGORITHM II (NSGA-II)

Novel non-dominated sorting genetic algorithm II (NSGA-II) is a powerful optimization algorithm used in a wide range of applications, including laser cladding.Figure 4.5 shows that output of the computational fluid

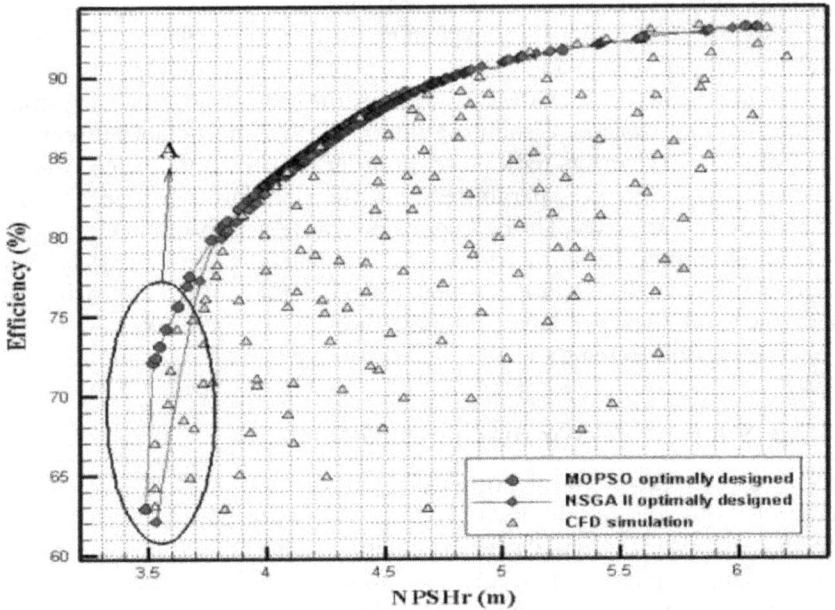

Figure 4.5 Data from the CFD simulation are superimposed on a graph showing the MOPSO and NSGA II's optimal Pareto front [29].

dynamics(CFD) simulation using NSGA II. This figure, which shows how well this article obtained the Pareto front, shows that the MOPSO Pareto front lies in the best feasible combination of the objective values of CFD data. The NSGA-II algorithm is particularly well-suited to LC optimization problems because it is capable of efficiently finding the Pareto-optimal solutions for optimization problems. In laser cladding, the objective is to optimize multiple parameters simultaneously to achieve a specific set of properties in the final product [29].

NSGA-II works by dividing the search space into several non-dominated fronts, where each front represents a set of solutions that are not dominated by any other solution in the set. The algorithm then assigns a fitness value to each solution based on its rank and crowding distance, where the rank indicates the front to which the solution belongs, and the crowding distance ensures diversity among solutions in the same front. In laser cladding, NSGA-II can be used to optimize the process parameters to achieve a set of properties, such as surface roughness, porosity, and hardness, while minimizing production costs and maximizing efficiency. The algorithm can handle multiple objectives simultaneously and can generate a set of optimal solutions that represent trade-offs between the objectives. LC process parameters are optimized using a hybrid algorithm (TS-GEP) and NSGA-II. They increased the rate of energy and material utilization in the LC process by employing this technique. Using the similar NSGA-II algorithm, two other researchers Jiang [30] and Lin et al. [31] optimized the laser cladding process parameters. The outcomes demonstrate the effectiveness of this technology in lowering energy loss during laser cladding while also enhancing the microhardness of the developed cladding layer. According to Zhao et al.'s [32] findings, using the optimized parameters obtained through NSGA-II algorithm could decrease the depth of the component heat-affected zone and increase cladding efficiency. Overall, the NSGA-II algorithm is a powerful tool for optimizing laser cladding parameters, and it can help manufacturers produce high-quality components with specific properties while minimizing costs and maximizing efficiency.

4.3.1 Particle swarm optimization model in laser cladding (PSO)

PSO is a metaheuristic optimization algorithm that is commonly used in engineering and optimization problems. In the context of laser cladding, PSO can be used to optimize the LC process parameters for achieving desirable material properties and minimizing defects. The process parameters can affect the quality of the coating and the properties of the resulting material. However, finding the optimal set of LC process parameters can be a challenging task, as it involves a large search space and multiple objectives, such as minimizing porosity, maximizing hardness, and minimizing residual stress.

PSO can help to address these challenges by searching for the optimal LC process parameters through a swarm of particles, which move in the search space and update their positions and velocities based on their own experience and the best experience of the swarm. In laser cladding, each particle represents a set of process parameters, and the fitness function evaluates the quality of the resulting coating based on the desired objectives. The PSO algorithm can be implemented in different ways, depending on the specific objectives and constraints of the problem. For example, the PSO algorithm can be modified to handle multi-objective optimization problems, where multiple conflicting objectives need to be optimized simultaneously. Additionally, PSO can be combined with other optimization techniques, like genetic algorithms or simulated annealing, to enhance the performance and robustness of the optimization process. The PSO algorithm is a cost-effective and efficient optimization technique compared to traditional optimization methods such as genetic algorithms. Due to their quick convergence, straightforward operation, and minimal parameter requirements, they are frequently utilized in multi-objective optimization problems [33]. In order to optimize the ANN, Pant et al. [34] created a prediction model taking into account the process parameters as well as the height and width of the clad. The outcomes demonstrated that the PSO-ANN model's predicted results were superior to those of the ANN model in terms of accuracy. A multiobjective quantum PSO method was utilized by Ma et al. [35] to identify the minimal dilution rate and residual stress. They discovered that the scanning speed had the biggest influence on the residual stress, while the defocus amount had the best impact on the dilution rate.

4.3.2 Random forest model

A random forest model is a type of ML algorithm that is used to classify and predict outcomes from data. It is based on the theory of decision trees, which are the building blocks of the model. The idea behind a random forest model is to take a large number of randomly selected decision trees and combine them to create a more accurate prediction. This is done by taking a subset of the data and using it to build a decision tree. Then, a second tree is created using a different subset of the data and the two trees are combined to create a more accurate prediction. This process is repeated until all of the trees in the random forest model are combined. Random forest models are used in LC to predict the properties of the material after the cladding process. The model takes into account factors such as the laser parameters, material properties, and the surface topology. The model can then predict the quality of the cladded part and can be used to help optimize the process. The LC process parameters, as well as a single-pass cladding layer size, were modeled using a random forest approach by Liang et al. [36].The findings demonstrate that this approach is capable of providing an accurate estimation of the LC process parameters needed to handle the cross-sectional geometry of a particular single-pass cladding layer.

4.3.3 Adaptive neuro-fuzzy inference model

The adaptive neuro-fuzzy inference model (ANFIS) is a powerful and popular data-driven modeling technique for predicting the output of LC. It is a hybrid system that combines the advantages of both ANNs and fuzzy logic systems. ANFIS is capable of improving its performance with the help of an adaptive learning algorithm, which enables it to learn from input–output data and modify its structure accordingly. ANFIS has been successfully used in several laser cladding applications, including process optimization, defect identification, and process monitoring. In addition, it has been used to identify defects in LC parts and to monitor the process in real time. ANFIS is a promising tool for predicting the output of LC and can be used to improve the performance and efficiency of the process [37].

4.4 CONCLUSION AND FUTURE SCOPE

In conclusion, laser cladding and AI are two emerging technologies that have the potential to revolutionize various industries. AI involves the use of algorithms and computer programs to simulate human intelligence and perform tasks such as learning, reasoning, and decision-making. When combined, laser cladding and AI can enhance the performance of LC processes by optimizing parameters such as laser power, scanning speed, and powder feed rate. AI can also be used to analyze the data generated during LC processes, enabling real-time monitoring and control. Overall, the integration of LC and AI has the potential to improve the efficiency and quality of manufacturing processes, leading to significant cost savings and better products. As these technologies continue to evolve, it will be interesting to see how they are applied in various industries and what new innovations they will bring as these optimization techniques still have few defects.

The future of AI in laser cladding is promising, and we can expect to see many new developments in this field in the coming years. One area where AI can have a significant impact on laser cladding is process optimization. By using AI algorithms to analyze data from the laser cladding process, engineers can identify the optimal parameters for each material and geometry combination. This can lead to improved process efficiency, higher-quality coatings, and reduced production costs. Another area where AI can be useful is in real-time process monitoring and control. By using ML algorithms, AI systems can monitor the laser cladding process and adjust the parameters in real time to optimize the quality of the coating. This can help reduce the risk of defects and ensure consistent coating quality, which is especially important for high-value components such as those used in aerospace or medical applications.

AI can also be used to develop new materials for laser cladding. By analyzing large amounts of data from experiments and simulations, AI systems can identify new material compositions and geometries that offer improved properties for laser cladding. This can lead to the development of new coatings with enhanced wear resistance, or other desirable properties. Finally, AI can play a crucial role in enabling automated LC systems. By combining AI algorithms with robotic systems, engineers can develop fully automated laser cladding processes that can run 24/7 with minimal human intervention. This can improve production efficiency, reduce labor costs, and increase throughput. Hence, the future of AI in laser cladding is bright, and we can expect to see many exciting new developments in this field in the years to come.

REFERENCES

[1]   Zhu, L., Xue, P., Lan, Q., Meng, G., Ren, Y., Yang, Z., Xu, P. and Liu, Z., 2021. Recent research and development status of laser cladding: A review. *Optics & Laser Technology, 138*, p.106915.

[2] Chakraborty, S.S. and Dutta, S., 2019. Estimation of dilution in laser cladding based on energy balance approach using regression analysis. *Sādhanā, 44*, pp.1–6.

[3] Nenadl, O., Ocelík, V., Palavra, A. and De Hosson, J.T.M., 2014. The prediction of coating geometry from main processing parameters in laser cladding. *Physics Procedia, 56*, pp.220–227.

[4] Wang, K., Du, D., Liu, G., Chang, B., Ju, J., Sun, S. and Fu, H., 2019. Microstructure and property of laser clad Fe-based composite layer containing Nb and B4C powders. *Journal of Alloys and Compounds, 802*, pp.373–384.

[5] Wang, K., Liu, W., Hong, Y., Sohan, H.S., Tong, Y., Hu, Y., Zhang, M., Zhang, J., Xiang, D., Fu, H. and Ju, J., 2023. An overview of technological parameter optimization in the case of laser cladding. *Coatings, 13*(3), p.496.

[6] Singh, J. and Singh, S. 2021. Neural network prediction of slurry erosion of heavy-duty pump impeller/casing materials 18Cr-8Ni, 16Cr-10Ni-2Mo, super duplex 24Cr-6Ni-3Mo-N, and grey cast iron. *Wear* [Internet], *476*, p.203741. Available from: https://doi.org/10.1016/j.wear.2021.203741

[7] Singh, J. 2019. Analysis on suitability of HVOF sprayed Ni-20Al, Ni-20Cr and Al-20Ti coatings in coal-ash slurry conditions using artificial neural network model. *Industrial Lubrication and Tribology,*;71, p.972–982.

[8] Bulgarevich, D.S., Tsukamoto, S., Kasuya, T., et al. 2018. Pattern recognition with machine learning on optical microscopy images of typical metallurgical microstructures. *Science Reports* [Internet], ;8, pp.3–9. Available from: http://dx.doi.org/10.1038/s41598-018-20438-6

[9] Shobha, G. and Rangaswamy, S. 2018. *Machine Learning* [Internet]. 1st ed. Elsevier B.V. Available from: http://dx.doi.org/10.1016/bs.host.2018.07.004

[10] Agatonovic-Kustrin, S. and Beresford R. 2000. Basic concepts of artificial neural network (ANN) modeling and its application in pharmaceutical research. *Journal of Pharmacy and Biomedical Analysis, 22*, pp.717–727.

[11] Singh, J. and Singh, S. 2022. A review on machine learning aspect in physics and mechanics of glasses. *Material Science and Engineering B* [Internet].*284*, p.115858. Available from: https://doi.org/10.1016/j.mseb.2022.115858

[12] Jin, G., Cai, Z., Guan, Y., Cui, X., Liu, Z., Li, Y. and Dong, M., 2018. High temperature wear performance of laser-cladded FeNiCoAlCu high-entropy alloy coating. *Applied Surface Science, 445*, pp.113–122.

[13] Abioye, T.E., McCartney, D.G. and Clare, A.T., 2015. Laser cladding of Inconel 625 wire for corrosion protection. *Journal of Materials Processing Technology, 217*, pp.232–240.

[14] Wang, G., Liu, X.B., Zhu, G.X., Zhu, Y., Liu, Y.F., Zhang, L. and Wang, J.L., 2022. Tribological study of Ti3SiC2/Cu5Si/TiC reinforced Co-based coatings on SUS304 steel by laser cladding. *Surface and Coatings Technology, 432*, p.128064.

[15] El Cheikh, H., Courant, B., Branchu, S., Hascoët, J.Y. and Guillén, R., 2012. Analysis and prediction of single laser tracks geometrical characteristics in

coaxial laser cladding process. *Optics and Lasers in Engineering, 50*(3), pp.413–422.

[16] Hamedi, M.J., Torkamany, M.J. and Sabbaghzadeh, J., 2011. Effect of pulsed laser parameters on in-situ TiC synthesis in laser surface treatment. *Optics and Lasers in Engineering, 49*(4), pp.557–563.

[17] Shan, B., Chen, J., Chen, S., Ma, M., Ni, L., Shang, F. and Zhou, L., 2022. Laser cladding of Fe-based corrosion and wear-resistant alloy: Genetic design, microstructure, and properties. *Surface and Coatings Technology, 433*, p.128117.

[18] Gan, Z., Yu, G., He, X. and Li, S., 2017. Numerical simulation of thermal behavior and multicomponent mass transfer in direct laser deposition of Co-base alloy on steel. *International Journal of Heat and Mass Transfer, 104*, pp.28–38.

[19] Muvvala, G., Karmakar, D.P. and Nath, A.K., 2017. Online monitoring of thermo-cycles and its correlation with microstructure in laser cladding of nickel based super alloy. *Optics and Lasers in Engineering, 88*, pp.139–152.

[20] Urbanic, R.J., Saqib, S.M. and Aggarwal, K., 2016. Using predictive modeling and classification methods for single and overlapping bead laser cladding to understand bead geometry to process parameter relationships. *Journal of Manufacturing Science and Engineering, 138*(5), p.051012.

[21] Pinkerton, A.J., 2015. Advances in the modeling of laser direct metal deposition. *Journal of Laser Applications, 27*(S1), p.S15001.

[22] Bakhtiyari, A.N., Wang, Z., Wang, L. and Zheng, H., 2021. A review on applications of artificial intelligence in modeling and optimization of laser beam machining. *Optics & Laser Technology, 135*, p.106721.

[23] Gao, J., Wang, C., Hao, Y., Liang, X. and Zhao, K., 2022. Prediction of TC11 single-track geometry in laser metal deposition based on back propagation neural network and random forest. *Journal of Mechanical Science and Technology, 36*(3), pp.1417–1425.

[24] Song, C., Liu, L., Yang, Y. and Weng, C., 2020. Prediction on geometrical characteristics of laser energy deposition based on regression equation and neural network. *IFAC-PapersOnLine, 53*(5), pp.89–96.

[25] Li, Y., Wang, K., Fu, H., Zhi, X., Guo, X. and Lin, J., 2021. Prediction for dilution rate of AlCoCrFeNi coatings by laser cladding based on a BP neural network. *Coatings, 11*(11), p.1402.

[26] Caiazzo, F. and Caggiano, A., 2018. Laser direct metal deposition of 2024 Al alloy: trace geometry prediction via machine learning. *Materials, 11*(3), p.444.

[27] Chen, T., Wu, W., Li, W. and Liu, D., 2019. Laser cladding of nanoparticle TiC ceramic powder: Effects of process parameters on the quality characteristics of the coatings and its prediction model. *Optics & Laser Technology, 116*, pp.345–355.

[28] Zhang, Y., Gong, B., Tang, Z. and Cao, W., 2022. Application of a bio-inspired algorithm in the process parameter optimization of laser cladding. *Machines, 10*(4), p.263.

[29] Nourbakhsh, A., Safikhani, H. and Derakhshan, S., 2011. The comparison of multi-objective particle swarm optimization and NSGA II

algorithm: applications in centrifugal pumps. *Engineering Optimization*, 43(10), pp.1095–1113.

[30] Jiang, X., Tian, Z., Liu, W., Tian, G., Gao, Y., Xing, F., Suo, Y. and Song, B., 2022. An energy-efficient method of laser remanufacturing process. *Sustainable Energy Technologies and Assessments*, 52, p.102201.

[31] Shu, L., Li, J., Wu, H. and Heng, Z., 2022. Optimization of multi-track laser-cladding process of titanium alloy based on RSM and NSGA-II algorithm. *Coatings*, 12(9), p.1301.

[32] Zhao, K., Liang, X., Wang, W., Yang, P., Hao, Y. and Zhu, Z. 2020. Multi-objective optimization of coaxial powder feeding laser cladding based on NSGA-II. *Chinese Journal of Lasers*, 47, 0102004.

[33] Xu, S.H., Mu, X.D., Chai, D. and Zhao, P., 2016. Multi-objective quantum-behaved particle swarm optimization algorithm with double-potential well and share-learning. *Optik*, 127(12), pp.4921–4927.

[34] Pant, P. and Chatterjee, D., 2020. Prediction of clad characteristics using ANN and combined PSO-ANN algorithms in laser metal deposition process. *Surfaces and Interfaces*, 21, p.100699.

[35] Ma, M., Xiong, W., Lian, Y., Han, D., Zhao, C. and Zhang, J., 2020. Modeling and optimization for laser cladding via multi-objective quantum-behaved particle swarm optimization algorithm. *Surface and Coatings Technology*, 381, p.125129.

[36] Liang, X., Wang, W., Zhao, K., Hao, Y., Yang, P. and Zhu, Z. 2020. Application of random forest re-gression analysis in trace geometry prediction of laser cladding. *Chinese Journal of Nonferrous Metals*, 30, 1644–1652.

[37] Sohrabpoor, H., 2016. Analysis of laser powder deposition parameters: ANFIS modeling and ICA optimization. *Optik*, 127(8), pp.4031–4038.

Chapter 5

Multi-objective optimisation of wire arc additive manufacturing deposition using genetic algorithm

Kashif Hasan Kazmi, Ajeet Kumar Bara and Sumit K. Sharma

5.1 INTRODUCTION

Additive manufacturing, sometimes called "3D printing", gained popularity in the manufacturing industry due to its high efficiency and low cost. It uses a computer-controlled technique to deposit the material layer by layer to create three-dimensional objects [1,2]. Additive manufacturing processes include powder bed fusion, sheet lamination, Vat polymerisation, binder jetting, directed energy deposition, and material extrusion as shown in Figure 5.1. Direct energy deposition (DED) is an additive manufacturing technique that precisely deposits material using a focused energy source (such as a laser, plasma, electron beam, or electric arc). DED gives the advantage of the deposition of multiple materials, which allows the creation of complex multi-material or multi-functional components [3]. Some famous techniques for direct energy deposition include WAAM, electron beam additive manufacturing (EBAM), and laser metal deposition (LMD), which give very high deposition rates as shown in Figure 5.2. WAAM is a developed manufacturing technique for the production of large-scale components at a reasonable cost. In WAAM, a wire electrode, typically formed of a metal or alloy, is heated up locally between the workpiece and the electrode using an electric arc [4]. The electrode is continuously fed into the arc, and the molten material is deposited into the substrate or previously deposited layers. A consumable electrode is commonly used with shielding gas (argon, helium, or a mixture) in the arc welding method known as gas metal arc welding (GMAW), sometimes known as metal inert gas welding (MIG). These shielding gases protect the weld pool and electrode from atmospheric contamination [5].

In wire arc additive manufacturing processes, parameters such as current, tool speed, wire feed rate, contact tip distance, and gas flow rate influence deposition quality [6]. Figure 5.3 shows the GMAW WAAM process. The welding current directly affects the heat input in the arc zone during deposition [7,8]. The heat input affects material deposition rate, depth of penetration, weld bead geometry, melting and fusion size, and the area of the

DOI: 10.1201/9781032713830-5

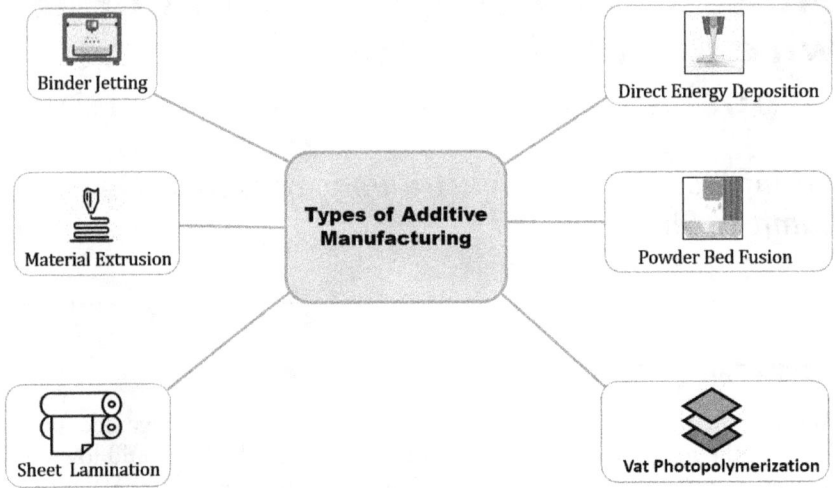

Figure 5.1 Different types of additive manufacturing process.

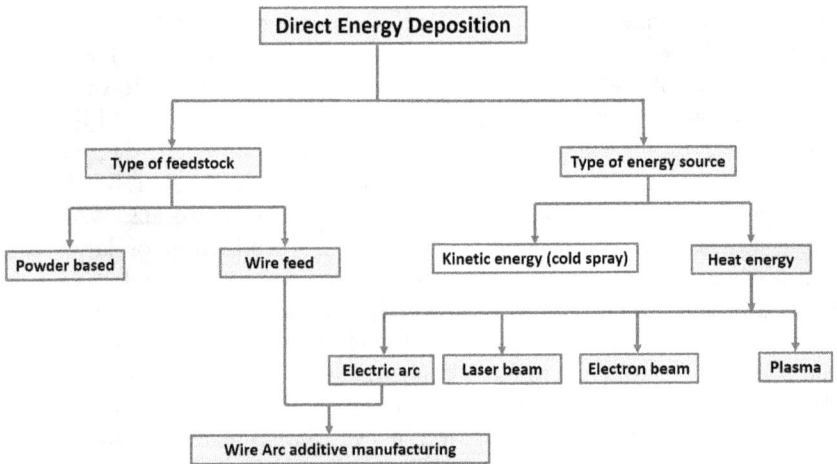

Figure 5.2 Classification of different DED techniques.

heat-affected zone around the weld [9]. Tool speed can impact the rate of deposition, heat input, fusion, and bonding, as well as material properties such as microstructure and mechanical properties [10]. The form and size of the melt pool, the weld bead, the deposition rate, and the overall surface finish of the fabricated item are all directly impacted by the wire feed rate.

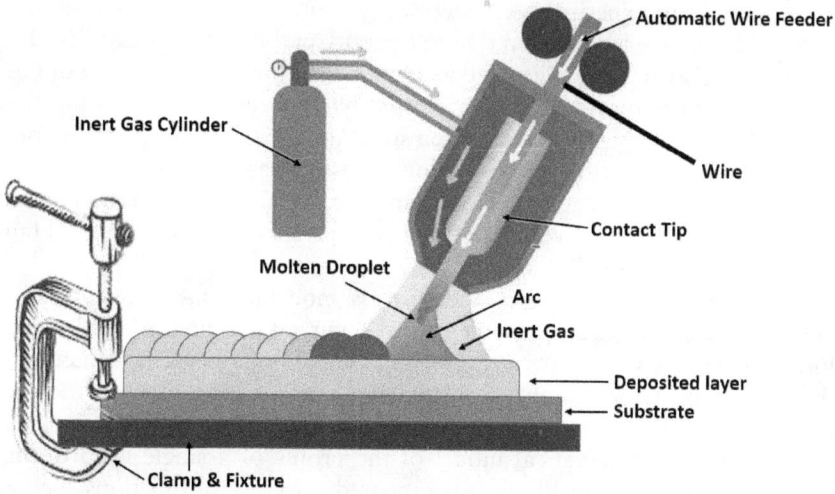

Figure 5.3 Schematic diagram of GMAW based WAAM.

Contact tip distance influences the depth of penetration into the workpiece, melt pool size and shape, stability of the arc, and metal transfer behaviour. The gas flow rate influences the efficiency of the shielding gas in protecting the weld pool and the whole deposition process. It determines the coverage and effectiveness of the shielding gas and controls the heat dissipation rate and stability of the welding arc [11].

Al5356 is a non-heat-treatable alloy with moderate tensile strength, high ductility, optimum electrical conductivity, excellent corrosion resistance, and weldability. Its ability to manufacture lightweight, high-strength parts with exceptional resistance to corrosion makes it applicable to numerous types of industries, such as aerospace, automotive, marine, general fabrication, and cryogenic applications [12]. These characteristics of Al 5356 make it an ideal material for additive manufacturing, especially WAAM. Optimisation of process parameters is crucial to achieving the desired shape, geometry, and quality of WAAM products. RSM is one of the most important tools for optimising process parameters and generating mathematical relationships between input variables and output responses[13,14]. Ideal conditions for a specific process or system can be identified through RSM. This approach is used to optimise product yield and adjust reaction conditions. The advantage of RSM is that it enables a person to examine multiple variables simultaneously and see how each variable affects the outcome. CCD uses statistical models to analyse experimental data and construct prediction models for process improvement without performing additional experiments [15]. CCD provides a large amount of data about the impacts of experiment variables and the total experimental error with a small number of experiments. Different types of CCDs can be used in multiple fields of experimentation due to their adaptability and versatility. The principles of natural selection and genetics constitute the foundation of GA. The notion of natural selection proposed by Darwin serves as inspiration for the GA family of stochastic search strategies [16]. Decoding, population generation, fitness evaluation, selection, crossover, and mutation are the six main components of a genetic algorithm. The genetic algorithm can be used to calculate the ideal or nearly optimal parameters in the WAAM to obtain the desired shape and quality [17,18].

Many experimental studies and process modelling have been done to enhance the surface quality, dimensional accuracy, mechanical, and metallurgical qualities of WAAM components. Suryakumar et al. [19] modelled the height and length of the deposited beads as an expression of the two input variables (tool speed and feed rate) using regression analysis. Cao et al. [20] created a mathematical model of the profile of a single bead during robotic metal arc gas welding and compared it to several functions such as parabola, gaussian, logistic, and sine, and the sine function provided the optimal fit. Xiong et al. [21] created an online monitoring and control system for bead geometry based on computer vision during GMAW-based thin wall deposition, resulting in significant material and energy savings. Xiong et al.'s study [22] was based on experimental data involving neural networks and regression-based algorithms built to predict bead shape in automated GMAW-based weld deposition. Xiong et al. [23] analysed the effects of varying processing parameters on three models of weld bead profiles (spherical arc, parabolic, and cosine functions) for estimating bead form. Ding et al. [24] found that the centre distance is an important factor in determining the flatness of the top surface and can be predicted using an overlapping tangent model. Ding et al. [25] also presented deposition route planning with no gaps in WAAM based on the notion of the medial axis transformation approach. Ding et al. [26] also proposed WAAM automated manufacturing technology, which uses CAD models to design deposition paths for creating any 3D functional item. Li et al. [27] created a model for improved bead overlapping using the weld bead spreading effect to regulate surface irregularity during overlapped deposition of many beads in WAAM. Youheng et al. [28] executed a multi-objective optimisation based on a response surface to get a defect-free surface in WAAM while controlling for factors including temperature, reinforcement, current, voltage, and bead width.

Some studies looked at the metallurgical and mechanical properties of WAAM-fabricated components manufactured from a variety of materials. Lu et al. [29] discovered that WAAM thick wall sections fabricated from copper-coated mild steel filament have anisotropic mechanical characteristics due to irregularly distributed granular features. Somashekara et al. [30] demonstrated the production of functionally graded materials through twin-wire-fed weld deposition, employing different materials, and found that the deposited hardness of a single bead was higher than that of overlapping beads. Samantaray et al. [31] discovered that 304L stainless steel's mechanical characteristics after semisolid processing enhance ductility and reduce the yield value and maximum tensile strength compared to its primarily processed counterpart. Haden et al. [32] observed that during the WAAM process of 304 stainless steel and ER70S mild steel, the wear rate decreased, and the microhardness values increased substantially with an increase in the length of the deposited component. Xing et al. [33] demonstrated that creating a bainite microstructure in a wall of GMAW-based AM increased its toughness. Some researchers explored double wire feed arc additive deposition of various materials to boost productivity and deposition efficiency while improving WAAM products' characteristics. Pan et al. [34] provided a comprehensive overview of research focused on the static mechanical properties of WAAM and summarised detailed information about the welding technology used and various processing conditions, such as heat treatment and interlayer cooling. Yet the techniques by which physical properties can be enhanced were not covered. Deposition of the beads using WAAM and their quality improvement in terms of desired geometry and surface finish are the multi-objective problems. When an objective function is in a trade-off relationship, minimising or maximising the objective function simultaneously is very difficult. In this research, experiments are designed based on RSM, and GA are used to optimise the objective.

5.2 EXPERIMENTAL PROCEDURE

Al5356 aluminium alloy beads have been deposited using a six-degree-of-freedom robot (Model: ABB 1520IRB), as shown in Figure 5.4. Al5083 plate with dimensions of $100 \times 100 \times 5$ mm has been used as a substrate plate, and a 0.8 mm diameter wire of Al5356 has been used as a filler wire, and their chemical composition is shown in Table 5.1. The robot is equipped with a synchronous welding machine (KempArc SYN 500) as a power source and an automatic wire feeder (KempArc DT 400) attached to the robot to feed the wire into the molten pool at the required speed. Argon was used as a shielding gas (20 l/min) to protect the molten pool from the outer environment and smoothen the deposition of the bead. All the robot movement, gas flow rate, wire feed rate, tool speed, current, and voltage were controlled through the robot controller (Model: IRC5). Beads were deposited according to CCD by changing the tool speed and current, and other deposition parameters were kept constant, as shown in Table 5.2. Bead height, bead width, and surface roughness were measured as output responses. Deposited beads were cut from the middle of deposited bead to eliminate the end effect, and bead height and width were measured after polishing the cross-section by using a toolmaker's microscope. Surface roughness was measured using a portable surface roughness tester (Mitutoyo SJ-210), and the average value was considered as an output response.

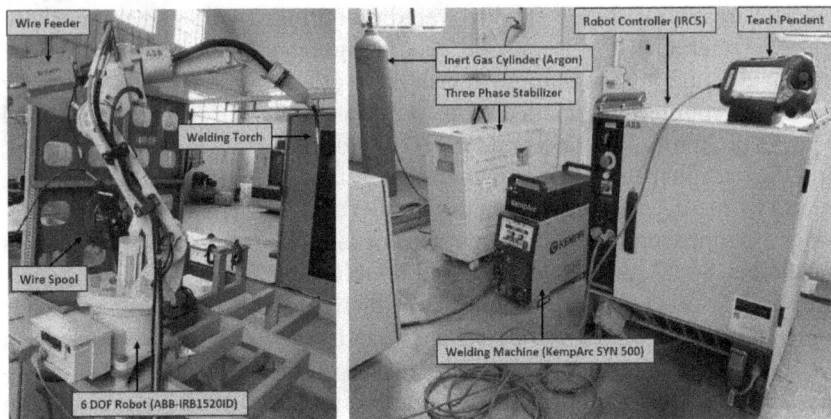

Figure 5.4 Experimental setup using for the deposition of Al5356 beads.

Table 5.1 Composition (wt. %) of Al5356 filler wire and Al5083 substrate plate

Alloys	Al	Si	Mg	Fe	Cu	Mn	Cr	Zn	Ti
Al5356	94.47	0.05	5.1	0.12	0.005	0.12	0.12	0.005	0.01
Al5083	94.5	0.09	4.4	0.27	0.01	0.62	0.07	0.02	0.02

Table 5.2 Fixed input process parameter used during deposition of beads

Process Parameters	Value
Filler wire	Al5356 (0.8 mm diameter)
Substrate plate	Al5083 (5 mm thickness)
Wire diameter	1.2 mm
Wire feeding rate (m/min)	7.5
Gas flow rate (lpm)	20 (Argon)
Standoff distance (mm)	13
Mode of operation	1 MIG

RSM is used to analyse the bead deposition, and experiments were carried out using a CCD.

Trial experiments were conducted to find out the input process parameter range during deposition, and two factors (current and tool speed) were considered for the deposition of beads. Fourteen sets of experiments were performed from the design matrix by varying the input process parameters, and their output response was noted using a toolmaker microscope and surface roughness tester, as shown in Table 5.3. As a result, the relationship between the input process parameter and the output response is generated in terms of the polynomial equation.

Table 5.3 Central composite response surface design and measured output responses

Std Order	Run Order	Pt Type	Blocks	Current, I (A)	Tool Speed, Vt (mm/min)	Bead Roughness (μm)	Bead Height (mm)	Bead Width (mm)
14	1	0	2	190	8.5	9.693	3.331	7.588
12	2	0	2	190	8.5	9.501	3.21	7.632
9	3	−1	2	232.426	8.5	17.624	4.021	12.301
10	4	−1	2	190	6.3787	11.596	3.708	7.776
8	5	−1	2	147.574	8.5	7.145	1.505	4.912
13	6	0	2	190	8.5	9.511	3.401	7.591
11	7	−1	2	190	10.6213	7.015	2.996	6.908
2	8	1	1	220	7	15.691	4.295	11.355
1	9	1	1	160	7	8.993	2.884	6.667
5	10	0	1	190	8.5	9.698	3.311	7.607
3	11	1	1	160	10	5.092	2.074	5.009
6	12	0	1	190	8.5	9.573	3.347	7.701
4	13	1	1	220	10	12.365	3.901	10.117
7	14	0	1	190	8.5	9.476	3.424	7.674

5.3 MATHEMATICAL MODELS

The equation $y = f(I, V_t)$ expresses the relationship between input process parameters (current and tool speed) and output responses (surface roughness, bead height and bead width). Response surface regression equation can be expressed as:

$$y = a_0 + \sum a_i x_i + \sum a_{ii} x_i^2 + \sum a_{ij} x_i x_j \qquad (5.1)\ [28]$$

where y is the response, a_0 the free term, a_i the linear term, x_i the number of factors, a_{ii} the quadratic term and a_{ij} an interaction term.

Surface roughness (SR) and bead geometry in terms of bead height (BH) and bead width (BW) can be expressed as a second-order response surface model as a function of input process parameters (current and tool speed) as shown in Equations 5.2, 5.3 and 5.4. This equation is obtained by using Minitab 17 statistical software.

$$SR = 48.06 - 0.4675\ C - 0.145\ TS + 0.001475\ C^2 \\ - 0.0944\ TS^2 + 0.00319\ C{*}TS \qquad (5.2)$$

$$BH = -3.86 + 0.1055\ C - 1.115\ TS - 0.000255\ C^2 \\ + 0.0289\ TS^2 + 0.00231\ C{*}TS \qquad (5.3)$$

$$BW = 18.82 - 0.1742\ C - 0.282\ TS + 0.000628\ C^2 \\ - 0.0297\ TS^2 + 0.00233\ C{*}TS \qquad (5.4)$$

where C denotes the current (A), TS denotes the tool speed (mm/sec), SR denotes the surface roughness (µm), BH denotes the bead height (mm), and BW denotes the bead width. Analysis of variance (ANOVA) is used to test the reliability of the developed mathematical model. Tables 5.4–5.7 show

Table 5.4 ANOVA table of surface roughness

Source	DF	Adj SS	Adj MS	F-Value	P-Value
Model	6	141.018	23.503	552.07	0.000
Block	1	0.102	0.102	2.40	0.165
Linear	2	127.092	63.546	1492.65	0.000
C	1	103.612	103.612	2433.77	0.000
TS	1	23.480	23.480	551.53	0.000
Square	2	13.741	6.841	161.38	0.000
C^2	1	13.007	13.007	305.52	0.000
TS^2	1	0.333	0.333	7.82	0.027
Two-way interaction	1	0.083	0.083	1.94	0.206
C*TS	1	0.083	0.083	1.94	0.206
Error	7	0.298	0.043		
Lack-of-fit	3	0.250	0.083	6.92	0.046
Pure error	4	0.048	0.012		
Total	13	141.316			
R-sq	0.9918				
R-sq (adj)	0.9848				
R-sq (pred)	0.9171				

Table 5.5 ANOVA table of bead height

Source	DF	Adj SS	Adj MS	F-Value	P-Value
Model	6	6.94814	1.15802	112.38	0.000
Block	1	0.08086	0.8086	7.85	0.026
Linear	2	6.38450	3.19225	309.79	0.000
C	1	5.77348	5.77348	560.29	0.000
TS	1	0.61102	0.61102	59.30	0.000
Square	2	0.43951	0.21976	21.33	0.001
C^2	1	0.38867	0.38867	37.72	0.000
TS^2	1	0.03128	0.03128	3.04	0.125
Two-way interaction	1	0.04326	0.04326	4.20	0.080
C*TS	1	0.04326	0.04326	4.20	0.080
Error	7	0.07213	0.01030		
Lack-of-fit	3	0.04679	0.01560	2.46	0.202
Pure error	4	0.02534	0.00633		
Total	13	7.02027			
R-sq	0.9897				
R-sq (adj)	0.9809				
R-sq (pred)	0.9431				

Table 5.6 ANOVA table of bead width

Source	DF	Adj SS	Adj MS	F-Value	P-Value
Model	6	56.0005	9.3334	141.01	0.000
Block	1	0.1444	0.1444	2.18	0.183
Linear	2	53.3611	26.6806	403.09	0.000
C	1	51.2357	51.2357	774.07	0.000
TS	1	2.1254	2.1254	32.11	0.001
Square	2	2.4508	1.2254	18.51	0.002
C^2	1	2.3603	2.3603	35.66	0.001
TS^2	1	0.0330	0.0330	0.50	0.503
Two-way-interaction	1	0.0441	0.0441	0.67	0.0441
C*TS	1	0.0441	0.0441	0.67	0.0441
Error	7	0.4633	0.0662		
Lack-of-fit	3	0.4574	0.1525	103.49	0.000
Pure error	4	0.0059	0.0015		
Total	13	56.4638			
R-sq	0.9918				
R-sq (adj)	0.9848				
R-sq (pred)	0.9171				

Table 5.7 Optimize process parameters and their experimental validation

Input Process Parameters		Output Responses		
Current (A)	Tool Speed (mm/s)		Experimental	Predicted
185	8.5	Surface Roughness (µm)	8.351	8.161
		Bead Height (mm)	3.317	3.201
		Bead Width (mm)	7.218	7.391

the ANVOVA table of output responses (surface roughness, bead height and bead width). The P-value (probability value) derived from the surface response is less than 5% for all output responses, and the R-sq (adj) and R-sq (pred) values are greater than 90%, indicating that the model is acceptable and significant.

5.4 PARAMETRIC EFFECTS

The interaction of current and tool speed on output response (surface roughness, bead height and bead width) is shown in Figure 5.5. It is observed from Figure 5.5(a) that an increase in the current increases the surface roughness value, but an increase in the tool speed decreases the surface roughness. The same pattern is observed in the case of bead height, which is shown in Figure 5.5(b), but from Figure 5.5(c), we can observe

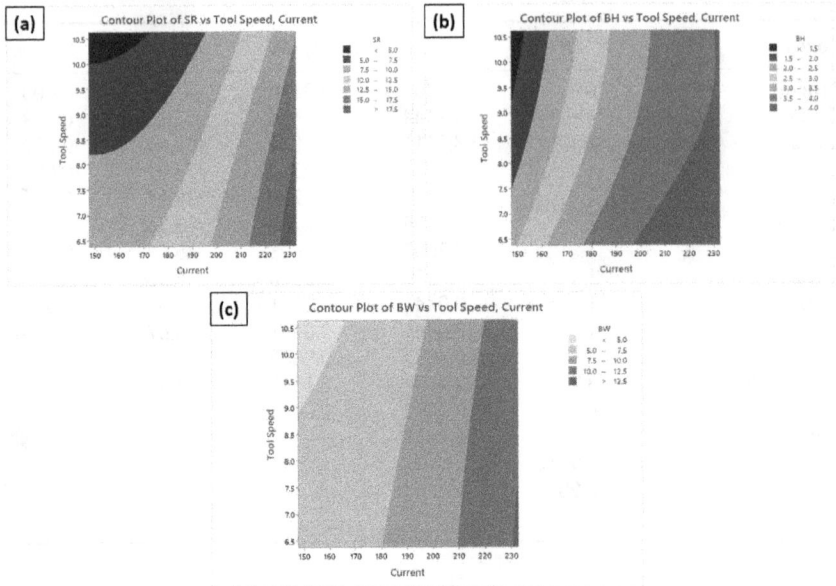

Figure 5.5 Effects of input process parameters (current and tool speed) on (a) surface roughness, (b) bead height, and (c) bead width.

that changes in current primarily affect bead width, and the influence of tool speed is not much as compared to other output responses. A higher current value increases the melting efficiency, leading to larger bead sizes and more convex shapes, while a lower current value deposits beads with smaller sizes and more concave sizes. A higher tool speed deposited beads with taller and narrower sizes, while a lower tool speed deposited wider and flatter beads. The contact time between the substrate plate and molten metal is largely affected by tool speed; higher tool speeds lower the contact time, resulting in a smaller heat-affected zone and vice versa. The cooling rate is largely affected by current and tool speed. The shape, size and surface roughness of the deposited beads affect the cooling rate of deposition and the contact time between the substrate plate and molten metal or dwell time between deposited layers.

5.5 MULTI-OBJECTIVE GENETIC ALGORITHM

Genetic algorithm (GA) is a non-convectional stochastic optimisation technique. The success of GA significantly lies in the correct calibration of GA parameters, which cover population size (N_p), crossover probability (p_c), and mutational probability (p_m) [16]. Through a detailed analysis of the GA, appropriate GA parameters were successfully obtained for application in the present research [35]. To achieve the most favourable deposition outcomes in the WAAM of three-dimensional components, a binary-coded GA was utilised with the intention of increasing the bead height and width and reducing surface roughness. The objective function is formulated as a maximisation problem for BH and BW and a minimisation problem for SA to optimise the parameters.

$$SR = 48.06 - 0.4675\ C - 0.145\ TS + 0.001475\ C^2 - 0.0944\ TS^2 + 0.00319\ C^*TS$$

$$BH = -3.86 + 0.1055\ C - 1.115\ TS - 0.000255\ C^2 + 0.0289\ TS^2 + 0.00231\ C^*TS$$

$$BW = 18.82 - 0.1742\ C - 0.282\ TS + 0.000628\ C^2 - 0.0297\ TS^2 + 0.00233\ C^*TS$$

subject to the constraints $160 \leq C \leq 220$, $7 \leq TS \leq 10$, where C and TS are taken as current and tool speed, respectively, these equations are used for GA-based optimisation. The optimisation was carried out using a binary-coded GA with a tournament selection approach, constraint crossover, and constraint-dependent mutation. Through a comprehensive exploration of various parameters, appropriate genetic algorithm parameters were identified and selected to effectively enhance the performance of the genetic algorithm. Single-pint crossover, tournament selection scheme and constraint-dependent mutation are used for optimisation. GA parameters were determined through a parametric study for better performance.

5.6 OPTIMISATION RESULTS

Figure 5.6 illustrates the fitness value's variation with the number of generations or iterations using the set of GA parameters ($N_p = 50$, $p_c = 0.8$ and $p_m = 0.01$). The optimised result of input process parameters and their output response obtained from genetic algorithm optimisation is shown in Table 5.5. Experiments were conducted for the deposition of beads on this optimum input process parameter to validate the GA optimisation results. There are some errors that occur due to the unavailability of the exact input process parameter option in the machine setup, the current of 185 A, and the tool speed of 8.5 mm/s (the nearest value) chosen for deposition. The results obtained from GA optimisation and experiments are shown in good correlation.

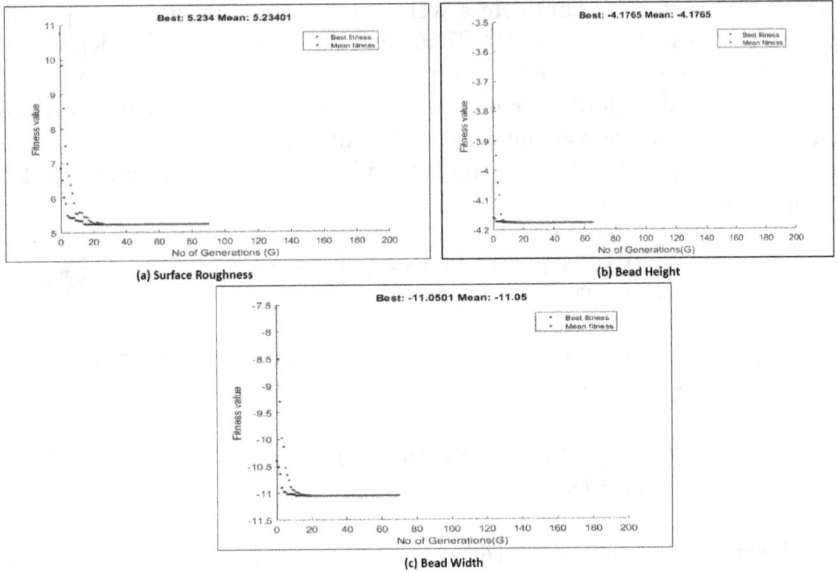

(a) Surface Roughness

(b) Bead Height

(c) Bead Width

Figure 5.6 Convergence plot of fitness value: (a) surface roughness, (b) bead height, and (c) bead width.

5.7 CONCLUSIONS

Bead geometry and surface roughness are important factors in deciding the quality of final 3D components, so optimising these outputs by selecting proper input parameters is crucial. This research proposes a multi-objective genetic algorithm to optimise the bead geometry and surface roughness using wire-arc additive manufacturing. RSM is used for the experimental design and a mathematical model has been developed. The P-value (probability value) is less than 5%, the R-sq (adj) and R-sq (pred) values are greater than 90%, and these values indicate the model is significant. Optimisation of the parameters using a genetic algorithm shows promising results. The proposed optimisation techniques were validated through experiments, and it is concluded that the combination of RSM and GA gives good results in optimising the shape and quality of WAAM-developed components.

REFERENCES

[1]   J. Gardan, "Additive manufacturing technologies: state of the art and trends," *Addit. Manuf. Handb.*, pp. 149–168, 2017.

[2]   S. K. Sharma, R. Mandal, and A. K. Shukla, "Processing techniques, principles, and applications of additive manufacturing," in *Additive*

Manufacturing: Advanced Materials and Design Techniques, CRC Press, Cambridge, 2023, p. 187.

[3] M. L. Dezaki et al., "A review on additive/subtractive hybrid manufacturing of directed energy deposition (DED) process," *Adv. Powder Mater.*, p. 100054, 2022.

[4] D. Ding, Z. Pan, D. Cuiuri, and H. Li, "Wire-feed additive manufacturing of metal components: technologies, developments and future interests," *Int. J. Adv. Manuf. Technol.*, vol. 81, no. 1, pp. 465–481, 2015.

[5] C. R. Cunningham, J. M. Flynn, A. Shokrani, V. Dhokia, and S. T. Newman, "Invited review article: Strategies and processes for high quality wire arc additive manufacturing," *Additive Manufacturing*, vol. 22, pp. 672–686, 2018. doi: 10.1016/j.addma.2018.06.020.

[6] M. D. Barath Kumar and M. Manikandan, "Assessment of process, parameters, residual stress mitigation, post treatments and finite element analysis simulations of wire arc additive manufacturing technique," The Korean Institute of Metals and Materials, vol. 28, no. 1., 2022. doi: 10.1007/s12540-021-01015-5.

[7] K. H. Kazmi, S. K. Sharma, A. K. Das, A. Mandal, and A. Shukla, "Development of wire arc additive manufactured Cu–Si alloy: Study of microstructure and wear behavior," *J. Mater. Eng. Perform.*, 2023, doi: 10.1007/s11665-023-07972-9.

[8] K. H. Kazmi, A. K. Das, S. K. Sharma, A. Mandal, and A. K. Shukla, "Wire arc additive manufacturing of ER-4043 aluminum alloy: Evaluation of bead profile, microstructure, and wear behavior," *Weld. World*, no. 0123456789, 2023, doi: 10.1007/s40194-023-01558-8.

[9] R. Warsi, K. H. Kazmi, and M. Chandra, "Mechanical properties of wire and arc additive manufactured component deposited by a CNC controlled GMAW," *Mater. Today Proc.*, vol. 56, no. November, pp. 2818–2825, 2022, doi: 10.1016/j.matpr.2021.10.114.

[10] K. H. Kazmi, S. K. Sharma, A. K. Das, A. Mandal, A. Kumar Shukla, and R. Mandal, "Wire arc additive manufacturing of ER-4043 aluminum alloy: Effect of tool speed on microstructure, mechanical properties and parameter optimisation," *J. Mater. Eng. Perform.*, vol. 20, 2023, doi: 10.1007/s11665-023-08309-2.

[11] M. Chaturvedi, E. Scutelnicu, C. C. Rusu, L. R. Mistodie, D. Mihailescu, and S. Arungalai Vendan, "Wire arc additive manufacturing: Review on recent findings and challenges in industrial applications and materials characterisation," *Metals (Basel).*, vol. 11, no. 6, 2021, doi: 10.3390/met11060939.

[12] M. Köhler, S. Fiebig, J. Hensel, and K. Dilger, "Wire and arc additive manufacturing of aluminum components," *Metals (Basel).*, vol. 9, no. 5, p. 608, 2019.

[13] H. Tian, Z. Lu, F. Li, and S. Chen, "Predictive modeling of surface roughness based on response surface methodology after WAAM," in *2019 International Conference on Electronical, Mechanical and Materials Engineering (ICE2ME 2019)*, 2019, pp. 47–50.

[14] M. Islam, A. Buijk, M. Rais-Rohani, and K. Motoyama, "Process parameter optimisation of lap joint fillet weld based on FEM–RSM–GA integration technique," *Adv. Eng. Softw.*, vol. 79, pp. 127–136, 2015.

[15] F. Veiga, A. Suárez, E. Aldalur, I. Goenaga, and J. Amondarain, "Wire arc additive manufacturing process for topologically optimised aeronautical fixtures," *3D Print. Addit. Manuf.*, vol. 10, no. 1, pp. 23–33, 2023.

[16] D. E. Goldberg, *Genetic Algorithms*. Pearson Education India, 2013.

[17] A. Kumar and K. Maji, "Selection of process parameters for near-net shape deposition in wire arc additive manufacturing by genetic algorithm," *J. Mater. Eng. Perform.*, vol. 29, no. 5, pp. 3334–3352, 2020, doi: 10.1007/s11665-020-04847-1.

[18] B. Panda, K. Shankhwar, A. Garg, and M. M. Savalani, "Evaluation of genetic programming-based models for simulating bead dimensions in wire and arc additive manufacturing," *J. Intell. Manuf.*, vol. 30, no. 2, pp. 809–820, Feb. 2019, doi: 10.1007/s10845-016-1282-2.

[19] S. Suryakumar, K. P. Karunakaran, A. Bernard, U. Chandrasekhar, N. Raghavender, and D. Sharma, "Weld bead modeling and process optimisation in hybrid layered manufacturing," *Comput. Des.*, vol. 43, no. 4, pp. 331–344, 2011.

[20] Y. Cao, S. Zhu, X. Liang, and W. Wang, "Overlapping model of beads and curve fitting of bead section for rapid manufacturing by robotic MAG welding process," *Robot. Comput. Integr. Manuf.*, vol. 27, no. 3, pp. 641–645, 2011.

[21] J. Xiong, G. Zhang, Z. Qiu, and Y. Li, "Vision-sensing and bead width control of a single-bead multi-layer part: material and energy savings in GMAW-based rapid manufacturing," *J. Clean. Prod.*, vol. 41, pp. 82–88, 2013.

[22] J. Xiong, G. Zhang, J. Hu, and L. Wu, "Bead geometry prediction for robotic GMAW-based rapid manufacturing through a neural network and a second-order regression analysis," *J. Intell. Manuf.*, vol. 25, no. 1, pp. 157–163, 2014.

[23] J. Xiong, G. Zhang, H. Gao, and L. Wu, "Modeling of bead section profile and overlapping beads with experimental validation for robotic GMAW-based rapid manufacturing," *Robot. Comput. Integr. Manuf.*, vol. 29, no. 2, pp. 417–423, 2013.

[24] D. Ding, Z. Pan, D. Cuiuri, and H. Li, "A multi-bead overlapping model for robotic wire and arc additive manufacturing (WAAM)," *Robot. Comput. Integr. Manuf.*, vol. 31, pp. 101–110, 2015.

[25] D. Ding, Z. Pan, D. Cuiuri, and H. Li, "A practical path planning methodology for wire and arc additive manufacturing of thin-walled structures," *Robot. Comput. Integr. Manuf.*, vol. 34, pp. 8–19, 2015.

[26] D. Ding et al., "Towards an automated robotic arc-welding-based additive manufacturing system from CAD to finished part," *Comput. Des.*, vol. 73, pp. 66–75, 2016.

[27] Y. Li, Y. Sun, Q. Han, G. Zhang, and I. Horváth, "Enhanced beads overlapping model for wire and arc additive manufacturing of multi-layer

multi-bead metallic parts," *J. Mater. Process. Technol.*, vol. 252, pp. 838–848, 2018.

[28] F. Youheng, W. Guilan, Z. Haiou, and L. Liye, "Optimisation of surface appearance for wire and arc additive manufacturing of Bainite steel," *Int. J. Adv. Manuf. Technol.*, vol. 91, no. 1, pp. 301–313, 2017.

[29] X. Lu, Y. F. Zhou, X. L. Xing, L. Y. Shao, Q. X. Yang, and S. Y. Gao, "Open-source wire and arc additive manufacturing system: Formability, microstructures, and mechanical properties," *Int. J. Adv. Manuf. Technol.*, vol. 93, pp. 2145–2154, 2017.

[30] M. A. Somashekara and S. Suryakumar, "Studies on dissimilar twin-wire weld-deposition for additive manufacturing applications," *Trans. Indian Inst. Met.*, vol. 70, pp. 2123–2135, 2017.

[31] D. Samantaray, V. Kumar, A. K. Bhaduri, and P. Dutta, "Microstructural evolution and mechanical properties of type 304 L stainless steel processed in semi-solid state," *Int. J. Metall. Eng.*, vol. 2, no. 2, pp. 149–153, 2013.

[32] C. V Haden, G. Zeng, F. M. Carter III, C. Ruhl, B. A. Krick, and D. G. Harlow, "Wire and arc additive manufactured steel: Tensile and wear properties," *Addit. Manuf.*, vol. 16, pp. 115–123, 2017.

[33] X. Xing et al., "Microstructure optimisation and cracking control of additive manufactured bainite steel by gas metal arc welding technology," *J. Mater. Eng. Perform.*, vol. 28, pp. 5138–5145, 2019.

[34] Z. Pan, D. Ding, B. Wu, D. Cuiuri, H. Li, and J. Norrish, "Arc welding processes for additive manufacturing: a review," *Trans. Intell. Weld. Manuf.*, pp. 3–24, 2018.

[35] K. Maji, D. K. Pratihar, and S. Patra, "Modelling of electrical discharge machining process using regression analysis, adaptive neuro-fuzzy inference system and genetic algorithm," *Int. J. Data Mining, Model. Manag.*, vol. 2, no. 1, pp. 75–94, 2010.

Chapter 6

Techniques for improving engineering material's tribological performance using laser cladding process in laser surface texturing

Hitesh Vasudev and Amrinder Mehta

6.1 INTRODUCTION

Friction and wear are significant factors in engineering applications involving surface interactions. The field of tribology, which investigates the interactions between surfaces in motion, holds significant significance in the realm of material and part design. Its primary objective is to mitigate wear and friction, hence augmenting performance and durability. Tribology additionally contributes to the mitigation of energy consumption in machinery and the enhancement of device safety. Engineers possess the ability to enhance the performance and durability of components by comprehending the intricate mechanisms of friction and wear. This comprehension moreover facilitates the advancement of improved lubricants and coatings aimed at mitigating friction and minimizing wear [1–3]. Engineers can effectively design engineered surfaces that exhibit optimal interaction through the application of tribology. Surface texturing has emerged as a prominent strategy for effectively regulating tribological performance, attracting considerable attention in recent times. Engineers possess the capability to alter the friction and wear characteristics of materials through the incorporation of distinct patterns, structures, or textures onto the surface. In the realm of tribology, the discipline pertaining to the study of friction, wear, and lubrication, the area of focus commonly denoted as "surface engineering" or "surface texturing" is recognized. Surface texturing techniques can be employed to mitigate friction and wear, enhance the effectiveness of lubricants, and optimize sealing functionalities [4–6]. Enhancing tribological performance in several applications can be achieved through the utilization of an efficient approach. Surface texturing can be employed as a means to enhance the contact characteristics of surfaces and increase the longevity of components. Additionally, it has the capability to enhance the surface characteristics of components, resulting in advantageous outcomes such as resistance against corrosion and prevention of ice formation [7–9]. Figure 6.1 depicts the schematic representation of the laser cladding process.

DOI: 10.1201/9781032713830-6

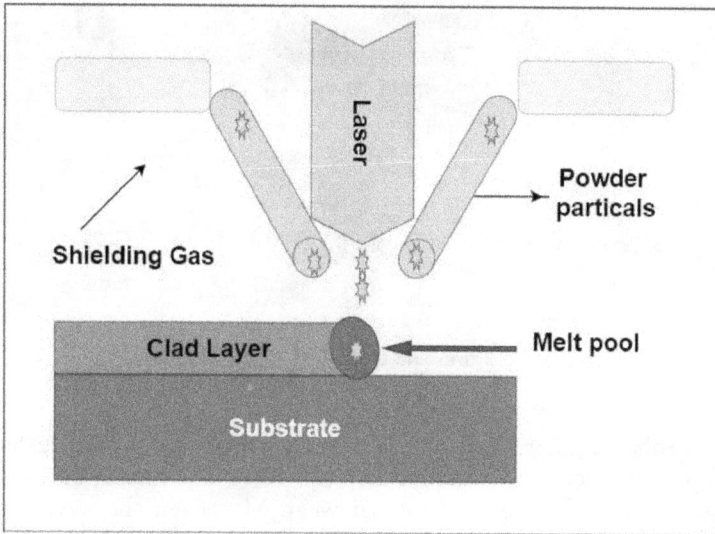

Figure 6.1 Laser cladding process.

In recent years, there has been a notable surge in scholarly inquiry and assessment pertaining to the application of cladded layers composed of superalloys on diverse substrates. This research has predominantly centred on investigating the tribological characteristics, encompassing friction, wear, and lubrication performance. Superalloys are a class of materials that exhibit remarkable mechanical strength, resistance to creep, and resistance to corrosion. These properties render them well-suited for many demanding applications characterized by elevated temperatures and significant mechanical stress [10–12]. The applications include aerospace engineering, power generation, and gas turbine engines. The process of cladding entails the application of a superalloy layer over a substrate material, with the aim of improving certain characteristics or safeguarding the substrate against adverse conditions. The presence of a cladded layer can have a substantial impact on the overall performance of the material, particularly with regards to its wear resistance and friction characteristics.

There are many different surface textures that can be used, as shown in (Figure 6.2), to improve tribological performance.

These minute surface characteristics are often measured in micrometers. Microtextures have the potential to enhance the ability to retain lubricants, minimize contact area, and alter the distribution of contact pressure. Consequently, these effects contribute to the decrease of friction and wear. The incorporation of microtextures inside tribological systems has been

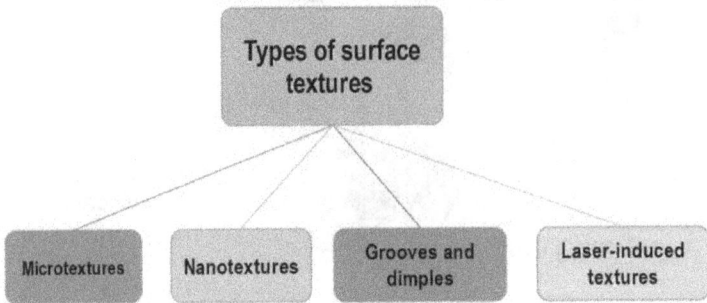

Figure 6.2 Variety of surface textures and types.

found to enhance their performance through the provision of a cushioning effect upon contact. Additionally, they can serve as a means of regulating the coefficient of friction and rate of wear. Moreover, the incorporation of microtextures has been shown to enhance the durability of surfaces by increasing their resistance to both wear and corrosion [13–15]. The use of microtextures can also be employed to generate a hydrodynamic phenomenon, hence diminishing the necessity for lubrication and enhancing the functionality of tribological systems. Furthermore, they can serve the purpose of mitigating noise and minimizing vibration. These properties are of an even smaller magnitude, typically measured on the nanoscale scale. The impact of nanotextures on friction and wear behavior is substantial due to their influence on surface energy and adhesion [16–18]. Nanotextures are present on the surfaces of various materials such as metals, polymers, and ceramics. Moreover, they have the capability to be manipulated in order to provide distinct textures that are tailored for particular applications. Nanotextures have the potential to mitigate frictional forces and enhance the longevity of various materials. The utilization of this technology has the potential to yield advantages in a diverse array of fields, encompassing the realm of automotive components as well as the domain of medical implants. Nanotextures have the potential to modify the wettability of surfaces, hence enhancing their resistance to water or dust particles [19–21].

The incorporation of grooves and dimples onto a surface has been found to have several beneficial benefits, including enhanced lubricant retention, the promotion of hydrodynamic phenomena, and the formation of protective tribofilms. These effects collectively contribute to the reduction of friction and wear. This phenomenon leads to a decrease in the necessity for regular lubrication and maintenance, thereby resulting in energy and cost savings. Additionally, it enhances the operating lifespan of the component, resulting in enhanced productivity and performance [22–24].

Moreover, these impacts can also contribute to the preservation of a more uniform performance, as the levels of friction and wear will exhibit greater consistency. Consequently, this will contribute to enhancing the dependability and security of the component. Lasers provide precise customization and manipulation of tribological characteristics through the generation of controlled surface textures. This facilitates a decrease in friction and the occurrence of wear between interacting surfaces. Laser cladding has gained significant traction across various industries, including automotive, aerospace, and medical devices. The utilization of lasers in generating textured surfaces also confers enhanced resistance to corrosion, hence augmenting the durability of the product and diminishing associated maintenance expenses. Laser cladding has the potential to enhance the operational effectiveness of components through the augmentation of their efficiency and mitigation of noise levels [25–27].

The integration of various scales of texture, such as micro and nanotextures, inside hierarchical structures can yield synergistic outcomes, resulting in enhanced tribological performance. Hierarchical architectures provide greater stability in comparison to individual textures. Furthermore, it has been observed that hierarchical textures exhibit superior self-lubrication and anti-wear characteristics in comparison to their smooth counterparts. In addition, the implementation of hierarchical textures has been shown to enhance the fatigue resistance of surfaces while concurrently mitigating friction [28–30]. This characteristic renders them well-suited for applications that necessitate a combination of wear resistance and strength. Surface texturing for tribology offers several advantages, including reduced friction coefficients, extended lifespan of components, enhanced energy economy, and greater reliability. In real-world engineering applications, the implementation of surface texturing requires a thorough examination of various factors including the choice of materials, manufacturing techniques, and the specific operational conditions of the system. This guarantees that the surface texturing is appropriate for the tribological system and achieves the intended performance. In addition, it is imperative to grasp the intricacies associated with surface texturing in order to achieve favorable outcomes in tribology applications [31–33]. Hence, it is vital to possess a thorough comprehension of the tribology system and the method of surface texturing. The establishment of a collaborative relationship between engineers and material scientists is important in order to guarantee the effectiveness and triumph of tribology applications. Surface texturing is an auspicious approach for enhancing the tribological characteristics of materials, with the capacity to significantly transform various technical domains, including automotive, aerospace, pharmaceutical, and energy applications [34–36]. The aforementioned technology possesses the capacity to diminish energy expenditures, enhance operational efficiency, and prolong the durability of constituent parts. Surface texturing has emerged as a viable and economically efficient approach in the field of tribology, warranting its consideration for use in various tribological applications. This technology exhibits considerable promise and has the potential to significantly influence numerous industries.

6.1.1 Laser cladding

Laser cladding (LC) is a cutting-edge process for modifying the surface characteristics of materials. It entails the use of laser beams to build controlled microstructures on the surface of a material, modifying its surface properties. Because of its potential to increase surface functions and performance, LC has found uses in a variety of industries, including automotive, aerospace, medical, and electronics. It is a fast and efficient method for producing high-precision surfaces with a wide range of characteristics [37–39]. Furthermore, as shown in Figure 6.3, it typically involves some important steps. LC is also a cost-effective process which can be used to modify existing components or create new ones.

The first step is to choose a suitable laser source. The laser used is determined by the material to be textured, the desired surface qualities, and the process precision required. For example, a short-pulse laser is best for texturing plastics, while a continuous-wave laser is better for metals. The power of the laser beam also needs to be considered, as it affects the depth of the texturing. The speed of the laser beam needs to be adjusted to achieve the desired results. Once the laser source is chosen, it can be adjusted to achieve the desired texturing. The laser can also be used to create patterns and shapes on the material's surface. Post-processing techniques can be used to refine the texture and achieve the desired finish [40–42].

The laser parameters, such as power, pulse duration, wavelength, and repetition rate, are then optimized to produce the appropriate microstructures on the material's surface after the laser source has been chosen. The optimal parameters will depend on the material's properties and the desired outcome of the process. The laser is then used to modify the material's surface in a precise and controlled manner. The microstructures on the surface are inspected to ensure that they meet the desired specifications. The material is then tested to verify that the desired results have been achieved [43–45].

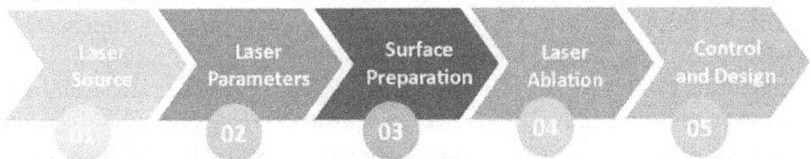

Figure 6.3 Some critical processes are typically involved in the laser cladding process.

The process is then repeated with different parameters, if needed, to ensure the best possible outcome. The surface of the material is normally cleaned and prepped before applying LC to ensure optimal adhesion and quality of the textured characteristics. The LC is then applied in multiple coats, with the thickness and texture of the finish determined by the number of coats applied. Once the final coat is applied, the surface is cured to set the texture into the material. The material is then sealed and polished to protect the surface from wear and tear. Finally, the textured finish is tested to ensure it meets the desired specifications. The laser beam is aimed at the surface of the material and focussed to achieve a particular spot size and intensity [46–48]. The material absorbs the energy from the laser, which leads to localized melting and vaporization of the substance. When the material is allowed to cool down and become solid, the microstructures that are wanted are formed. The mobility of the laser as well as the intensity of its power can be manipulated to produce a variety of distinct patterns, forms, and densities of microstructures on the surface of the material. This enables accurate manipulation of the surface qualities of the finished product. This makes it possible to customize the surface of the material for specific applications. The surface can be roughened or smoothed in order to improve grip or reduce drag. It can also be used to create patterns that can aid in the performance of a product [49–51]. As shown in Figure 6.4, there are many advantages to using laser surface texturing, including some important points.

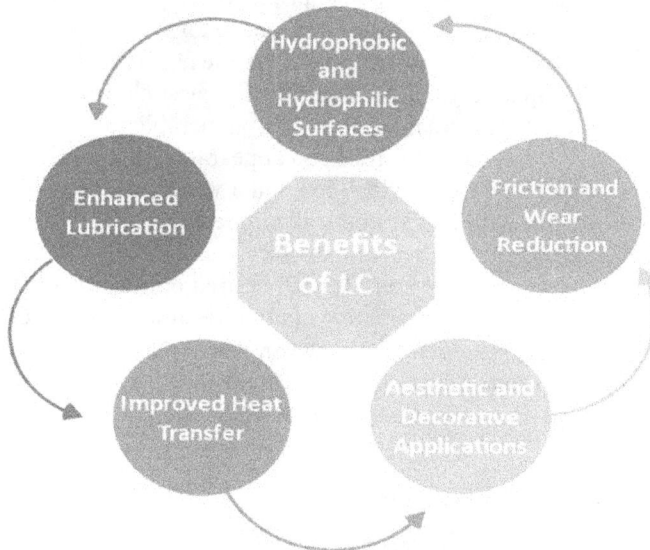

Figure 6.4 Among the benefits of laser cladding (LC).

LC can produce patterns that reduce friction and wear between surfaces, making it useful for applications such as engine components and cutting instruments. LC is also used to improve fuel efficiency in automotive engines, as well as to reduce noise and vibration in machines. Furthermore, LC can be used to improve the performance of cutting tools and other precision instruments. LC can also be used to increase the efficiency of manufacturing processes, reduce material waste and improve product quality [52]. Depending on the application, LC can be used to create surfaces that are either water-repellent (hydrophobic) or water-attractive (hydrophilic). These surfaces have many applications, from medical implants to self-cleaning windows and car surfaces. LC can also be used to modify the surface of a material to make it more durable, increasing its lifetime and reducing its maintenance costs. LC can also be used to modify the surface of a material to increase its adhesion properties, making it easy to paint or coat with other materials. LC can be used to create surfaces with specific electrical properties, such as electrical conductivity or insulating properties [53–55].

Textured surfaces can help mechanical components retain lubricants more effectively, enhancing their performance and durability. This reduces friction between components, which in turn reduces wear and tear, and helps extend the life of the parts. Additionally, textured surfaces help to disperse heat, increasing the efficiency of the parts [56]. Most importantly, textured surfaces help improve the safety of mechanical components by reducing the chances of them slipping and malfunctioning. Laser cladding can improve heat dissipation in high-temperature applications and electronic devices. This is achieved by creating a specific pattern on the surface of the material, which increases its surface area and allows heat to be dissipated more quickly. Laser cladding can also increase the friction between surfaces, providing better grip and stability. Additionally, laser cladding can increase the durability of materials, making them less vulnerable to wear and tear. It can also be used to improve the aesthetic appearance of materials, creating unique and eye-catching designs [57–59]. On a variety of materials, LC can be used to create intricate and aesthetically appealing surface patterns. This technique is used in fields such as architecture, automotive, and fashion. LC can also be used to produce materials with desired optical properties. It has become a popular choice for creating unique decorative surfaces. This is particularly useful for products like eyewear, where the optical properties of the material are of utmost importance. It can also be used to create surfaces with anti-reflective and anti-glare properties [60–62].

6.1.1.1 Biomedical Applications

In the medical field, LC can be used to generate bioactive surfaces on implants, thereby facilitating their integration with the host tissue. Additionally, LC can be used to create 3D structures for tissue engineering, enabling the development of artificial organs. Furthermore, LC can be applied to create biocompatible coatings for medical devices, helping to reduce their potential to induce inflammation. The success of laser cladding is highly dependent on optimizing the laser parameters, comprehending the material properties and customizing the process to satisfy the requirements of each application. As technology continues to advance, laser cladding will persist. With the development of laser surface texturing, precision and efficiency of machining processes have improved significantly. This has enabled numerous industries to benefit from the cost-effectiveness of laser surface texturing. Its advantages have ensured that the technology will remain in use for years to come. Despite this, laser cladding does have some drawbacks [63–65]. The tooling costs associated with laser cladding are high and the process itself can be complex. Additionally, laser cladding is not suitable for certain materials and geometries.

6.1.2 Periodic surface structures induced by lasers, fundamentals, and theory

Laser surface structuring is a technique used to modify the surface of a material by irradiating it with a laser. The formation of laser-induced periodic surface structures (LIPSS), also known as ripples or laser-induced surface gratings, is an intriguing byproduct of this process. LIPSS are periodic patterns that can form on the surface of a material when laser radiation is applied [66]. The periodicities and orientations of these structures are dependent on the laser parameters and material properties. LIPSS can be used to improve the surface properties of a material, such as increasing its roughness or enhancing its reflectivity [67]. These structures are also used in the fabrication of optical components, such as waveguides and diffractive optical elements. LIPSS are also used for medical applications, such as controlling cell adhesion and improving the biocompatibility of materials. This is because of the high aspect ratio and nanoscale features of LIPSS, which enable them to interact with biological systems uniquely [68].

As shown in Figure 6.5, the fundamentals of laser surface structuring are essential to understanding the principles of laser–material interaction and the effects of laser structuring on the surface of materials. It is also important to understand the properties and characteristics of the materials that are being laser structured. It is important to understand the potential applications of laser surface structuring.

Depending on the laser characteristics and material qualities, different interaction mechanisms can occur when a material is treated with a laser beam. Interference between input laser light and surface-scattered waves, as well as self-organization due to nonlinear processes and material

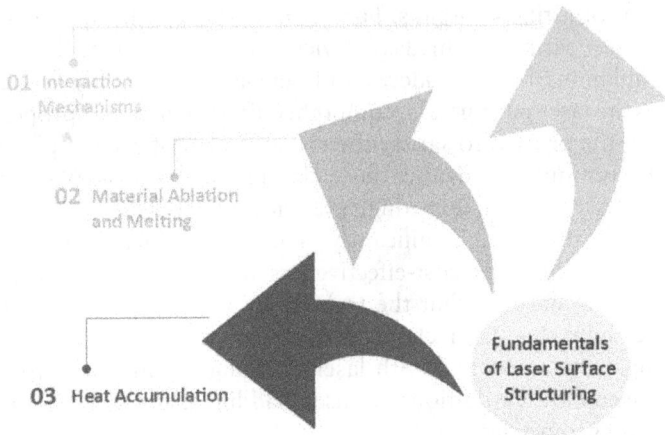

Figure 6.5 Fundamentals of laser surface structuring is crucial to comprehending laser–material principles.

redistribution caused by laser-induced temperature gradients, are the two fundamental mechanisms for the creation of LIPSS. These LIPSS can affect the materials' optical, electrical, and tribological properties. This makes LIPSS a useful tool for creating surfaces with desired qualities. LIPSS can be used in many applications where surface properties are important, such as in biomedical materials, optical coatings, and tribological surfaces. Thus, LIPSS are a promising tool for creating surfaces with superior qualities [69]. Depending on the laser's intensity and pulse duration, it can cause material ablation (removal) or dissolving. At high laser intensities and brief pulse durations, it is possible to rapidly vaporize (ablate) the material. The material may undergo localized melting and resolidification at lesser intensities and longer pulses. This process is known as laser ablation and is used for a wide range of applications. In some cases, it can be used to shape materials as well as create microscopic features. In other cases, it can be used to create precisely controlled chemical reactions. This makes laser ablation an ideal tool for micro- and nanofabrication, particularly for materials that are difficult to machine or chemically etch. Laser ablation can also be used to modify surfaces to improve adhesion, reduce friction, and increase corrosion resistance [70]. Multiple laser pulses striking the same location on the material can lead to heat accumulation. This phenomenon can facilitate the formation of LIPSS and lead to novel interactions. This can be used to modify the material's surface properties, such as its optical reflectance. The use of LIPSS can also enable the production of MEMS and NEMS structures, which can be used in a variety of applications.

6.1.3 Theory of laser-induced periodic surface structures

The formation of LIPSS is a complex phenomenon, and their theoretical origin is still the subject of active research. Several theories and models have been proposed to explain the formation of LIPSS, as depicted in Figure 6.6. Despite the numerous models and theories, the exact mechanism for the formation of LIPSS is still not completely understood. Research is ongoing better to understand this process and its implications for various technologies. Further research is needed to gain a better understanding of LIPSS and to develop new theories to explain their formation [71–73]. This could lead to more efficient and effective methods for fabricating surfaces with LIPSS.

According to the interference theory, periodic structures result from the interference between the incident laser beam and surface-scattered waves. This interference produces periodic patterns consisting of regions of constructive and destructive interference. This interference pattern depends on the wavelength of the incident laser, the geometry of the surface features, and the refractive index of the material. The periodic structures can be used to manipulate light in various ways, such as for optical communications and data storage. The light manipulation also has applications in photonics, sensing, and microscopy. In addition, this interference phenomenon can be used to study the physical properties of materials [74–76].

SPPs are collective electron oscillations at the surface of a material. LIPSS can be produced by the interaction of incident laser light with

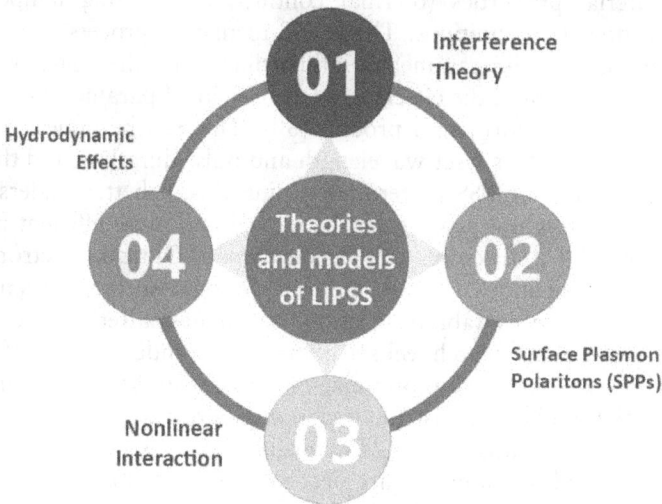

Figure 6.6 Theories and models presented the LIPSS.

SPPs. These oscillations enhance the absorption of light, and the resulting energy is converted into LIPSS. This process is known as surface plasmon-polariton enhanced laser induced periodic surface structures (SPPE-LIPSS). SPPE-LIPSS can be used to modify the physical and optical properties of materials and can be used for applications like nanolithography, biosensing, and solar cell fabrication [77]. At high laser intensities, nonlinear effects occur, resulting in a spatial redistribution of energy and matter on the surface of the material. This redistribution can lead to periodic structure self-organization. These nanostructures have numerous applications, such as for surface-enhanced Raman spectroscopy, improving corrosion resistance, and increasing the efficiency of solar cells. These nanostructures can also be used as templates for further nanostructure fabrication, and as catalysts for chemical reactions [78]. Additionally, these nanostructures can be used to improve the thermal, electrical, and optical properties of the material. Similar to the Rayleigh–Taylor instability in fluid dynamics, laser-induced temperature gradients can contribute to the formation of surface instabilities, which result in the formation of ripples. These ripples can act as a surface for the laser to propagate along, leading to an increased intensity at the surface and further instabilities. This can lead to the formation of more complex structures such as bubbles and cavities. These structures can be used for a variety of applications, including the fabrication of micro- and nanostructures. Additionally, such instabilities can also be used to study the fundamental properties of laser–material interactions [79]. It is important to note that the formation of LIPSS is affected by several parameters, including laser wavelength, pulse duration, fluence (energy per unit area), incident angle, material properties (thermal conductivity, melting temperature), and environmental conditions. The LIPSS formation process is a complex phenomenon, with the parameters influencing each other. Therefore, it is essential to understand the effects of each individual parameter in order to optimize the LIPSS formation process [80]. This can be done by varying the parameters, such as laser wavelength and pulse duration, and then analyzing the resulting LIPSS pattern. By doing this, a better understanding of the LIPSS formation process can be gained and more efficient methods for creating LIPSS structures can be developed. Optical, electronic, and biomaterial applications of laser-induced periodic surface structures for modulating surface wettability, friction, and light–matter interactions are promising. Ongoing research seeks to gain a deeper understanding of the fundamental mechanisms and to optimize the process for specific applications. The potential applications of this technology are vast and wide-ranging, from medical and industrial uses to consumer devices [81]. However, further research and development are needed to fully realize the potential of these technologies. To this end, scientists and engineers are working hard to push the boundaries of what is possible and to bring these technologies to market. It is an exciting time for the field, and the possibilities are endless.

6.2 MOST RECENT DEVELOPMENTS IN LASER-INDUCED PERIODIC SURFACE PATTERNS ON DIFFERENT MATERIALS

This technology allows for the creation of complex topographical structures on materials using laser pulses. The structures have a wide range of applications, from anti-reflection coatings to wettability enhancement and biomedical applications. These structures can be used to create optical devices, such as lenses, mirrors, and filters, as well as biomedical implants, such as stents and catheters [82–84]. In addition, these structures can also be used to enhance the performance of materials, such as improving the corrosion resistance of metals or increasing the heat dissipation of electronics.

6.2.1 LC by direct laser ablation

Laser ablation is a process used to create surface textures on materials by removing material from the surface. This method is frequently used to enhance materials' functionality and performance in a variety of applications. The laser used in LC is typically a high-powered, focused beam that can precisely extract material from the material's surface. Laser ablation produces a surface with small pits and grooves, resulting in a textured surface. The LC-created texture can have varied effects on the properties of the material. It can increase the coefficient of friction, decrease wear and tear, improve adhesion, and increase wettability. These enhancements can be advantageous in a variety of applications, including automotive components, aerospace components, medical devices, and even consumer electronics [85–87].

Laser cladding has numerous advantages over conventional surface treatment techniques. Since there is no physical contact between the laser and the material, the danger of damage or contamination is reduced. It is also a precise and highly controllable technique that permits the creation of complex and customized surface textures. In addition, LC is a relatively quick and effective method, making it appropriate for mass production. In addition to being environmentally sustainable, the method does not require the use of chemicals or other potentially hazardous substances [88–90]. The LC by direct laser ablation is a versatile and efficient technique that can considerably enhance the performance and functionality of a wide range of materials and applications. This procedure is anticipated to play a crucial role in shaping the future of material processing and manufacturing as technology advances. It is an ideal choice for many industries, especially those that prioritize safety and sustainability, and it has the potential to revolutionize the way materials are processed and manufactured [91–93]. This procedure is cost-effective, efficient, and fast, and it is expected to have a positive environmental impact due to reduced energy consumption. As such, it is a promising choice for many industries, and its potential impact on material processing and manufacturing should not be underestimated.

6.2.2 LC by laser shock processing

LC is a method for modifying the surface properties of materials via laser shock processing. In this procedure, high-energy laser pulses are directed onto the surface of the material to generate shockwaves under control. These shockwaves induce surface modifications, such as texturing, melting, and hardening, which enhance the efficacy of the material. This method is especially useful for improving the wear resistance and anti-corrosion properties of the material. Additionally, the process is fast and cost-effective, making it a popular choice for many industries. The process is also eco-friendly, as it does not require the use of any hazardous chemicals. Moreover, the process can be used on a variety of materials, including metals, ceramics, and polymers [94–96]. LC is primarily used to enhance the wear resistance, fatigue strength, and friction coefficient of materials. By creating micropatterns or textures on the surface, LC can increase the contact area and provide greater lubrication, thereby reducing wear and extending the lifespan of the material. This technique also helps to reduce noise and vibration and can be used to improve the surface finish of components. LC can also be used to increase the wettability of a surface, making it more resistant to dirt and dust. This improved wettability helps reduce friction and wear, which can further extend the lifespan of the components. It can also reduce energy consumption and lead to improved performance [97–99].

In addition, the compressive residual stress produced by laser shock processing can increase the material's fatigue resistance by reducing the initiation and propagation of cracks. This makes LC an appropriate method for improving the performance of components subject to cyclic loading, such as turbine blades, gearboxes, and bearings. Laser cladding has advantages over conventional methods of surface treatment because it is noncontact, precise, and readily adaptable to a variety of materials and geometries. In addition, surface patterns can be tailored to specific application requirements. In general, laser cladding by laser shock processing is a promising method for enhancing the performance and durability of materials in a variety of industries [100–102]. It is a subject of ongoing research and development due to its prospective applications in the aerospace, automotive, and biomedical industries. It also offers a viable alternative to traditional machining and grinding processes, as it does not require complex tooling. Additionally, it can be used to create intricate surface patterns for improved functionality and design.

6.3 EFFECTS OF LASER PARAMETERS ON SURFACE TEXTURING

Significant effects of LC parameters on the surface properties of a material may include surface roughness and laser intensity, scanning speed, and pulse frequency can influence the surface's roughness. Greater laser power and slower scanning pace can result in a surface with a coarser texture. LC can induce compressive stresses on the surface of a material, resulting in increased surface hardness and erosion resistance. LC can also improve tribological properties such as wear resistance and friction coefficient [103–105]. It can also be used to increase the adhesion strength of coatings on certain materials. Textured surfaces can alter the material's friction and wear properties, reducing wear and enhancing lubrication efficiency. This makes them ideal for applications such as automotive parts, bearings, and seals. Textured surfaces offer an economical alternative to traditional lubrication methods, such as the use of oil or grease [106–108].

As shown in Figure 6.7, the material and laser parameters, LC can alter the surface's moisture properties to make it more hydrophobic or hydrophilic. This property can be used to improve the material's performance in

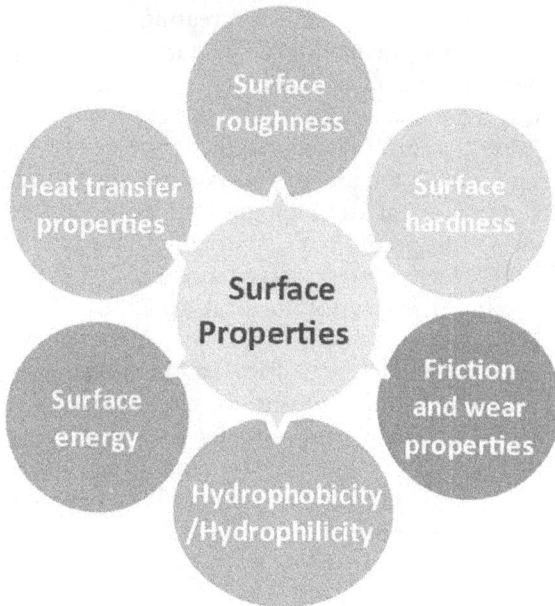

Figure 6.7 Effect of surface properties.

a variety of applications. It also helps to reduce the risk of corrosion and contamination, making the material more durable and reliable [109,110]. This also increases the material's resistance to wear and tear, making it more suitable for long-term use. The laser surface treatment also helps to reduce costs, as fewer replacements are needed due to improved durability. Laser texturing can alter the surface energy of the material, thereby altering its adhesion properties and enhancing bonding in specific applications. This technique can be used for a variety of purposes, such as improving the performance of adhesives and coatings, as well as providing a non-slip surface. It also has the potential to reduce costs and eliminate the need for manual processes. Laser texturing can also be used to create decorative patterns on a surface, as well as improve product appearance [111–113]. LC can increase a material's fatigue life by minimizing stress concentrations and enhancing crack resistance. This can help reduce the risk of unexpected failure, resulting in improved reliability and safety. LC can also be used to extend the service life of components, thus reducing maintenance costs.

Texturing the surface can improve its heat conduction capabilities, allowing it to dissipate heat more efficiently. This can be especially helpful for materials that are exposed to high temperatures or require precise temperature regulation. Heat conduction is an important factor to consider when selecting materials for these applications. Laser cladding can increase the corrosion resistance of a material by creating a protective oxide layer or by modifying the microstructure of the surface. This can be done by creating micro-pits and micro-channels that can trap corrosion products and prevent them from spreading to the bulk material [114–116]. It can also be used to alter the surface energy, which can increase the adhesion of coatings. LC can improve a material's surface biocompatibility, making it more acceptable for medical and biological applications. This is achieved by modifying surface properties such as roughness, wettability, and hydrophobicity. LC also reduces the risk of bacterial and fungal infection, which is an important consideration for medical applications. To achieve the necessary surface qualities while avoiding any negative impacts, the LC parameters must be carefully controlled. To optimize the surface properties of various materials and applications, certain laser parameters may be required [117–119]. The power, pulse length, overlap, speed, and scan pattern may all need to be adjusted. To obtain the desired result, it is important to understand the basic principles of laser surface treatment and adjust the parameters accordingly.

6.3.1 Pulse duration types

Commercial pulse lasers can be recognized by the duration of their laser pulses. The duration of laser pulses is an important component in determining the laser's use and performance in a variety of sectors. As illustrated

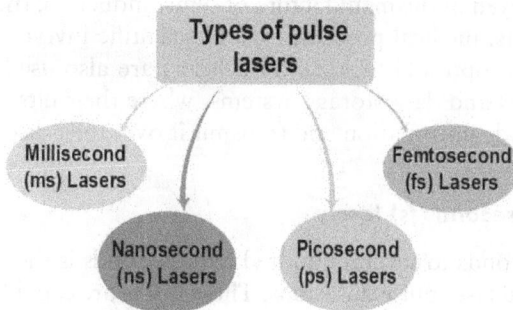

Figure 6.8 Pulse duration and their applications.

in Figure 6.8, there is a brief review of the various types of pulse lasers based on their pulse duration.

6.3.1.1 Millisecond (ms) lasers

The pulse durations of millisecond lasers typically range from a few to tens of milliseconds. These lasers generate rather lengthy pulses, and macro processing activities are frequently connected to their use. They are frequently employed in the processing of materials, including the drilling, welding, and cutting of thicker materials [120,121]. They are also used in the medical field for a range of treatments, including eye surgery, skin resurfacing, and laser therapy. Additionally, they are used in the development of microprocessors and other semiconductors.

6.3.1.2 Nanosecond (ns) lasers

The pulse durations of nanosecond lasers typically range from a few to a few hundred nanoseconds. These lasers are used in many different industries, such as the processing of materials, branding, engraving, and even medical procedures like tattoo removal. They are adaptable for a variety of applications because they establish a balance between pulse energy and duration. Nanosecond lasers are also more efficient than other lasers due to their shorter pulses, making them an ideal choice for a variety of applications [122,123]. They are also more cost-effective than other types of lasers, making them a popular choice for many businesses.

6.3.1.3 Picosecond (ps) lasers

The pulse durations of picosecond lasers are typically in the range of a few picoseconds to a few tens of picoseconds. These lasers are renowned for their capacity to produce incredibly brief pulses with high peak powers. They are employed in the manufacture of semiconductors, the processing of precise materials, medical procedures, and scientific investigation, particularly in ultrafast optics [124,125]. Such lasers are also used in laser-based communications and data storage systems, where their ultrafast pulses can be used to encode information and transmit it over long distances.

6.3.1.4 Femtosecond (fs) lasers

A few femtoseconds to several hundred femtoseconds is the common range for femtosecond laser pulse durations. These lasers are capable of producing ultrafast pulses with enormous peak energies. They are extensively employed in disciplines like ophthalmology, micromachining, ultrafast spectroscopy, and nonlinear optics. Femtosecond lasers are also widely used in the medical field for laser surgery, as they can precisely target individual cells without damaging surrounding tissue [126,127]. In addition, they can be used to etch extremely small features into materials, such as in the manufacture of microchips.

Each variety of pulse laser has advantages and uses in particular fields. The needs of the specific application, as well as the desired level of accuracy and energy delivery, will determine the laser pulse duration. For industrial applications, short-pulse durations are often necessary for precision cutting and welding [128–131]. In medical applications, longer pulse durations may be used to reduce the risk of tissue damage. For research applications, ultrashort pulses may be used to measure extremely fast processes.

6.3.2 Laser power intensity

In LC operations, laser power intensity is a crucial variable that profoundly affects the surface texturing result. The amount of energy provided per unit area of the surface of the target material is referred to as the laser power intensity, also known as laser fluence. The higher the laser power intensity, the deeper and wider the texturing. However, too high of a laser intensity can cause overheating and adverse effects on the surface of the material. Therefore, it is important to find the right balance between laser power intensity and the desired surface texturing [132–134]. Laser power intensity has an impact on the surface texturing process, as illustrated in Figure 6.9. High laser power intensity can result in a high-quality surface texture. However, too much power can cause a decrease in the surface quality due to thermal damage. Therefore, the laser power intensity needs to be carefully adjusted for optimal results. Higher laser power intensities give more energy to the target material per unit area, increasing material removal and ablation. Deeper and larger texture patterns may result as a result on the surface. Increased energy levels also cause higher

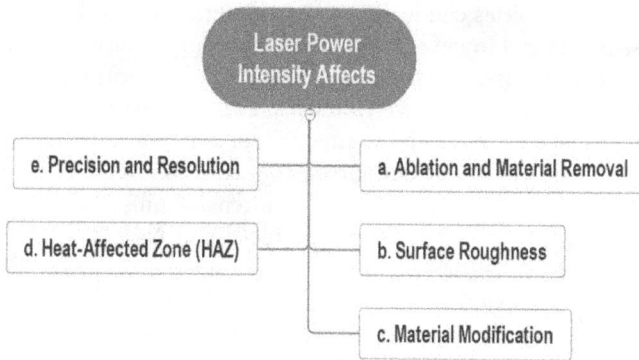

Figure 6.9 Surface texturing is affected by the laser's power level.

temperatures in the target material, resulting in increased stress and strain on the surface. This can lead to increased material damage and reduced surface quality.

The surface roughness of the textured region is directly influenced by the laser power intensity [135–137]. Due to increased material removal and the development of microstructures on the surface, fluence levels that higher tend to produce textures that are rougher. The texture depth also increases as the laser power intensity increases. However, if the laser power intensity is too high, it can lead to the formation of pores and cracks in the material. The quality of the texture heavily depends on the laser power intensity and the material properties. It is therefore important to consider these factors to get the desired results. Moderate laser power levels can modify a material without completely ablating it. For some applications, these changes in material characteristics, such as hardness, crystallinity, and surface chemistry, may be beneficial [138–140]. Laser treatments can be used to tailor the properties of a material to meet the requirements of a specific application. This is especially useful for materials that are difficult to machine or process using conventional methods.

Greater heat-affected zones around the textured area can be produced by using more power, which can change the material's properties there. To prevent undesirable thermal consequences, power intensity must be carefully managed. Overheating can cause warping, cracking, and other damage to the material. The speed of the laser should also be adjusted accordingly to prevent any potential problems. It is important to use the correct settings when laser texturing to achieve the desired results. Typically, lower power intensities are employed to produce finer and more precise surface textures. They enable greater resolution and control during the texturing process. This allows for improved surface finish, greater accuracy, and higher quality results. It also reduces the risk of material distortion or damage [141,142]. Higher power intensities can lead to faster material removal rates, reducing the processing time. However, this may also compromise the quality and precision of the textured surface. As a result, it is important to determine the right balance between power intensity and the quality of the textured surface. This balance should be adjusted depending on the application and the desired output. Depending on the precise needs of the surface texturing application, the right laser power intensity must be chosen. When optimizing the laser power intensity for successful laser surface texturing, factors including material type, desired texture pattern, processing speed, and material heat sensitivity must be considered. The laser power intensity must be high enough to ensure sufficient material ablation, but low enough to avoid thermal damage. The correct setting should produce the desired texture in the shortest possible time [143–145]. It is important to remember that the laser power intensity should be adjusted depending on the material being processed.

6.4 APPLICATIONS

Due to its effectiveness in changing the surface properties of materials, LC has a wide range of applications in a variety of industries. There are some of the main uses for laser surface texturing, as shown in Figure 6.10) [146]. LC is widely utilized in the automobile industry to raise the effectiveness and performance of engine parts. To decrease friction, wear, and fuel consumption, it is applied to piston rings, cylinder liners, and other crucial elements. Additionally, LC can improve the tribological characteristics of transmission gears, lowering noise and extending component life. LC is employed in aerospace applications to improve the efficiency of wing surfaces, landing gear, and engine components. It enhances fuel efficiency, lessens wear and tear, and helps reduce overall weight, which enhances aircraft performance [147,148]. In many industrial processes, LC is used to create precise surface structures that improve lubrication, lessen tool wear, and increase overall machining efficiency. Cutting tools, dies, molds, and other crucial components all use it. LC is used in steam and gas turbines, among other power generation machinery, to increase efficiency and lower maintenance costs [149]. In the printing and packaging sectors, laser cladding is used to produce precise surface patterns on rollers, molds, and printing plates. It enhances printing quality and lowers the possibility of production-related material sticking. LC is used to alter the surface of weaving and knitting components throughout the textile manufacturing process. To increase the effectiveness of the manufacturing process, it can produce anti-slip textures and decrease friction. LC is used to make microstructures on the surfaces of semiconductors, optical lenses, and displays in the electronics and optics sectors [150]. This improves these components' optical

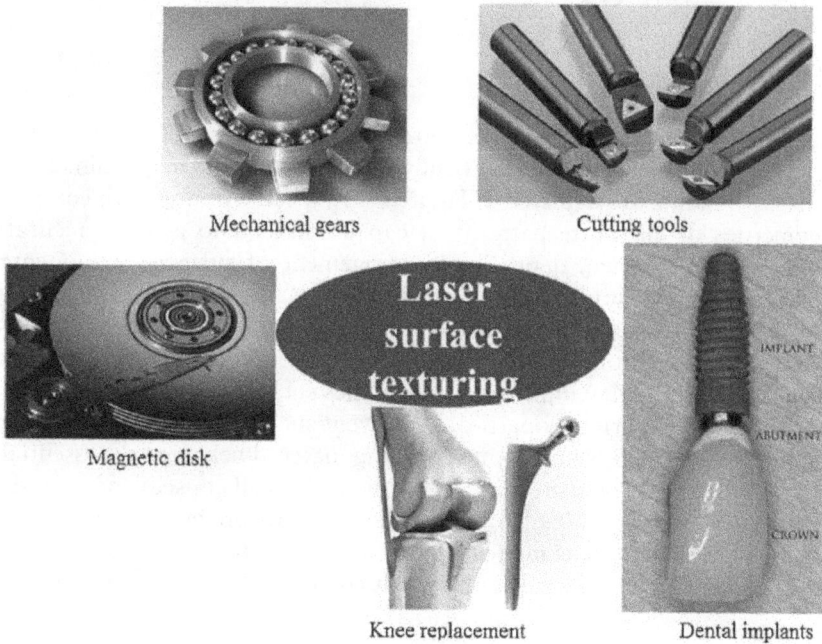

Figure 6.10 Some important applications of the laser surface texturing [146].

qualities and functionality. LC is used in solar panels and wind turbine components to enhance light absorption and lower friction, increasing the efficiency and durability of energy conversion. LC is employed in maritime applications to lessen friction and fouling on ship hulls and offshore structures, improving fuel economy and lowering maintenance costs [151]. These high-temperature and high-stress components operate better and last longer when the surface qualities are optimized. LC is employed in biomedical applications to provide textured surfaces on surgical equipment, prosthetics, and implants. This promotes better tissue integration, lowers friction, and increases biocompatibility.

These examples demonstrate how versatile laser cladding is as a potent method for boosting the effectiveness, dependability, and efficiency of numerous engineering materials and components in many sectors.

6.5 CHALLENGES AND FUTURE DIRECTIONS

Despite its advantages, LC has some challenges such as the need for a precise control of the laser parameters, the heat-affected zone, and the high cost of the equipment. Research is ongoing to improve the efficiency and accuracy of the process in order to make it more accessible to a wider range of industries [152–154].

6.5.1 Current challenges

While LC has many benefits, there are a number of obstacles that must be overcome before it can be successfully used. Some of the major difficulties in laser cladding are depicted in Figure 6.11. When working with complex geometries or elaborate patterns, it can be difficult to achieve accurate control over the size, depth, and arrangement of surface components. Complex laser scanning and control systems are required to maintain consistency and uniformity across a vast region [155]. Laser cladding doesn't work well on all kinds of materials. Due to laser treatment, some materials may display unfavorable outcomes such heat damage, cracking, or changes in material properties. To prevent such problems, choosing the right laser parameters and comprehending material behavior are essential. Heat produced by laser irradiation can have thermal consequences on the material, such as melting or recrystallization. To prevent distortions or changes in the material's microstructure and characteristics, it is crucial to control these heat impacts [156, 157]. It could be expensive to implement

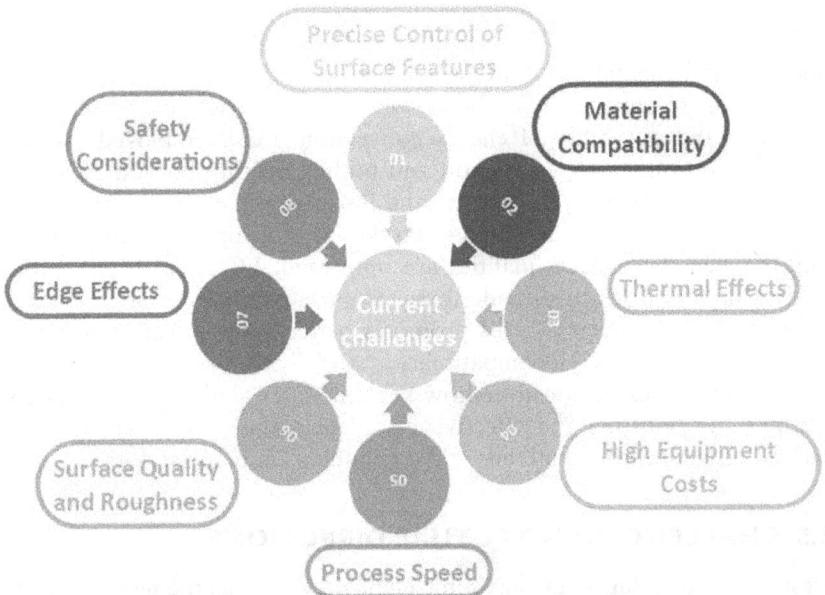

Figure 6.11 Major obstacles in laser surface texturing.

laser cladding if you need specialist laser systems and processing gear. Small-scale applications or enterprises with tight budgets may find it difficult to afford such equipment. When producing intricate patterns or tiny textures, laser cladding can be a time-consuming procedure. To ensure timely and cost-effective output, it is essential to strike a balance between processing speed and accuracy. Poor process control or insufficient laser parameters might provide irregular or rough surface textures, which might not fulfill the standards for some applications. It might be difficult to achieve consistent and homogeneous surface textures up to the material's edges [158–160]. The beam profile and the processing approach may cause edge effects, such as tapering or abnormalities. Potential hazards associated with laser systems include laser radiation and gases. When using laser cladding equipment, it is crucial to ensure operator security and adherence to safety requirements. The transition from laboratory- to industrial-scale LC may be complicated by issues with process stability, cost efficiency, and upholding high standards.

It might be challenging to integrate laser cladding into current manufacturing or production procedures. It can be a big task to integrate with other surface treatment techniques and make sure that downstream procedures are compatible. Advanced laser technologies, material scientific expertise, process optimization, and tight cooperation between researchers, engineers, and industry professionals are all necessary to meet these problems [161–163]. Despite these difficulties, laser cladding is still a valuable method for improving the functionality and performance of engineered materials in a variety of applications.

6.6 FUTURE DIRECTIONS

As new opportunities and uses for this adaptable method are continually made possible by continuous research and technological breakthroughs, the future potential of LC is encouraging. The development of additive manufacturing (3D printing) technologies will be greatly aided by LC. It will be feasible to add intricate and practical surface characteristics directly to 3D-printed components, enhancing their performance and functionality, by adding LC into the printing process. The development of "smart surfaces" that can react to environmental factors like humidity, pressure, or temperature may be made possible through LC. Materials that can display adaptability to friction, wettability, or other qualities because of surface customization could find creative uses in industries including robots, sensors, and biomedical equipment. LC shows considerable promise for the creation of biomedical materials with improved tissue integration and biocompatibility [164]. LC has the potential to enhance the functionality and durability of medical implants and devices by accurately manipulating surface textures.

The use of LC at the nanoscale level will be made possible by developments in laser technology, resulting in ultrafine surface features with distinctive optical, mechanical, and electrical properties. Nanophotonics, nanoelectronics, and other cutting-edge fields will be able to use this. By lowering energy use, chemical use, and material waste, LC can support environmentally responsible manufacturing operations [165]. It can be used with eco-friendly lubricants and surface coatings to provide a more environmentally friendly option for a variety of industrial applications. LC can support the creation of cutting-edge tribological coatings. It will be feasible to make surfaces that are self-lubricating or wear-resistant and have specialized qualities by combining laser cladding with coating materials. Improved light management and optical performance can be achieved by fabricating accurate micro-optical elements on optical components via laser surface texturing. Imaging systems, lasers, and other optical devices will all enjoy the benefit of LC [166–168]. LC can aid in energy-saving efforts by increasing machinery effectiveness and minimizing friction-related energy losses. Several businesses, including manufacturing, energy production, and transportation, may be significantly impacted by this. Adaptive surface texturing depending on application requirements, real-time quality assurance, and effective process management will all be made possible by the integration of LC with automation and artificial intelligence (AI) technologies [169]. LC can be utilized in space exploration to increase component performance and durability in hostile space settings, resulting in the creation of more dependable and long-lasting space systems. The future application of laser cladding will develop as research goes on and laser technology progresses, spurring innovation across numerous industries and creating new opportunities for improving the performance and functionality of materials.

6.7 CONCLUSION

Tribology is crucial in engineering applications involving surface interaction, with a focus on studying friction and wear phenomena to optimize performance and increase longevity. Surface engineering, also known as surface texturing, is a methodology used to control tribological performance through the incorporation of distinct patterns, structures, or textures. Surface texturing treatments can mitigate friction and wear, enhance lubricant effectiveness, and optimize sealing properties. It can also enhance contact characteristics, longevity of components, and yield favorable outcomes such as corrosion resistance and ice prevention. A range of surface textures can be employed to enhance tribological performance, including microtextures, nanotextures, grooves, dimples, and hierarchical textures. These textures contribute to reduced friction and wear, reducing the need for regular lubrication and maintenance, resulting in energy and cost savings. Laser cladding (LC) is a growing trend in industries such as automotive, aerospace, medicine, and electronics, offering a cost-effective approach to tribology and warranting consideration for implementation in various tribology applications. LC involves selecting an appropriate laser source, fine-tuning the laser's operational characteristics, and adjusting the material's surface properties. This process is iterated with varying parameters to optimize the final result. Laser cladding can generate hydrophobic or hydrophilic surfaces suitable for various applications, such as medical implants, self-cleaning windows, and automotive surfaces. Laser-induced periodic surface structures (LIPSS) are periodic patterns that can manifest on a material's surface upon exposure to laser light. These structures have been demonstrated to enhance surface characteristics, augment surface roughness, improve reflectivity, and are employed in the manufacturing of optical components and medical contexts. The production of LIPSS remains incompletely comprehended, necessitating more investigation to enhance our understanding and formulate novel explanatory frameworks for their genesis. Surface plasmon-polariton enhanced laser induced periodic surface structures (SPPE-LIPSS) are laser-induced periodic surface structures that can alter the physical and optical characteristics of many materials, with applications in nanolithography, biosensing, and solar cell production. Laser-induced periodic surface structures (LIPSS) are a rapid, cost-efficient, and environmentally conscious technique used to enhance materials' characteristics, including wear resistance, fatigue strength, and friction coefficient. This technology is advantageous in various industrial sectors, such as optical devices, biomedical implants, and materials aimed at enhancing corrosion resistance and heat dissipation. LC, using direct laser ablation, is a highly adaptable and effective method that can greatly augment the performance and usefulness of diverse materials and applications. The impact of laser parameters on surface texturing encompasses various factors such as surface roughness, compressive stresses, tribological properties, lubrication efficiency, moisture properties, fatigue life, and biocompatibility. Enhancing surface roughness can be achieved through manipulation of the microstructure of the surface, while enhancing corrosion resistance can be achieved by generating a protective oxide layer or changing the microstructure. LC also has the capability to enhance surface biocompatibility, making it more suitable for use in medical and biological applications.

To achieve desired surface characteristics while mitigating adverse effects, it is crucial to exercise meticulous control over the parameters of LC. Specific laser parameters, including power, pulse length, overlap, speed, and scan pattern, may be necessary to enhance the surface characteristics of different materials and applications. Pulse durations play a critical role in determining the application and efficacy of lasers across different industries. The intensity of laser power plays a vital role in laser cladding procedures. An elevated laser power intensity can yield a superior surface texture, but excessive power can lead to surface overheating and detrimental consequences.

The use of laser power at moderate levels can alter a material's characteristics without causing complete ablation, allowing laser treatments to customize features for specific application demands. Lower power intensities provide enhanced resolution and control during the texturing procedure, while increased power intensities may result in accelerated material removal rates, reducing overall processing time. The determination of the optimal laser power intensity for effective laser cladding depends on the specific application and desired outcome. Factors such as material type, preferred texture pattern, processing pace, and material sensitivity play a crucial role in optimizing laser power intensity.

REFERENCES

[1] H.A. Abdullah, R.A. Anaee, Characteristics and Morphological Studies of Nd Doped Titanium Thin Film Coating on SS 316L by DC Sputtering, DIYALA *Journal of Engineering Sciences* 15(3) (2022).

[2] M. Abdulwahab, V. Aigbodion, M. Enechukwu, Anti-corrosion of isothermally treated Ti-6Al-4V alloy as dental biomedical implant using non-toxic bitter leaf extract, *Chemical Data Collections* 24 (2019) 100271.

[3] N. Abhijith, D. Kumar, D. Kalyansundaram, Development of single-stage TiNbMoMnFe high-entropy alloy coating on 304L stainless steel using HVOF thermal spray, *Journal of Thermal Spray Technology* 31(4) (2022) 1032–1044.

[4] A.R. Ahmady, A. Ekhlasi, A. Nouri, M.H. Nazarpak, P. Gong, A. Solouk, High entropy alloy coatings for biomedical applications: A review, *Smart Materials in Manufacturing* 1 (2023) 100009.

[5] A. Amer, A. Al-Shehri, V. Cunningham, H. Saiari, A. Alshamrany, A. Meshaikhis, Artificial Intelligence to Enhance Corrosion Under Insulation Inspection, Abu Dhabi International Petroleum Exhibition and Conference, SPE, 2020, p. D012S116R121.

[6] Q. An, L. Huang, S. Jiang, Y. Bao, M. Ji, R. Zhang, L. Geng, Two-scale TiB/Ti64 composite coating fabricated by two-step process, *Journal of Alloys Compounds* 755 (2018) 29–40.

[7] S. Bajda, Y. Liu, R. Tosi, K. Cholewa-Kowalska, M. Krzyzanowski, M. Dziadek, M. Kopyscianski, S. Dymek, A.V. Polyakov, I.P. Semenova, Laser cladding of bioactive glass coating on pure titanium substrate with highly refined grain structure, *Journal of the Mechanical Behavior of Biomedical Materials* 119 (2021) 104519.

[8] R.R. Behera, A. Hasan, M.R. Sankar, L.M. Pandey, Laser cladding with HA and functionally graded TiO2-HA precursors on Ti–6Al–4V alloy for enhancing bioactivity and cyto-compatibility, *Surface Coatings Technology* 352 (2018) 420–436.

[9] V. Bhojak, J.K. Jain, T.S. Singhal, K.K. Saxena, C. Prakash, M.K. Agrawal, V. Malik, Friction stir processing and cladding: an innovative surface engineering technique to tailor magnesium-based alloys for biomedical implants, *Surface Review Letters* (2023) 2340007.

[10] N. Hua, Z. Qian, B. Lin, Z. Liao, Q. Wang, P. Dai, H. Fang, P.K. Liaw, Formation of a protective oxide layer with enhanced wear and corrosion resistance by heating the TiZrHfNbFe0. 5 refractory multi-principal element alloy at 1,000° C, *Scripta Materialia* 225 (2023) 115165.

[11] M.Z. Ibrahim, A. Halilu, A.A. Sarhan, T. Kuo, F. Yusuf, M. Shaikh, M. Hamdi, In-vitro viability of laser cladded Fe-based metallic glass as a promising bioactive material for improved osseointegration of orthopedic implants, *Medical Engineering Physics* 102 (2022) 103782.

[12] M.Z. Ibrahim, A.A. Sarhan, T. Kuo, F. Yusof, M. Hamdi, Characterization and hardness enhancement of amorphous Fe-based metallic glass laser cladded on nickel-free stainless steel for biomedical implant application, *Materials Chemistry Physics* 235 (2019) 121745.

[13] S. Bose, D. Ke, H. Sahasrabudhe, A. Bandyopadhyay, Additive manufacturing of biomaterials, *Progress in Materials Science* 93 (2018) 45–111.

[14] M. Braik, H. Al-Zoubi, H. Al-Hiary, Artificial neural networks training via bio-inspired optimisation algorithms: modelling industrial winding process, case study, *Soft Computing* 25 (2021) 4545–4569.

[15] D. Castro, P. Jaeger, A.C. Baptista, J.P. Oliveira, An overview of high-entropy alloys as biomaterials, *Metals* 11(4) (2021) 648.

[16] O.N. Çelik, Microstructure and wear properties of WC particle reinforced composite coating on Ti6Al4V alloy produced by the plasma transferred arc method, *Applied Surface Science* 274 (2013) 334–340.

[17] U. Chadha, S.K. Selvaraj, A.S. Lamsal, Y. Maddini, A.K. Ravinuthala, B. Choudhary, A. Mishra, D. Padala, V. Lahoti, A. Adefris, Directed energy deposition via artificial intelligence-enabled approaches, *Complexity* 2022 (2022).

[18] S. Chatterjee, S.S. Mahapatra, V. Bharadwaj, B.N. Upadhyay, K.S. Bindra, Prediction of quality characteristics of laser drilled holes using artificial intelligence techniques, *Engineering with Computers* 37 (2021) 1181–1204.

[19] A. Mehta, G. Singh, Consequences of hydroxyapatite doping using plasma spray to implant biomaterials, *Journal of Electrochemical Science Engineering Failure Analysis* 13(1) (2023) 5–23.

[20] Q. Chen, G.A. Thouas, Metallic implant biomaterials, *Materials Science Engineering: R: Reports* 87 (2015) 1–57.

[21] T. Chen, W. Li, D. Liu, Y. Xiong, X. Zhu, Effects of heat treatment on microstructure and mechanical properties of TiC/TiB composite bioinert ceramic coatings in-situ synthesized by laser cladding on Ti6Al4V, *Ceramics International* 47(1) (2021) 755–768.

[22] Y. Chen, Y. Hu, S. Zhang, X. Mei, Q. Shi, Optimized erosion prediction with MAGA algorithm based on BP neural network for submerged low-pressure water jet, *Applied Sciences* 10(8) (2020) 2926.

[23] Y. Cheng, H. Yang, Y. Yang, J. Huang, K. Wu, Z. Chen, X. Wang, C. Lin, Y. Lai, Progress in TiO 2 nanotube coatings for biomedical applications: a review, *Journal of Materials Chemistry B* 6(13) (2018) 1862–1886.

[24] G. Chryssolouris, K. Alexopoulos, Z. Arkouli, Artificial Intelligence in Manufacturing Processes, A Perspective on Artificial Intelligence in Manufacturing, Springer 2023, pp. 15–39.

[25] A. Mehta, H. Vasudev, N. Jeyaprakash, Role of sustainable manufacturing approach: microwave processing of materials, *International Journal on Interactive Design Manufacturing* (2023) 1–17.

[26] R. Comesaña, Special Issue on Surface Treatment by Laser-Assisted Techniques, MDPI Coatings, 2020, p. 580.

[27] M. Dehghan Manshadi, N. Alafchi, A. Tat, M. Mousavi, A. Mosavi, Comparative analysis of machine learning and numerical modeling for combined heat transfer in Polymethylmethacrylate, *Polymers* 14(10) (2022) 1996.

[28] Z. Deng, D. Liu, G. Liu, Y. Xiong, S. Xin, S. Li, C. Li, T. Chen, Study on the corrosion resistance of SiC particle-reinforced hydroxyapatite-silver gradient bioactive ceramic coatings prepared by laser cladding, *Surface Coatings Technology* (2023) 129734.

[29] A.R. Dhar, D. Gupta, S.S. Roy, A.K. Lohar, Forward and backward modeling of direct metal deposition using metaheuristic algorithms tuned artificial neural network and extreme gradient boost, *Progress in Additive Manufacturing* (2022) 1–15.

[30] C. Domínguez-Trujillo, F. Ternero, J.A. Rodríguez-Ortiz, J.J. Pavón, I. Montealegre-Meléndez, C. Arévalo, F. García-Moreno, Y. Torres, Improvement of the balance between a reduced stress shielding and bone ingrowth by bioactive coatings onto porous titanium substrates, *Surface Coatings Technology* 338 (2018) 32–37.

[31] A.B. Edathazhe, Investigation of properties, corrosion and bioactivity of novel BaO added phosphate glasses and glass-ceramic coating on biomedical metallic implant materials, National Institute of Technology Karnataka, Surathkal, 2018.

[32] G. Ertugrul, A. Hälsig, J. Hensel, J. Buhl, S. Härtel, Efficient multi-material and high deposition coating including additive manufacturing by tandem plasma transferred arc welding for functionally graded structures, *Metals* 12(8) (2022) 1336.

[33] V. Eyupoglu, E. Polat, B. Eren, R.A. Kumbasar, Two-dimensional assessment of cobalt transport and separation through ionic polymer inclusion membrane: experimental optimization and artificial neural network modeling, *Journal of Dispersion Science Technology* 44(5) (2023) 763–778.

[34] A. Mehta, H. Vasudev, S. Singh, Sustainable manufacturing approach with novel thermal barrier coatings in lowering CO2 emissions: Performance analysis with probable solutions, *International Journal on Interactive Design Manufacturing* (2023) 1–13.

[35] L. Fan, H. CHEN, Y. Dong, X. LI, L. DONG, Y. YIN, Corrosion behavior of Fe-based laser cladding coating in hydrochloric acid solutions, *Acta Metall Sin* 54(7) (2018) 1019–1030.

[36] X.a. Fan, X. Gao, G. Liu, N. Ma, Y. Zhang, Research and prospect of welding monitoring technology based on machine vision, *The International Journal of Advanced Manufacturing Technology* 115 (2021) 3365–3391.

[37] A. Farazin, C. Zhang, A. Gheisizadeh, A. Shahbazi, 3D bio-printing for use as bone replacement tissues: A review of biomedical application, *Biomedical Engineering Advances* (2023) 100075.

[38] R. Fathi, H. Wei, B. Saleh, N. Radhika, J. Jiang, A. Ma, M.H. Ahmed, Q. Li, K.K.J.A.M.T. Ostrikov, Past and present of functionally graded coatings: Advancements and future challenges, *Applied Materials Today* 26 (2022) 101373.

[39] O.S. Fatoba, O.S. Adesina, A. Popoola, Evaluation of microstructure, microhardness, and electrochemical properties of laser-deposited Ti-Co coatings on Ti-6Al-4V Alloy, *The International Journal of Advanced Manufacturing Technology* 97 (2018) 2341–2350.

[40] Y. Sharma, A. Mehta, H. Vasudev, N. Jeyaprakash, G. Prashar, C. Prakash, Analysis of friction stir welds using numerical modelling approach: a comprehensive review, *International Journal on Interactive Design Manufacturing* (2023) 1–14.

[41] J. Feng, Y. Tang, J. Liu, P. Zhang, C. Liu, L. Wang, Bio-high entropy alloys: Progress, challenges, and opportunities, *Frontiers in Bioengineering Biotechnology* 10 (2022) 977282.

[42] M. Feng, W. Yao, J. An, Z. Huang, Y. Yuan, Y. Yao, Prediction Method of Graphene Defect Modification Based on Neural Network, *Mobile Information Systems* 2022 (2022).

[43] G. Singh, A. Mehta, A. Bansal, Electrochemical behaviour and bio-compatibility of claddings developed using microwave route, *Journal of Electrochemical Science Engineering Failure Analysis* 13(1) (2023) 173–192.

[44] S.F.S. Ferreiro, J. Larreina, M. Tena, J. Leunda, I. Garmendia, A. Arnaiz, Artificial intelligence methodology for smart and sustainable manufacturing industry, *IFAC-PapersOnLine* 54(1) (2021) 1041–1046.

[45] S. Fetni, Q.D.T. Pham, V.X. Tran, L. Duchêne, H.S. Tran, A.M. Habraken, Thermal field prediction in DED manufacturing process using Artificial Neural Network, *Soft Computing* 24 (2021) 73–78.

[46] H. Singh, A. Mehta, Y. Sharma, H. Vasudev, Role of expert systems to optimize the friction stir welding process parameters using numerical modelling: a review, *International Journal on Interactive Design Manufacturing* (2023) 1–17.

[47] F.J.S.E. Findik, Innovation, Laser cladding and applications, *Sustainable Engineering Innovation* 5(1) (2023) 1–14.

[48] Q. Fu, W. Liang, J. Huang, W. Jin, B. Guo, P. Li, S. Xu, P.K. Chu, Z. Yu, Research perspective and prospective of additive manufacturing of bio-degradable magnesium-based materials, *Journal of Magnesium Alloys* (2023) 1032–1044.

[49] N. Singh, A. Mehta, H. Vasudev, P.S. Samra, A review on the design and analysis for the application of Wear and corrosion resistance coatings, *International Journal on Interactive Design Manufacturing* (2023) 1–25.

[50] M. Ganjali, M. Ganjali, S. Sadrnezhaad, Y. Pakzad, Laser cladding of Ti alloys for biomedical applications, *Laser Cladding of Metals* (2021) 265–292.

[51] M. Ganjali, A. Yazdanpanah, M. Mozafari, Laser deposition of nano coatings on biomedical implants, Emerging Applications of Nanoparticles and Architecture Nanostructures, Elsevier 2018, pp. 235–254.

[52] F. Ghadami, A.S.R. Aghdam, Improvement of high velocity oxy-fuel spray coatings by thermal post-treatments: A critical review, Thin Solid Films 678 (2019) 42–52.

[53] P.K. Verma, A. Mehta, H. Vasudev, V.J.S.R. Kumar, Performance of thermal spray coated metallic materials for bio-implant applications, *Surface Review Letters* (2023) 2250017.

[54] J. Garcia-Herrera, J. Henao, D. Espinosa-Arbelaez, J. Gonzalez-Carmona, C. Felix-Martinez, R. Santos-Fernandez, J. Corona-Castuera, C. Poblano-Salas, J. Alvarado-Orozco, Laser cladding deposition of a Fe-based metallic glass on 304 stainless steel substrates, *Journal of Thermal Spray Technology* 31(4) (2022) 968–979.

[55] S. Gaytan, L. Murr, E. Martinez, J. Martinez, B. Machado, D. Ramirez, F. Medina, S. Collins, R. Wicker, Comparison of microstructures and mechanical properties for solid and mesh cobalt-base alloy prototypes fabricated by electron beam melting, *Metallurgical Materials Transactions A* 41 (2010) 3216–3227.

[56] T. Hanawa, Reconstruction and regeneration of surface oxide film on metallic materials in biological environments, *Corrosion Reviews* 21(2–3) (2003) 161–182.

[57] V.S. Yedida, A. Mehta, H. Vasudev, S. Singh, Role of numerical modeling in predicting the oxidation behavior of thermal barrier coatings, *International Journal on Interactive Design Manufacturing* (2023) 1–10.

[58] Y. Gomez Taborda, M. Gómez Botero, J.G. Castaño-González, A. Bermúdez-Castañeda, Assessment of physical, chemical, and tribochemical properties of biomedical alloys used in explanted modular hip prostheses: A review, *Proceedings of the Institution of Mechanical Engineers, Part H: Journal of Engineering in Medicine* 236(4) (2022) 457–468.

[59] C. Han, Y. Li, Q. Wang, D. Cai, Q. Wei, L. Yang, S. Wen, J. Liu, Y. Shi, Titanium/hydroxyapatite (Ti/HA) gradient materials with quasi-continuous ratios fabricated by SLM: material interface and fracture toughness, *Materials Design* 141 (2018) 256–266.

[60] W. Harun, R. Asri, J. Alias, F. Zulkifli, K. Kadirgama, S. Ghani, J. Shariffuddin, A comprehensive review of hydroxyapatite-based coatings adhesion on metallic biomaterials, *Ceramics International* 44(2) (2018) 1250–1268.

[61] G. He, Y. Du, Q. Liang, Z. Zhou, L. Shu, Modeling and Optimization Method of Laser Cladding Based on GA-ACO-RFR and GNSGA-II, *International Journal of Precision Engineering Manufacturing-Green Technology* (2022) 1–16.

[62] B. Heer, A. Bandyopadhyay, Silica coated titanium using Laser Engineered Net Shaping for enhanced wear resistance, *Additive Manufacturing* 23 (2018) 303–311.

[63] W. Hu, J. Fang, Z. Liu, J. Tan, Intelligent design and optimization of wind turbines, Wind Energy Engineering, Elsevier 2023, pp. 315–325.

[64] Y. Hu, W. Cong, A review on laser deposition-additive manufacturing of ceramics and ceramic reinforced metal matrix composites, *Ceramics International* 44(17) (2018) 20599–20612.

[65] N. Hua, H. Huang, X. Zhang, Investigating the working efficiency of typical work in high-altitude alpine metal mining areas based on a SeqGAN-GABP mixed algorithm, *Advances in Civil Engineering* 2021 (2021) 1–12.

[66] J. Bonse, S. Gräf, Maxwell meets Marangoni—A review of theories on laser-induced periodic surface structures, *Laser Photonics Reviews* 14(10) (2020) 2000215.

[67] S. Höhm, A. Rosenfeld, J. Krüger, J. Bonse, Femtosecond laser-induced periodic surface structures on silica, *Journal of Applied Physics* 112(1) (2012).

[68] M.V. Shugaev, I. Gnilitskyi, N.M. Bulgakova, L.V. Zhigilei, Mechanism of single-pulse ablative generation of laser-induced periodic surface structures, *Physical Review B* 96(20) (2017) 205429.

[69] M. Garcia-Lechuga, D. Puerto, Y. Fuentes-Edfuf, J. Solis, J. Siegel, Ultrafast moving-spot microscopy: Birth and growth of laser-induced periodic surface structures, *Acs Photonics* 3(10) (2016) 1961–1967.

[70] I. Gnilitskyi, T.J.-Y. Derrien, Y. Levy, N.M. Bulgakova, T. Mocek, L. Orazi, High-speed manufacturing of highly regular femtosecond laser-induced periodic surface structures: Physical origin of regularity, *Scientific reports* 7(1) (2017) 8485.

[71] J. Bonse, R. Koter, M. Hartelt, D. Spaltmann, S. Pentzien, S. Höhm, A. Rosenfeld, J. Krüger, Tribological performance of femtosecond laser-induced periodic surface structures on titanium and a high toughness bearing steel, *Applied Surface Science* 336 (2015) 21–27.

[72] C. Kunz, J. Bonse, D. Spaltmann, C. Neumann, A. Turchanin, J.F. Bartolomé, F.A. Mueller, S. Graef, Tribological performance of metal-reinforced ceramic composites selectively structured with femtosecond laser-induced periodic surface structures, *Applied Surface Science* 499 (2020) 143917.

[73] C. Florian, S.V. Kirner, J. Krüger, J. Bonse, Surface functionalization by laser-induced periodic surface structures, *Journal of Laser Applications* 32(2) (2020).

[74] J. Bonse, R. Koter, M. Hartelt, D. Spaltmann, S. Pentzien, S. Höhm, A. Rosenfeld, J. Krüger, Femtosecond laser-induced periodic surface structures on steel and titanium alloy for tribological applications, *Applied physics A* 117 (2014) 103–110.

[75] J.J. Ayerdi, N. Slachciak, I. Llavori, A. Zabala, A. Aginagalde, J. Bonse, D. Spaltmann, On the role of a ZDDP in the tribological performance of femtosecond laser-induced periodic surface structures on titanium alloy against different counterbody materials, *Lubricants* 7(9) (2019) 79.

[76] J. Bonse, S.V. Kirner, S. Höhm, N. Epperlein, D. Spaltmann, A. Rosenfeld, J. Krüger, Applications of laser-induced periodic surface structures (LIPSS), Laser-based Micro-and Nanoprocessing XI, *SPIE*, 2017, pp. 114–122.

[77] P. Onufrijevs, L. Grase, J. Padgurskas, M. Rukanskis, R. Durena, D. Willer, M. Iesalnieks, J. Lungevics, J. Kaupuzs, R. Rukuiža, Anisotropy of the Tribological Performance of Periodically Oxidized Laser-Induced Periodic Surface Structures, *Coatings* 13(7) (2023) 1199.

[78] J. Bonse, S. Höhm, S.V. Kirner, A. Rosenfeld, J. Krüger, Laser-induced periodic surface structures—A scientific evergreen, IEEE *Journal of Selected Topics in Quantum Electronics* 23(3) (2016).

[79] J. Bonse, S.J.N. Gräf, Ten open questions about laser-induced periodic surface structures, *Nanomaterials* 11(12) (2021) 3326.

[80] J. Bonse, S.V. Kirner, J. Krüger, Laser-induced periodic surface structures (LIPSS), Handbook of laser micro-and nano-engineering (2020) 1–59.

[81] F.A. Müller, C. Kunz, S. Gräf, Bio-inspired functional surfaces based on laser-induced periodic surface structures, *Materials* 9(6) (2016) 476.

[82] M.Z. Ibrahim, A.A. Sarhan, M. Shaikh, T. Kuo, F. Yusuf, M. Hamdi, Investigate the effects of the laser cladding parameters on the microstructure, phases formation, mechanical and corrosion properties of metallic glasses coatings for biomedical implant application, *Additive Manufacturing of Emerging Materials* (2019) 299–323.

[83] H.Z. Imam, Y. Zheng, P. Martinez, R. Ahmad, Vision-based damage localization method for an autonomous robotic laser cladding process, *Procedia CIRP* 104 (2021) 827–832.

[84] M.N. Jahangir, M.A.H. Mamun, M.P. Sealy, A review of additive manufacturing of magnesium alloys, AIP conference proceedings, AIP Publishing, 2018.

[85] X. Ji, C. Luo, J. Jin, Y. Zhang, Y. Sun, L. Fu, Tribocorrosion performance of 316L stainless steel enhanced by laser clad 2-layer coating using Fe-based amorphous powder, *Journal of Materials Research Technology* 17 (2022) 612–621.

[86] W. JIA, X. LIN, Numerical microstructure simulation of laser rapid forming 316L stainless steel, *Acta Metall Sin* 46(2) (2010) 135–140.

[87] N.O. Joy-anne, Y. Su, X. Lu, P.-H. Kuo, J. Du, D. Zhu, Bioactive glass coatings on metallic implants for biomedical applications, *Bioactive Materials* 4 (2019) 261–270.

[88] K. Kanishka, B.J.J.o.M.P. Acherjee, A systematic review of additive manufacturing-based remanufacturing techniques for component repair and restoration, *Journal of Manufacturing Processes* 89 (2023) 220–283.

[89] A. Kansal, A. Dvivedi, P. Kumar, Development and performance study of biomedical porous zinc scaffold manufactured by using additive manufacturing and microwave sintering, *Materials Manufacturing Processes* 38(8) (2023) 1020–1032.

[90] N. Kaushik, A. Meena, H.S. Mali, High entropy alloy synthesis, characterisation, manufacturing & potential applications: a review, *Materials Manufacturing Processes* 37(10) (2022) 1085–1109.

[91] V. Kaushik, N. Kumar, M. Vignesh, Magnesium role in additive manufacturing of biomedical implants–challenges and opportunities, *Additive Manufacturing* 55 (2022) 102802.

[92] M. Khanzadeh, W. Tian, A. Yadollahi, H.R. Doude, M.A. Tschopp, L.J.A.M. Bian, Dual process monitoring of metal-based additive manufacturing using tensor decomposition of thermal image streams, *Additive Manufacturing* 23 (2018) 443–456.

[93] A.M. Khorasani, M. Goldberg, E.H. Doeven, G. Littlefair, Titanium in biomedical applications—properties and fabrication: a review, *Journal of Biomaterials Tissue Engineering* 5(8) (2015) 593–619.

[94] M. Krzyzanowski, D. Svyetlichnyy, S. Bajda, Additive manufacturing of multi layered bioactive materials with improved mechanical properties: modelling aspects, Materials Science Forum, *Trans Tech Publ*, 2021, pp. 888–893.

[95] M. Kumar, S. Kumar, K. Jha, A. Mandal, Composite coating by TIG cladding with different rare earth oxides, *Surface Engineering* 38(3) (2022) 271–287.

[96] P. Kumar, N.K. Jain, A. Tiwari, Sustainable Polishing of Directed Energy Deposition–Based Cladding Using Micro-Plasma Transferred Arc, Advances in Sustainable Machining and Manufacturing Processes, CRC Press 2022, pp. 289–302.

[97] S. Kumar, P. Katyal, Factors affecting biocompatibility and biodegradation of magnesium based alloys, *Materials Today: Proceedings* 52 (2022) 1092–1107.

[98] S.S. Kumar, V. Tripathi, R. Sharma, G. Puthilibai, M. Sudhakar, K. Negash, Study on developments in protection coating techniques for steel, *Advances in Materials Science Engineering* 2022 (2022) 331–344.

[99] W. Li, P. Xu, Y. Wang, Y. Zou, H. Gong, F. Lu, Laser synthesis and microstructure of micro-and nano-structured WC reinforced Co-based cladding layers on titanium alloy, *Journal of Alloys Compounds* 749 (2018) 10–22.

[100] W.Y.S. Lim, J. Cao, A. Suwardi, T.L. Meng, C.K.I. Tan, H. Liu, Recent advances in laser-cladding of metal alloys for protective coating and additive manufacturing, *Journal of Adhesion Science Technology* 36(23–24) (2022) 2482–2504.

[101] J. Liu, D. Liu, S. Li, Z. Deng, Z. Pan, C. Li, T. Chen, The effects of graphene oxide doping on the friction and wear properties of TiN bioinert ceramic coatings prepared using wide-band laser cladding, *Surface Coatings Technology* 458 (2023) 129354.

[102] W. Liu, S. Liu, L. Wang, Surface modification of biomedical titanium alloy: micromorphology, microstructure evolution and biomedical applications, *Coatings* 9(4) (2019) 249.

[103] Z. Liu, K.C. Chan, L. Liu, S. Guo, Bioactive calcium titanate coatings on a Zr-based bulk metallic glass by laser cladding, *Materials Letters* 82 (2012) 67–70.

[104] M. Lu, P. McCormick, Y. Zhao, Z. Fan, H. Huang, Laser deposition of compositionally graded titanium oxide on Ti6Al4V alloy, *Ceramics International* 44(17) (2018) 20851–20861.

[105] N. Ma, S. Liu, W. Liu, L. Xie, D. Wei, L. Wang, L. Li, B. Zhao, Y. Wang, Research progress of titanium-based high entropy alloy: methods, properties, and applications, *Frontiers in Bioengineering Biotechnology* 8 (2020) 603522.

[106] A. Mahajan, S. Devgan, D.J.M. Kalyanasundaram, M. Processes, Surface alteration of Cobalt-Chromium and duplex stainless steel alloys for biomedical applications: a concise review, *Materials Manufacturing Processes* 38(3) (2023) 260–270.

[107] M.A. Mahmood, A.C. Popescu, I.N. Mihailescu, Metal matrix composites synthesized by laser-melting deposition: a review, *Materials* 13(11) (2020) 2593.

[108] M.A. Mahmood, A.C. Popescu, M. Oane, A. Channa, S. Mihai, C. Ristoscu, I.N. Mihailescu, Bridging the analytical and artificial neural

network models for keyhole formation with experimental verification in laser melting deposition: A novel approach, *Results in Physics* 26 (2021) 104440.

[109] A.I. Mahmoud Zakaria Alsayed, Laser cladding of FeCrMoCB metallic glass on nickel-free stainless-steel to develop durable and cost-effective biomedical implants/Mahmoud Zakaria Alsayed Abdalfattah Ibrahim, PhD diss, Universiti Malaya, 2019.

[110] J.D. Majumdar, A. Kumar, S. Pityana, I. Manna, Laser surface melting of AISI 316L stainless steel for bio-implant application, *Proceedings of the National Academy of Sciences, India Section A: Physical Sciences* 88 (2018) 387–403.

[111] A.S.H. Makhlouf, A. Barhoum, Emerging applications of nanoparticles and architectural nanostructures: current prospects and future trends, *Metals* 37 (2018).

[112] J. Mesquita-Guimarães, B. Henriques, F. Silva, Bioactive glass coatings, Bioactive glasses, Elsevier 2018, pp. 103–118.

[113] M.H. Miah, D. Singh Chand, G.S. Malhi, S. Khan, Influence of laser scanning power on microstructure and tribological behavior of NI-composite claddings fabricated on TC4 titanium alloy, *Aircraft Engineering Aerospace Technology* (2023) 103–118.

[114] A.-C. Mocanu, F. Miculescu, G.E. Stan, I. Pasuk, T. Tite, A. Pascu, T.M. Butte, L.-T. Ciocan, Modulated Laser Cladding of Implant-Type Coatings by Bovine-Bone-Derived Hydroxyapatite Powder Injection on Ti6Al4V Substrates—Part I: Fabrication and Physico-Chemical Characterization, *Materials* 15(22) (2022) 7971.

[115] A.-C. Mocanu, F. Miculescu, G.E. Stan, T. Tite, M. Miculescu, M.H. Țierean, A. Pascu, R.-C. Ciocoiu, T.M. Butte, L.-T. Ciocan, Development of ceramic coatings on titanium alloy substrate by laser cladding with pre-placed natural derived-slurry: Influence of hydroxyapatite ratio and beam power, *Ceramics International* 49(7) (2023) 10445–10454.

[116] B. Mohajernia, S.E. Mirazimzadeh, A. Pasha, R.J. Urbanic, Machine learning approaches for predicting geometric and mechanical characteristics for single P420 laser beads clad onto an AISI 1018 substrate, *The International Journal of Advanced Manufacturing Technology* (2022) 1–20.

[117] N.P. Msweli, S.O. Akinwamide, P.A. Olubambi, B.A. Obadele, Microstructure and biocorrosion studies of spark plasma sintered yttria stabilized zirconia reinforced Ti6Al7Nb alloy in Hanks' solution, *Materials Chemistry Physics* 293 (2023) 126940.

[118] A. Mthisi, A. Popoola, D. Adebiyi, O. Popoola, Laser Cladding of Ti-6Al-4V Alloy with Ti-Al2O3 Coating for Biomedical Applications, IOP Conference Series: Materials Science and Engineering, IOP Publishing, 2018, p. 012005.

[119] S. Mukherjee, S. Dhara, P. Saha, Laser surface remelting of Ti and its alloys for improving surface biocompatibility of orthopaedic implants, *Materials Technology* 33(2) (2018) 106–118.

[120] F. Mumali, Artificial neural network-based decision support systems in manufacturing processes: A systematic literature review, *Computers Industrial Engineering* 165 (2022) 107964.

[121] J. Škamat, O. Černašėjus, G. Zhetessova, T. Nikonova, O. Zharkevich, N. Višniakov, Effect of laser processing parameters on microstructure, hardness and tribology of NiCrCoFeCBSi/WC coatings, *Materials* 14(20) (2021) 6034.

[122] K. Munir, A. Biesiekierski, C. Wen, Y. Li, Surface modifications of metallic biomaterials, Metallic Biomaterials Processing and Medical Device Manufacturing, Elsevier 2020, pp. 387–424.

[123] V. Kumar, R. Verma, S. Kango, V.S. Sharma, Recent progresses and applications in laser-based surface texturing systems, *Materials Today Communications* 26 (2021) 101736.

[124] L. Murr, Metallurgy principles applied to powder bed fusion 3D printing/ additive manufacturing of personalized and optimized metal and alloy biomedical implants: An overview, *Journal of Materials Research Technology* 9(1) (2020) 1087–1103.

[125] P. Pou-Álvarez, A. Riveiro, X. Nóvoa, M. Fernández-Arias, J. Del Val, R. Comesaña, M. Boutinguiza, F. Lusquiños, J. Pou, Nanosecond, picosecond and femtosecond laser surface treatment of magnesium alloy: Role of pulse length, *Surface Coatings Technology* 427 (2021) 127802.

[126] R. Nazempour, Q. Zhang, R. Fu, X. Sheng, Biocompatible and implantable optical fibers and waveguides for biomedicine, *Materials* 11(8) (2018) 1283.

[127] A. Michalek, S. Qi, A. Batal, P. Penchev, H. Dong, T.L. See, S. Dimov, Submicron structuring/texturing of diamond-like carbon-coated replication masters with a femtosecond laser, *Applied Physics A* 126 (2020) 1–12.

[128] T.D. Ngo, A. Kashani, G. Imbalzano, K.T. Nguyen, D.J.C.P.B.E. Hui, Additive manufacturing (3D printing): A review of materials, methods, applications and challenges, *Composites Part B: Engineering* 143 (2018) 172–196.

[129] M. Niinomi, Recent metallic materials for biomedical applications, *Metallurgical Materials Transactions* A 33 (2002) 477–486.

[130] M.P. Nikolova, M.D. Apostolova, Advances in Multifunctional Bioactive Coatings for Metallic Bone Implants, *Materials* 16(1) (2022) 183.

[131] Z. Wu, H. Bao, Y. Xing, L. Liu, Tribological characteristics and advanced processing methods of textured surfaces: A review, *The International Journal of Advanced Manufacturing Technology* 114 (2021) 1241–1277.

[132] S. Omarov, N. Nauryz, D. Talamona, A. Perveen, Surface Modification Techniques for Metallic Biomedical Alloys: A Concise Review, *Metals* 13(1) (2022) 82.

[133] A.A. Oudah, M.A. Hassan, N. Almuramady, Materials manufacturing processes: Feature and trends, AIP Conference Proceedings, AIP Publishing, 2023.

[134] G. Padmanabham, R. Bathe, Laser materials processing for industrial applications, Proceedings of the National Academy of Sciences, *India Section A: Physical Sciences* 88 (2018) 359–374.

[135] U.M.R. Paturi, D.G. Vanga, S. Cheruku, S.T. Palakurthy, N.K.J.M.T.P. Jha, Estimation of abrasive wear of nanostructured WC-10Co-4Cr TIG weld cladding using neural network and fuzzy logic approach, *Materials Today: Proceedings* 78 (2023) 449–457.

[136] Y. Qu, T. Nguyen-Dang, A.G. Page, W. Yan, T. Das Gupta, G.M. Rotaru, R.M. Rossi, V.D. Favrod, N. Bartolomei, F. Sorin, Superelastic multimaterial electronic and photonic fibers and devices via thermal drawing, *Advanced Materials* 30(27) (2018) 1707251.

[137] Y. Qu, T. Nguyen-Dang, A.G. Page, W. Yan, T.D. Gupta, G.M. Rotaru, R.M. Rossi, V.D. Favrod, N. Bartolomei, F. Sorin, Stretchable optical and electronic fibers via thermal drawing, International Flexible Electronics Technology Conference (IFETC), IEEE, 2018, pp. 1–1.

[138] M.A. Rahman, T. Saleh, M.P. Jahan, C. McGarry, A. Chaudhari, R. Huang, M. Tauhiduzzaman, A. Ahmed, A.A. Mahmud, M.S. Bhuiyan, Review of Intelligence for Additive and Subtractive Manufacturing: Current Status and Future Prospects, *Micromachines* 14(3) (2023) 508.

[139] R. Ranjan, A.K. Das, Improving the Resistance to Wear and Mechanical Characteristics of Cladding Layers on Titanium and its Alloys: A Review, *Tribology in Industry* 44(1) (2023) 136.

[140] J. Reddy, M. Chamanzar, Parylene photonic waveguide arrays: a platform for implantable optical neural implants, CLEO: Applications and Technology, Optica Publishing Group, 2018, p. AM3P. 6.

[141] J.W. Reddy, M. Chamanzar, Low-loss flexible Parylene photonic waveguides for optical implants, *Optics Letters* 43(17) (2018) 4112–4115.

[142] R. Rojo, M. Prados-Privado, A.J. Reinoso, J.C. Prados-Frutos, Evaluation of fatigue behavior in dental implants from in vitro clinical tests: a systematic review, *Metals* 8(5) (2018) 313.

[143] D. Romanov, K. Sosnin, S.Y. Pronin, Y.F. Ivanov, V. Gromov, Structure and Properties of Electroexplosion Molybdenum Coating Deposited on Titanium Alloy VT6, *Metal Science Heat Treatment* 64(11) (2023) 639–647.

[144] W. Sallehhudin, A. Diab, Using machine learning to predict the fuel peak cladding temperature for a large break loss of coolant accident, *Frontiers in Energy Research* 9 (2021) 755638.

[145] M. Salonvaara, A. Desjarlais, A.J. Aldykiewicz Jr, E. Iffa, P. Boudreaux, J. Dong, B. Liu, G. Accawi, D. Hun, E. Werling, Application of Machine Learning to Assist a Moisture Durability Tool, *Energies* 16(4) (2023) 2033.

[146] B. Mao, A. Siddaiah, Y. Liao, P.L.J.J.o.M.P. Menezes, Laser surface texturing and related techniques for enhancing tribological performance of engineering materials: A review, *Journal of Manufacturing Processes* 53 (2020) 153–173.

[147] A. Santos, J. Teixeira, C. Fonzar, E. Rangel, N. Cruz, P.N. Lisboa-Filho, A Tribological Investigation of the Titanium Oxide and Calcium Phosphate Coating Electrochemical Deposited on Titanium, *Metals* 13(2) (2023) 410.

[148] A. Sharma, A. Singh, V. Chawla, J. Grewal, A. Bansal, Microwave processing and characterization of alumina reinforced HA cladding for biomedical applications, *Materials Today: Proceedings* 57 (2022) 650–656.

[149] R.K. Sharma, G.P.S. Sodhi, V. Bhakar, R. Kaur, S. Pallakonda, P. Sarkar, H. Singh, Sustainability in manufacturing processes: Finding the environmental impacts of friction stir processing of pure magnesium, *CIRP Journal of Manufacturing Science Technology* 30 (2020) 25–35.

[150] A. Shearer, M. Montazerian, J.J. Sly, R.G. Hill, J.C. Mauro, Trends and perspectives on the commercialization of bioactive glasses, *Acta Biomaterialia* (2023).

[151] J. Singh, S. Singh, A.J.J.o.E.S. Verma, Engineering, Artificial intelligence in use of ZrO2 material in biomedical science, *Journal of Electrochemical Science Engineering* 13(1) (2023) 83–97.

[152] K. Singh, S. Mohan, S. Konovalov, M. Graf, Effect of nano-hydroxyapatite and post heat treatment on biomedical implants by sol-gel and HVOF spraying, Nanomaterials for Sustainable Tribology, CRC Press 2023, pp. 257–285.

[153] P. Singh, A. Bansal, V.K. Verma, Hydroxyapatite reinforced surface modification of SS-316L by microwave processing, *Surfaces Interfaces* 28 (2022) 101701.

[154] P. Singh, P. Kumar, An overview of biomedical materials and techniques for better functional performance, life, sustainability and biocompatibility of orthopedic implants, *Indian J. Sci. Tech* 11 (2018) 1–7.

[155] P. Singh, H. Vasudev, A. Bansal, Effect of post-heat treatment on the microstructural, mechanical, and bioactivity behavior of the microwave-assisted alumina-reinforced hydroxyapatite cladding, Proceedings of the Institution of Mechanical Engineers, *Part E: Journal of Process Mechanical Engineering* (2022) 09544089221116168.

[156] T.P. Singh, H. Singh, H. Singh, Characterization of thermal sprayed hydroxyapatite coatings on some biomedical implant materials, *Journal of Applied Biomaterials Functional Materials* 12(1) (2014) 48–56.

[157] R. Soni, S. Pande, S. Kumar, S. Salunkhe, H. Natu, H.M.A.M. Hussein, Wear Characterization of Laser Cladded Ti-Nb-Ta Alloy for Biomedical Applications, *Crystals* 12(12) (2022) 1716.

[158] R. Soni, S. Pande, Laser Cladded Ti-Alloys in Biomedical Applications: A Review, *Trends in Biomaterials Artificial Organs* 36(2) (2022).

[159] J. Sousa, R. Darabi, A. Reis, M. Parente, L.P. Reis, J.C. de Sa, An Adaptive Thermal Finite Element Simulation of Direct Energy Deposition With Reinforcement Learning: A Conceptual Framework, ASME International Mechanical Engineering Congress and Exposition, American Society of Mechanical Engineers, 2022, p. V02BT02A037.

[160] P. Sun, N. Yan, S. Wei, D. Wang, W. Song, C. Tang, J. Yang, Z. Xu, Q. Hu, X. Zeng, Microstructural evolution and strengthening mechanisms of Inconel 718 alloy with different W addition fabricated by laser cladding, *Materials Science Engineering: A* 868 (2023) 144535.

[161] B.R. Sunil, A.S.K. Kiran, S. Ramakrishna, Surface functionalized titanium with enhanced bioactivity and antimicrobial properties through surface engineering strategies for bone implant applications, *Current Opinion in Biomedical Engineering* 23 (2022) 100398.

[162] D. Svetlizky, M. Das, B. Zheng, A.L. Vyatskikh, S. Bose, A. Bandyopadhyay, J.M. Schoenung, E.J. Lavernia, N. Eliaz, Directed energy deposition (DED) additive manufacturing: Physical characteristics, defects, challenges and applications, *Materials Today* 49 (2021) 271–295.

[163] S. Tayebati, K.T. Cho, A hybrid machine learning framework for clad characteristics prediction in metal additive manufacturing, arXiv preprint arXiv:.01872 (2023).

[164] S. Teoh, Fatigue of biomaterials: a review, *International Journal of Fatigue* 22(10) (2000) 825–837.

[165] I.d.V. Tomaz, F.H.G. Colaço, S. Sarfraz, D.Y. Pimenov, M.K. Gupta, G. Pintaude, Investigations on quality characteristics in gas tungsten arc welding process using artificial neural network integrated with genetic algorithm, *The International Journal of Advanced Manufacturing Technology* 113(11–12) (2021) 3569–3583.

[166] T.S. Tshephe, S.O. Akinwamide, E. Olevsky, P.A. Olubambi, Additive manufacturing of titanium-based alloys-A review of methods, properties, challenges, and prospects, Heliyon (2022).

[167] R. Wandra, C. Prakash, S. Singh, Investigation on surface roughness and hardness of β-Ti alloy by ball burnishing assisted electrical discharge cladding for bio-medical applications, *Materials Today: Proceedings* 50 (2022) 848–854.

[168] R. Wandra, C. Prakash, S. Singh, Experimental investigation and optimization of surface roughness of β-Phase titanium alloy by ball burnishing assisted electrical discharge cladding for implant applications, *Materials Today: Proceedings* 48 (2022) 975–980.

[169] D.R. Unune, G.R. Brown, G.C. Reilly, Thermal based surface modification techniques for enhancing the corrosion and wear resistance of metallic implants: A review, *Vacuum* (2022) 111298.

Chapter 7

Prediction and performance of thermal cladding using artificial intelligence and machine learning
Design analysis and simulation

Hitesh Vasudev and Amrinder Mehta

7.1 INTRODUCTION

Thermal cladding, also known as thermal spray cladding or thermal spray coating, is a method for applying a protective layer to a substrate. The process entails the application of a molten or semi-molten cladding material to the surface of a substrate. This technique is frequently used in a variety of industries to improve the surface properties of the substrate, including corrosion resistance, abrasion resistance, thermal insulation, and even aesthetic appearance. Thermal cladding is often used in industries such as aerospace, automotive, petrochemical, oil and gas, and power generation. It is also commonly used in special applications, such as turbine blades, exhaust manifolds, and fuel injectors. Utilising a thermal spray cannon or torch, a process known as thermal spraying is typically used to apply thermal cladding [1–4]. The cladding material is delivered into the gun or torch in the form of powder, wire, or rod. The molten or semi-molten material is then accelerated towards the substrate using a high-velocity gas stream, typically compressed air or nitrogen. The cladding material is deposited onto the substrate, forming a bond. The process can be used to repair or protect a variety of components from wear and corrosion. It can also be used to create a decorative finish. The cladding material's molten particles stick to the substrate and solidify, providing a protective layer. The coating mechanically bonds with the substrate, improving characteristics and giving protection against environmental causes or other forms of degradation [5]. Thermal cladding can use a variety of cladding materials depending on the desired qualities and application requirements. Metals such as aluminium, zinc, stainless steel, or nickel-based alloys, as well as ceramics, carbides, and polymers, are commonly utilised materials. Thermal cladding is an effective way of protecting components from wear, corrosion, and other forms of damage. It also provides an additional layer of insulation to help reduce the transfer of heat. Furthermore, it can also improve the aesthetics of components [6–9].

DOI: 10.1201/9781032713830-7

Thermal cladding has various advantages over other types of covering. It enables the deposition of thick coatings with thicknesses ranging from a few micrometres to several millimetres, offering great protection and durability. It also provides freedom in terms of cladding material selection, allowing for customised coverings with specific qualities [10]. Furthermore, because thermal spray may be sprayed to complex shapes and huge surfaces, it is suited for a wide range of applications. Thermal cladding is also a relatively cost-effective option, making it a popular choice for many industries. Additionally, it is a sustainable solution, as it helps reduce waste by reusing materials [11–13]. Thermal cladding is a valuable method for depositing protective layers onto substrates, resulting in improved properties and a longer longevity for a variety of industrial components and structures.

7.1.1 Challenges in the thermal cladding

Thermal cladding, the act of depositing a layer of material onto a substrate using a heat source, offers various obstacles that must be solved for successful application. Some of the major issues in thermal cladding are depicted in Figure 7.1. It can be difficult to choose a cladding material that is compatible with the substrate and meets the intended functional requirements [14–16]. To reduce the risk of cracking or delamination during the chilling

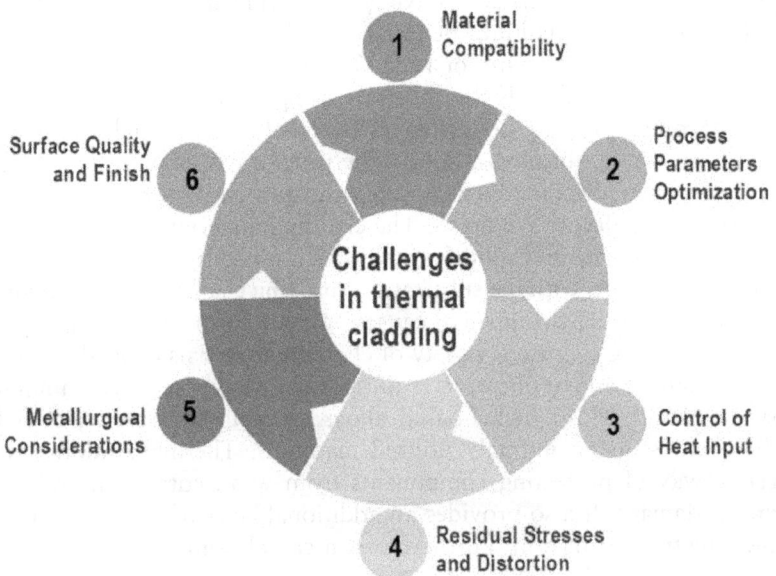

Figure 7.1 Challenges faced in the thermal cladding process.

procedure, the cladding material should have similar thermal expansion characteristics to the substrate. It is also important to consider the aesthetic requirements, such as colour, texture, and gloss, and the cost of the material. The chosen material must also be able to withstand any environmental conditions it may be exposed to. Identifying the optimal process parameters, such as laser power, scanning speed, powder input rate, and beam diameter, is essential for attaining the desired cladding characteristics. Due to the interactions between different parameters and their effects on the resulting microstructure and properties, determining the optimal parameter combination for a particular application can be difficult and time-consuming [17–19]. Therefore, numerical simulations can be a helpful tool in predicting the effects of different parameters on the microstructure and properties. Additionally, experimental studies can provide a more accurate depiction of the process and help refine the numerical models. To accomplish uniform melting and solidification of the cladding material, it is essential to maintain precise control over the heat input during thermal cladding. Variations in heat input can result in flaws such as porosity, incomplete fusion, and excessive dilution, jeopardising the quality and integrity of the clad layer. This can be achieved through careful monitoring and regulation of the welding parameters [20–22]. The welding speed, the preheat temperature, and the size of the cladding material all have to be taken into account to ensure the final product is of the highest quality. Thermal cladding generates significant temperature gradients and thermal cycles, causing residual stresses and distortion [23]. These stresses and distortions can impact the dimensional accuracy and mechanical properties of the clad component, potentially resulting in performance issues. These issues can be mitigated through careful selection of materials, careful design, and effective heat treatment.

Additionally, proper process control and monitoring can help to ensure that the cladding process is successful. Quality assurance practices should also be implemented throughout the process to ensure that the end product is of the highest quality. Additionally, regular testing for dimensional accuracy and mechanical properties should be conducted to guarantee the performance of the clad component. The thermal history of the cladding material during the manufacturing process can affect its microstructure and properties. Controlling factors such as cooling rate, solidification behaviour, and phase transformations are essential for achieving the desired microstructural characteristics and avoiding problems such as solidification fractures or undesirable phase formations. These factors can be controlled by adjusting the temperature profile of the heat treatment [24–26]. The desired properties can also be achieved by controlling the temperature during the annealing process. This includes controlling the rate of heating and cooling, as well as the duration of the process. This allows the material to be processed with a minimum of distortion and ensures that the final product has the desired properties. The process can also be optimised to reduce energy costs and improve the efficiency of the process. For many applications, achieving a uniform and defect-free cladded surface is essential [27–29]. However, the presence of flaws such as spatter, pores, or fractures can degrade the surface quality, impacting functionality, aesthetics, and subsequent machining or finishing processes. Therefore, cladding processes must be closely monitored and controlled to ensure a quality end product. Quality control measures should be taken to identify and eliminate any flaws in the cladding process. These measures may include visual inspections and using specialised measuring instruments to detect imperfections. Proper training of personnel and review of production parameters can help in minimising the occurrence of surface defects. Monitoring and controlling the thermal cladding process in real time are essential for ensuring quality and performance consistency [30–32]. Due to the harsh process conditions and restricted access to the cladding zone, it can be difficult to employ effective monitoring techniques such as temperature monitoring, thermal imaging, and in-situ sensing [33–35]. Automated sensors can be used to monitor and control the cladding process. These sensors can be implemented remotely and provide real-time data that can be used to ensure quality and performance consistency. Additionally, automated sensors can be used to detect and report any potential anomalies in the process. This drastically reduces the risk of manufacturing defects and can help to significantly reduce wastage due to errors. Automated sensors can also be used to optimise the process and ensure the highest quality products are produced [36–38].

Frequently, experimental investigations, process optimisation techniques, advanced modelling and simulation, and continuous development efforts are required to address these challenges. Collaboration between material scientists, process engineers, and equipment manufacturers is essential for overcoming these obstacles and advancing thermal cladding technologies. Through close partnerships between these stakeholders, the development of new materials and processes can be accelerated [39–41]. This ensures the best possible results in terms of cost, quality, and safety. Ultimately, this can lead to significant progress in thermal cladding technologies. Such progress can open up new applications of thermal cladding, such as in areas like aerospace and automotive engineering. This could lead to faster, safer, and more efficient production processes. Ultimately, this could result in better products and services for consumers around the world.

7.1.2 Importance of the artificial intelligence (AI) and machine learning (ML) techniques

Through the use of artificial intelligence (AI) and machine learning (ML) techniques, thermal cladding prediction and performance can be improved. As shown in Figure 7.2, AI and ML can improve thermal cladding procedures

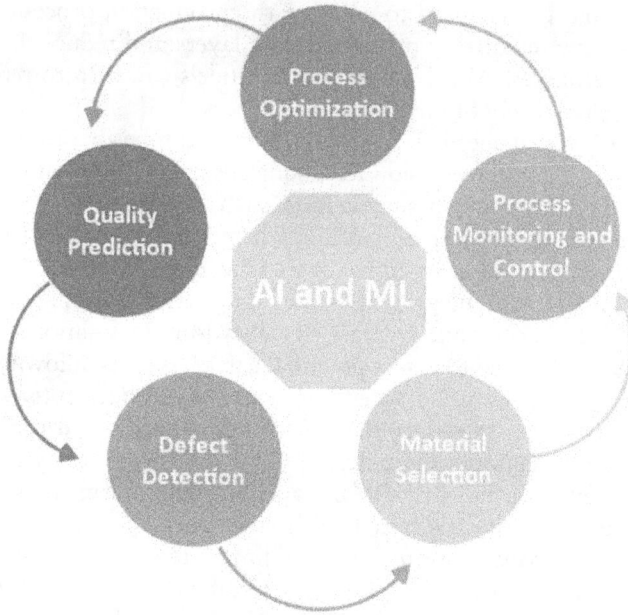

Figure 7.2 Importance of AI and ML approach.

in a variety of ways. Artificial intelligence and machine learning algorithms can analyse vast quantities of data from thermal cladding processes, such as material properties, process parameters, and performance metrics. These algorithms can identify patterns, correlations, and optimal process conditions for enhanced cladding performance by employing advanced data analytics techniques [42]. This can result in increased productivity, fewer defects, and enhanced material properties. This can lead to significant cost savings and improved product quality. It also has the potential to reduce environmental impact by improving energy efficiency and reducing waste and emissions. These cost savings and quality improvements can also result in increased customer satisfaction and loyalty [43]. This can lead to increased sales and profits for companies. To predict the quality and performance of cladding layers based on input parameters such as laser power, scanning speed, material composition, and substrate properties, machine learning models can be trained using historical data. Using these models, engineers can optimise the cladding process and accomplish desired results by making informed decisions [44]. Furthermore, AI-driven predictive models can help to anticipate any potential process deviations and alert the engineers to take preventive actions. This will enable companies to reduce costs, save time, and increase efficiency in the production of cladding layers.

Moreover, AI-driven predictive models can provide insights into patterns and trends that can be used to enhance the production process. This can help to improve the quality of the cladding layers and reduce the need for additional resources. AI-driven predictive models can help to reduce risks and increase the safety of the production process [45].

Cladding layer images can be analysed in real-time or processed retrospectively using computer vision techniques based on artificial intelligence. By identifying and categorising defects such as fractures, porosity, and incomplete fusion, AI systems can immediately provide feedback for process adjustments or reject defective parts. This contributes to the preservation of the quality and integrity of the cladding layers. This in turn can help reduce production costs and increase efficiency [46]. In addition, AI-assisted inspections can be used to detect potential safety hazards, allowing for proactive action and preventing potentially catastrophic accidents. AI-assisted inspections can help to reduce the risk of human error and provide more accurate results. This can ultimately result in improved safety and quality for a wide range of industries. ML algorithms can aid in the selection of suitable cladding materials based on performance specifications. By analysing material properties, previous cladding outcomes, and intended performance characteristics, AI models can recommend optimal materials for optimising the thermal cladding process and achieving the desired protective layer properties [47]. This automated selection process helps to reduce the time and cost of selecting the right material while also increasing the overall efficiency of the cladding process. Additionally, AI models can be trained to identify any potential problems before they occur, helping to minimise disruption and increase safety. AI can help to create a more accurate and reliable cladding process, with fewer errors and fewer delays. AI and ML can be utilised for real-time process parameter monitoring and feedback control. By perpetually analysing sensor data from the cladding process, AI algorithms can detect anomalies, predict potential issues, and adjust process parameters in real time to ensure optimal cladding performance [48]. This enables manufacturers to improve quality, reduce waste, and increase product consistency and reliability. AI and ML can also be used to automate processes, reducing the need for manual labour and increasing operational efficiency. AI and ML can also be used to improve safety by monitoring workers and alerting them when they are in danger. Additionally, AI can be used to identify maintenance issues before they become costly problems [49].

Using AI and ML, thermal cladding processes can be optimised, defects can be reduced, and the overall performance of the cladding layers can be substantially enhanced. This ultimately results in more efficient and reliable implementations for cladding in a variety of industries. AI and ML can create a significant cost savings in terms of time and resources [50–52]. Additionally, the improved precision can result in improved safety due to the reduced risk of catastrophic failure. By using AI and ML in thermal cladding processes, the quality and reliability of the cladding layers are improved, leading to cost savings and increased safety.

7.2 THERMAL CLADDING TECHNIQUES

Thermal cladding techniques involve the thermal deposition of a protective layer onto a substrate. Various industries, including aerospace, automotive, electronics, and energy production, employ these techniques to improve the surface properties of materials and safeguard them from environmental factors. As shown in Figure 7.3, three thermal cladding techniques are commonly employed [53–55]. Thermal spray coating includes heating and driving feedstock materials onto the substrate with a high-velocity gas or plasma stream. The feedstock materials could be powders, wires, or rods. When molten or semi-molten particles collide with a substrate, they solidify and produce a coating. Flame spraying, plasma spraying, high-velocity oxy-fuel (HVOF) spraying, and electric arc spraying are all thermal spray techniques. These methods offer good coating adherence, thickness control, and a diverse range of material alternatives. Flame spraying is the most common thermal spray technique and is used for a wide range of applications [56–58]. It is cost-effective and can be used to coat materials with a range of properties, such as corrosion resistance and wear resistance. It is also capable of producing high-quality coatings, with good bond strength and a uniform surface finish. Additionally, flame spraying is capable of spraying onto complex shapes and can be used to apply multiple layers of coatings. Physical vapour deposition (PVD) is a method in which a target material is evaporated or sputtered in a vacuum chamber. The vaporised or spewed atoms or molecules condense on the substrate, resulting in the formation of

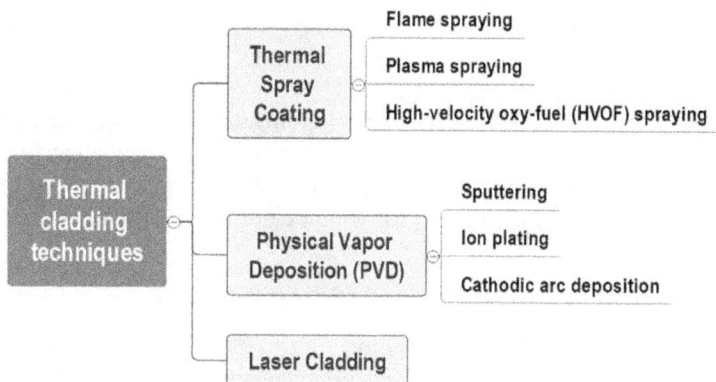

Figure 7.3 Thermal cladding processes.

a thin film covering. Evaporation, sputtering, ion plating, and cathodic arc deposition are all PVD processes [59–62].

PVD coatings provide excellent adhesion, a high density, and perfect control over film composition and thickness. They are frequently utilised in applications that need wear resistance, corrosion resistance, and enhanced surface qualities. PVD coatings are also used in various industries such as automotive, aerospace, electronics, and medical. They can be applied to a variety of materials such as metals, ceramics, and even plastics. Additionally, PVD coatings can be applied in extremely thin layers, making them ideal for many applications. A high-energy laser beam is used to melt a powdered or wire feedstock material, which is then deposited onto the substrate [63–65]. The laser beam scans the surface of the substrate, forming a molten pool into which the feedstock material is inserted. The applied material solidifies quickly, forming a clad layer. The use of laser cladding allows for fine control over coating thickness, composition, and localised deposition. It is commonly employed in the repair of damaged components, the addition of wear-resistant layers, and the creation of complex geometries. Laser cladding is a cost-effective and efficient method for a wide range of applications. It is also an environmentally friendly process as it does not require the use of hazardous chemicals. Furthermore, its high precision and repeatability make it ideal for creating complex parts with intricate geometries. Additionally, laser cladding is a fast process and can be used to add a range of materials with varying properties. The resulting parts are durable and long-lasting, making it a reliable choice for many repair and manufacturing applications [66–68].

Depending on the application requirements, these thermal cladding techniques provide distinct advantages. Material qualities, coating thickness, adhesion, surface preparation, and cost factors should all be considered when determining the best procedure for a given substrate and protective layer [69]. Different processes can be used for different substrates, such as aluminium, steel, and titanium. Additionally, the choice of thermal cladding technique will depend on the desired properties of the substrate and protective layer, such as corrosion resistance, heat resistance, and wear resistance [70]. The choice of the thermal cladding technique should be made carefully, as the wrong choice can lead to costly problems, such as unanticipated wear or corrosion. It is important to consult an experienced professional when making decisions about which cladding technique to use.

7.2.1 Overview of the laser cladding process

In manufacturing and materials engineering, laser cladding is a precise and efficient method for applying a layer of material to a base metal or substrate. Utilising a high-powered laser beam, a powdered or wire substrate is melted and fused to the surface, resulting in a metallurgical bond. The layer

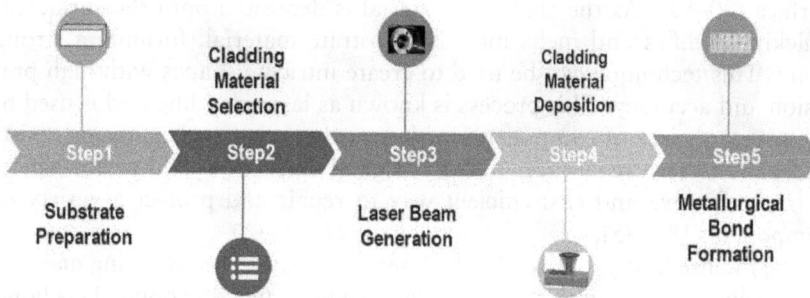

Figure 7.4 Process stages involved in laser cladding.

of material applied can be any metal, alloy, or ceramic and can be used to repair or reinforce components. Laser cladding also offers a low-cost alternative to welding and other machining processes. The steps involved in the laser cladding process are depicted in Figure 7.4.

To assure proper adhesion of the cladding material, the base metal or substrate is cleaned and prepared thoroughly. This may entail surface cleaning, grinding, or other forms of pre-treatment. The substrate must then be dried and heated to the appropriate temperature for the cladding material to adhere [71,72]. Finally, the cladding material is applied and allowed to cool. The cladding material, in powder or filament form, is selected based on its desired properties, such as composition, hardness, wear resistance, corrosion resistance, or other application-specific characteristics. After the material is selected, it is applied to the surface of the object to be cladded. It is usually done by thermal spraying, electric arc spraying, or other specialised processes [73–75]. The cladding material is then fused to the substrate, forming a hard, protective layer. This cladding layer helps to protect the substrate from corrosion or wear, as well as improving its aesthetic properties and increasing its lifetime [76].

On the substrate's surface is focused a high-power laser beam, which is typically produced by a carbon dioxide (CO_2) or fibre laser. The laser beam provides the required energy to melt the cladding material and create a reservoir of molten material. The molten material is then pushed out of the reservoir, and a new layer of cladding material is added. This process is repeated until the three-dimensional shape is achieved. The finished product is then cooled and removed from the substrate [77–79]. The molten material can be manipulated to achieve the desired shape, which can be complex. The 3D shape can then be used for a variety of applications, such as creating prototype parts for testing or manufacturing components for industrial use. The cladding material is inserted into the laser-generated molten pool, either as a powder or as wire. The laser beam and controlled movement of the substrate or cladding nozzle ensure exact material deposition onto the surface [80–82]. As the cladding material is deposited onto the surface, it quickly solidifies and melts into the substrate material, forming a strong bond. This technique can be used to create intricate shapes with high precision and accuracy. This process is known as laser cladding and is used to improve a variety of components with regard to wear and corrosion resistance, and to restore surfaces to their original dimensions and performance. It is an effective and cost-efficient way to repair and protect a variety of components [83–85].

The intense heat produced by the laser beam causes the cladding material to dissolve and fuse with the substrate, forming a metallic bond. This bond between the cladding layer and the substrate assures superior adhesion and mechanical properties. Laser cladding is a specialised form of welding that is used to add a layer of material onto a substrate. It can be used to improve the performance of certain components, such as those that are exposed to harsh environments or used in high-wear applications [86–88]. This process is advantageous because it can be performed quickly and efficiently with minimal waste. In addition, it does not require the use of filler materials, allowing for a stronger bond and a more uniform coating. The molten pool quickly solidifies as the laser beam travels along the surface, creating a solid layer of cladding material [89]. The microstructure and resulting material qualities can be influenced by altering the cooling rate. This process is known as laser cladding and is often used to repair and protect components in a variety of industries. It can also be used to manufacture components from scratch, allowing for the creation of complex geometries with tight tolerances. Laser cladding can also be used to modify the surface properties of existing components, such as increasing corrosion resistance, wear characteristics, and fatigue strength [90].

The clad layer may need post-processing processes after the cladding process is finished in order to produce the appropriate surface finish, dimensional accuracy, or final shape. This post-processing may involve re-machining, grinding, polishing, heat treatment, or annealing. The post-processing step is important to ensure the cladding meets the required specifications. The post-processing step should be planned carefully to minimise the cost and time required for the process. Any modifications needed should also be taken into consideration in order to produce the desired output. Finally, the cladding should be thoroughly inspected to ensure the highest quality of the final product. In comparison to conventional cladding methods, laser cladding has a number of benefits [91–93]. It gives the deposition process fine control, enabling complicated shapes, thin coatings, and targeted repairs. Additionally, the method results in little heat input, little deformation, and little dilution between the substrate and the cladding material. This makes laser cladding useful for a variety of applications, including surface protection, repair, and additive manufacturing. It also makes it possible to use different materials and to create customised material properties. The process is also fast, energy-efficient, and cost-effective, making it an attractive option for many industries [94–96]. Laser cladding can be used on a variety of materials, including metals, ceramics, and composites. Additionally, it can be used to join dissimilar materials, allowing for more complex designs. Laser cladding is also highly precise, allowing for intricate details to be added to components. It is also a non-contact process, which reduces the risk of contamination and damage to the material [97].

7.2.2 Advantages laser cladding process characteristics

Advanced laser surface cladding is a very adaptable technology with several uses in the mechanical fields. The method of using a laser beam to melt and fuse a coating material onto a substrate surface is known as laser surface cladding, also known as laser cladding or laser deposition. The laser beam is used to precisely deposit the coating material, which is often in the form of a powder or wire, onto the substrate surface [98–100]. The coating material is melted by the high-energy laser before solidifying to create a layer that is both functional and protective. Compared to conventional cladding techniques like thermal spray or welding, this procedure has a number of benefits [101]. It is much faster and produces very little waste, meaning the process is more cost-effective. It also yields high-quality results with excellent adhesion and uniformity of the coating. Some of the numerous benefits of advanced laser surface cladding are depicted in Figure 7.5.

Wear-resistant and corrosion-resistant coatings can be applied to components exposed to harsh operating conditions via laser cladding. It contributes to the increased durability and longevity of mechanical elements such as engine components, valves, shafts, and moulds. Laser cladding also allows for customisation of the components, as the coatings can be applied in specific areas [102–104]. Laser surface cladding is an effective method for restoring deteriorated or broken components. It can be used to reinforce the material and restore the original dimensions of parts, thereby reducing the need for costly replacements. The process is also fast and cost-effective, and it can be used to coat a variety of materials, including metals, ceramics, and polymers. Additionally, laser cladding can be used to introduce functional properties to a material, such as wear, corrosion, and oxidation resistance. Laser cladding is a reliable and efficient method for repairing and resurfacing components. It produces a high-quality and consistent finish that is strong and durable. In the production of moulds, dies, and tools, laser cladding is commonly employed. By applying wear-resistant coatings to these components, their service life can be substantially increased, thereby decreasing downtime and maintenance expenses [105–107]. Additionally, laser cladding has the advantage of being a fast

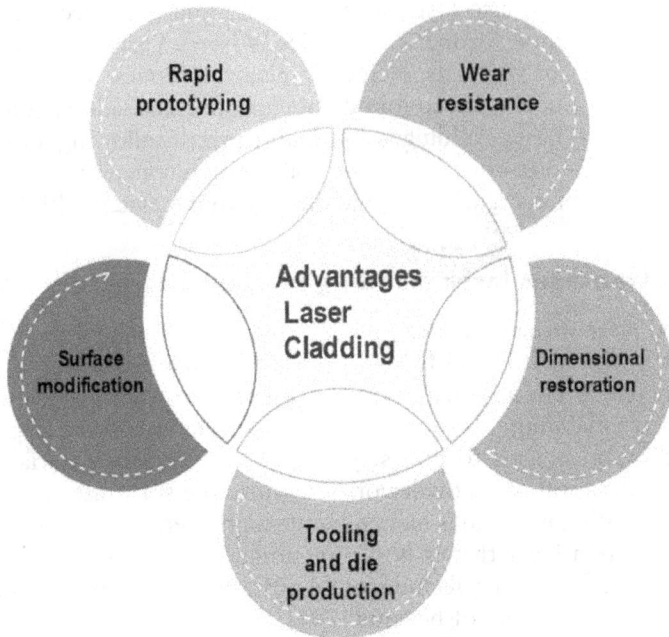

Figure 7.5 Several benefits of advanced laser surface cladding.

and efficient process. It is also a cost-effective solution, as it can be used to repair and restore components, as opposed to having to replace them. Laser cladding can also be used to improve the properties of components, such as corrosion resistance, hardness, and wear resistance. This makes it a great choice for a variety of applications in different industries [108–110].

Laser cladding can alter mechanical components' surface properties. This includes increasing friction, decreasing the coefficient of friction, enhancing heat transfer, and enhancing electrical conductivity. These modifications can enhance the functionality of components in a variety of applications [111]. Laser cladding also allows for the introduction of new materials and elements that can further improve the performance of components. Additionally, it can increase the wear resistance and extend the service life of components, reducing maintenance costs. Additionally, laser cladding can be utilised as an additive manufacturing method. It enables the production of complex geometries and the deposition of multiple materials, thereby enabling rapid prototyping and the production of customised parts. Laser cladding is particularly attractive for applications that require the integration of multiple materials, due to its ability to deposit multiple materials in the same part [112–114]. This makes it ideal for applications that require the integration of multiple materials with different properties. Advanced laser surface cladding provides precise control over the coating process, exceptional coating-to-substrate adhesion, and the ability to customise material properties. This makes it a valuable technology in the mechanical industries for augmenting performance, extending component life, and decreasing maintenance costs [115–117]. It is also used for repairing worn or damaged components, eliminating the need for costly replacements. Additionally, it can be used to create new materials with unique properties.

7.3 ROLE OF MACHINE LEARNING TECHNIQUES IN THERMAL CLADDING

Thermal cladding can utilise machine learning techniques to effectively resolve various challenges and improve process optimisation, control, and quality. As depicted in Figure 7.6, there are several advantages and applications of machine learning in thermal cladding. To determine the best parameter combinations for desired cladding features, machine learning techniques can be employed to analyse massive datasets of process parameters and corresponding cladding results. Machine learning models can assist with process optimisation for increased efficiency and quality by understanding the intricate correlations between process variables and cladding results [118,119]. This approach can lead to a reduction of trial and error and an increase in productivity. Additionally, it can help to avoid costly mistakes by predicting the outcome of the process ahead of time. Machine learning models can also be used to identify and diagnose potential problems, allowing for quick and effective troubleshooting.

This can help to reduce downtime and improve the overall performance of a process. Using the input process parameters as input, machine learning

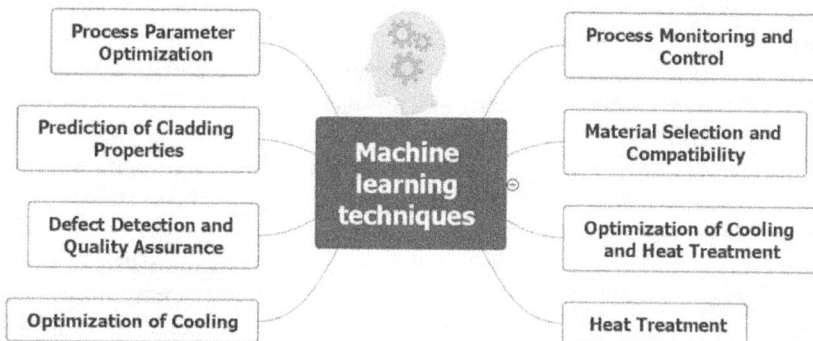

Figure 7.6 Machine learning's techniques in thermal cladding benefits.

may be used to create prediction models that calculate cladding attributes such as cladding layer thickness, hardness, or microstructure. With the help of these models, cladding conditions may be quickly evaluated and optimised without the need for protracted experimental trials. This significantly reduces the cost and time associated with cladding processes while ensuring optimal performance. Moreover, it helps to minimise the risk of defects in cladding layers [120]. This makes the cladding process more efficient and reliable and helps to improve the quality of the final product. It also reduces waste and environmental impact, as less materials are needed for production.

To detect problems like as pores, fractures, or splatter, machine learning algorithms can analyse photos or sensor data collected during the cladding process. Machine learning models can find patterns and abnormalities in data by training on annotated data and providing real-time feedback and quality assurance. This improves process repeatability and reduces time-consuming manual inspections [121]. This technology can be used to ensure that welds meet strict quality and safety standards, as well as reduce costs associated with manufacturing. Additionally, machine learning can be used to predict faults and anticipate failure points before they occur. This helps to reduce the risk of accidents or product failures, improving safety and reliability. Machine learning algorithms can be used to monitor and alter the cladding process in real time to maintain consistent quality [122]. Machine learning models can recognise process variations, predict possible faults, and trigger necessary corrective actions by analysing sensor data such as temperature, pressure, or energy input. This can significantly reduce production costs and optimise the overall efficiency of the cladding process. Machine learning can also be used to store and analyse data from previous cladding runs, allowing for better control of the process and further optimisation. By analysing material attributes, substrate compatibility, and performance requirements, machine learning can aid in the selection of acceptable cladding materials [123]. Machine learning algorithms can help find the optimum material solutions for certain applications by leveraging historical data and experience, hence optimising the whole cladding process. This can help to minimise material waste, reduce costs, and improve energy efficiency. Machine learning can also assist with predicting future material requirements and helping to anticipate changes in the cladding industry. Machine learning algorithms can enable faster and more accurate material selection and can help to automate the cladding process, leading to improved productivity and cost savings [124].

Following thermal cladding, machine learning algorithms can optimise the cooling and heat treatment procedures. Machine learning models can offer ideal cooling rates, heat treatment cycles, or stress relief techniques to decrease residual stresses, distortion, and increase material performance by analysing material attributes and thermal history data. Machine learning models can also detect and diagnose abnormal thermal treatments, such as overheating or undercooling, which can lead to material degradation [125]. Moreover, they can also provide solutions to prevent such occurrences in the future. Machine learning models can also be used to optimise machining processes, such as cutting speed, feed rate, and tool selection to improve surface finish and reduce machining time. Additionally, they can be used to detect potential tool wear and provide feedback for tool maintenance. Machine learning approaches enable data-driven process optimisation by continuously analysing process data and finding trends and patterns. Machine learning models can suggest process alterations or parameter adjustments for enhanced process control and performance by utilising past data and feedback. These models can also be used to predict future outcomes and help to identify areas for improvement. This can lead to significant cost savings and improved efficiency. Additionally, machine learning models can be used to detect abnormalities in process data, such as unexpected outliers or anomalies, which could otherwise go unnoticed [126]. This can help to identify potential risks or problems that could lead to significant losses or damage. It is vital to highlight that the successful application of machine learning approaches in thermal cladding necessitates the use of high-quality and representative datasets, domain expertise, and coordination between materials scientists, process engineers, and data scientists. Furthermore, it is important to define effective performance metrics and analyse the results to identify areas for improvement. Finally, the continual optimisation of the thermal cladding process will lead to improved safety, quality, and efficiency [127]. This can be achieved by making use of data-driven decision making and predictive analytics. These data-driven insights can be used to inform the decisions made by process engineers and operators, driving the desired outcomes.

7.4 INTELLIGENT OPTIMISATION METHODS

To minimise the difference between the predicted outputs and the actual outputs, optimisation methods adjust the internal parameters (weights and biases) of the network. By minimising the difference, the network can improve its accuracy and performance. The optimisation process can be done manually or with automated algorithms. Ultimately, it helps the network to learn and adapt to new data [128]. This allows the network to generalise better and improve its accuracy on unseen data. The optimisation process is an important part of deep learning and helps to ensure the network is able to accurately classify data.

7.4.1 Artificial neural network model

The link between laser cladding process parameters and cladding layer size can be studied using an artificial neural network (ANN). Neural networks are effective tools for modelling complicated interactions and can be trained to predict or analyse data patterns [129]. To create an ANN for this purpose, you would normally follow the procedures indicated in Figure 7.7.

Collect a dataset containing various laser cladding process parameters (such as laser power, scanning speed, powder supply rate, etc.) and the associated cladding layer sizes. To capture the complete relationship, the dataset must contain a broad range of parameter values. The dataset can be used to analyse the effect of each parameter on the cladding layer size. This will enable us to develop models that can predict the cladding layer size for any given set of parameter values [130]. The dataset can then be used to optimise the laser cladding process parameters to achieve the desired cladding layer size. The models can also be used to develop a closed-loop control system to dynamically adjust the parameters to maintain the cladding layer size within a certain range. The collected data must be preprocessed to assure its quality and suitability for neural network training. This phase may entail removing outliers, normalising input parameters and separating data into training and testing sets. Preprocessing helps the neural network to learn more accurately and quickly [131]. It ensures that the data fed into the network is compatible with the network's structure and does not contain any irrelevant information. Preprocessing also helps the network avoid overfitting and provides a generalised understanding of the data. It can also reduce the training time and optimise the performance of the model [132].

Determine the network architecture, including the number and type of layers, the number of neurons in each layer, and the activation functions. You can begin with a basic architecture and then modify it as necessary. Once you have determined the architecture, it's time to start training the model. Start with a small number of training epochs to assess the performance, and

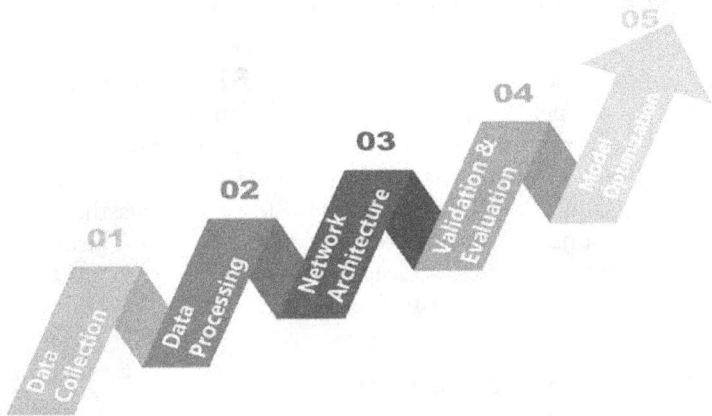

Figure 7.7 Steps followed by the artificial neural network model.

then increase the number of epochs if needed. Evaluate the performance of the model and fine-tune the architecture if needed. Once you are satisfied with the performance, deploy the model and monitor it in production. Make adjustments as needed to ensure that the model is performing as expected [133]. Train the neural network with the help of the training dataset. During training, the network modifies its internal parameters (weights and biases) according to the data's input–output relationships. For training neural networks, numerous optimisation algorithms are available, including gradient descent and its variants. The goal of the optimisation process is to minimise a predefined cost function, which measures the difference between the expected and the actual output of the network for the given input. The optimisation process stops when the cost function reaches a predefined threshold. This minimisation of the cost function is done by adjusting the weights and parameters of the network. The resulting model is then used to make predictions on unseen data [134]. Using the testing dataset, validate the trained network to ensure that it generalises well to unobserved data. Evaluate the network's efficacy using relevant metrics, such as mean squared error (MSE) and R-squared value. Once the validation is complete, the network can be deployed in the production environment. Monitor the network's performance over time and adjust hyperparameters as needed to ensure optimal performance [135]. After deployment, continue to use the testing dataset to evaluate the network's performance. Make sure to also monitor the network for any bias or errors that can arise due to changing input data or environment. Finally, document the process and results for future reference. Based on the validation results, fine-tune the architecture and hyperparameters of the network. This step may entail modifying the number of layers or neurons or experimenting with various activation functions. By doing this, it is possible to optimise the performance of the network and achieve the desired accuracy [136,137]. This process of fine-tuning the network is essential in order to produce a model that is able to generalise well, and can make accurate predictions on unseen data. After the architecture and hyperparameters of the network have been fine-tuned, the model can then be tested on unseen data to evaluate its accuracy. Finally, the model can then be deployed for practical use.

After training and validating the neural network, it can be used to predict cladding layer sizes based on new sets of process parameters. Examine the network's predictions and compare them to actual measurements to determine its precision and dependability. If the network's predictions are accurate, it can be used in production process optimisation. This could lead to improved efficiency and cost savings for the company. The company can then use the optimised production process to produce more with less, leading to increased productivity and profitability [122]. Remember that developing an ANN is an iterative process, and you may need to revisit certain stages to improve the performance of the network. In addition, a large and diverse dataset is essential for training an accurate model, so be sure to capture enough data for meaningful outcomes. Regularly check for overfitting and adjust the network structure accordingly. Finally, be sure to monitor the network in real time to detect changes in the environment and adapt quickly. Test your model with different scenarios and parameters.

7.4.2 Genetic algorithm optimises BP neural network (GABP)

The Genetic algorithm optimises BP neural network (GABP) is a method that combines the genetic algorithm (GA) and a backpropagation (BP) neural network in order to optimise the network parameters for the laser cladding procedure. This strategy seeks to improve the neural network's performance by identifying optimal weights and biases. As illustrated in Figure 7.8, GABP can be utilised to optimise a BP neural network for the laser cladding procedure [138]. Design the BP neural network's architecture, including the number of layers, the number of neurons in each layer, and the activation functions. The input layer would contain the laser cladding process parameters, while the output layer would represent the predicted cladding layer thickness. The network should be trained using a set of measured cladding layer thickness data [139]. The number of neurons in the hidden layers should be adjusted to minimise the error between the predicted and measured data.

The activation functions should be chosen to ensure the network is able to learn the non-linear relationships involved in the laser cladding process. The output layer should be linear to ensure the output values remain within the same range as the input values. The network should be tested and validated to ensure that it is able to accurately predict the cladding layer thickness.

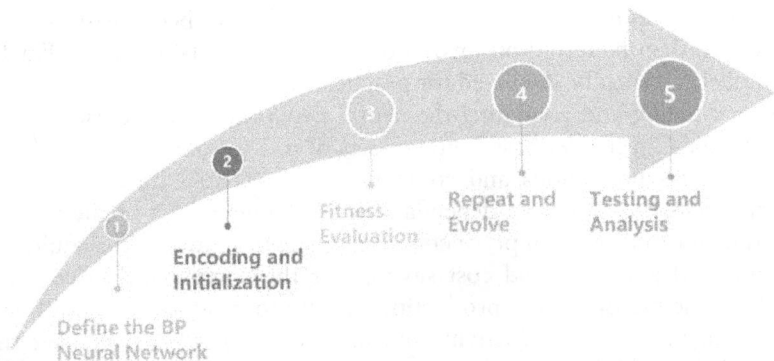

Figure 7.8 Steps followed by the genetic algorithm optimises BP neural network (GABP).

Finally, the network should be deployed in the production environment. Define an appropriate chromosome encoding scheme for neural network parameters (weights and biases). Create a population of individuals, each of which represents a set of network parameters. Initialise the population with randomly generated chromosomes [140]. Then run a genetic algorithm to optimise the fitness function and find the best set of network parameters. Evaluate the fitness of each individual in a population by training the respective neural network with the BP algorithm and measure its performance on a validation dataset. The fitness function could be determined by metrics like mean squared error (MSE) or precision [141].

Utilise genetic operators (selection, crossover, and mutation) to generate a new generation of individuals from the population. Select individuals with higher fitness values in order to increase their chances of reproduction. In order to produce progeny, crossover combines the genetic material of two parents, whereas mutation introduces random changes to the parameters of the offspring [142]. Perform multiple generations of fitness evaluations and genetic operations. The objective is to iteratively enhance the population by selecting fitter individuals and producing progeny with more favourable parameter combinations. Define a termination condition, such as attaining a maximum number of generations or reaching an acceptable level of physical fitness. Once the termination condition is met, the algorithm terminates, and the optimal parameters for the BP neural network are represented by the finest individual (with the highest fitness). Evaluate the efficacy of the optimised BP neural network using a separate test dataset [143]. Assess the accuracy and generalisability of the model by comparing its predictions to the actual cladding layer sizes. Analyse the results and, if necessary, adjust the network. GABP can effectively optimise the network parameters for the laser cladding process by integrating the GA's ability to explore a large search space and the BP neural network's ability to learn. It can enhance the neural network's accuracy and predictive ability when estimating cladding layer sizes based on process parameters.

7.5 CONCLUSION

Thermal cladding, also known as thermal spray cladding or thermal spray coating, is a process that applies a protective layer to a substrate, enhancing its properties such as corrosion resistance, abrasion resistance, thermal insulation, and aesthetic appearance. It finds extensive applications in industries like aerospace, automotive, petrochemical, oil and gas, and power generation. The process involves applying molten or semi-molten cladding material to the surface using a high-velocity gas stream, forming a strong bond with the substrate, providing protection against wear, corrosion, and creating a decorative finish. Thermal cladding offers several advantages over other coatings, such as being highly durable and versatile. However, it presents challenges such as selecting compatible materials, precise control over heat input, monitoring factors like cooling rate, solidification behaviour, and phase transformations, and real-time process control. Artificial intelligence (AI) and machine learning (ML) techniques have emerged as valuable tools to optimise thermal cladding performance. AI and ML algorithms can analyse vast amounts of data from thermal cladding processes, identifying patterns, correlations, and optimal process conditions for improved cladding performance. This leads to increased productivity, fewer defects, and enhanced material properties, resulting in significant cost savings and improved product quality. AI-driven predictive models can anticipate process deviations and prompt engineers to take preventive actions, reducing costs and saving time. Real-time monitoring using AI-assisted computer vision techniques can analyse cladding layer images, identifying and categorising defects, enabling immediate feedback for process adjustments or rejecting defective parts. ML algorithms also aid in selecting appropriate cladding materials based on performance specifications, minimising human error and providing more accurate results. In conclusion, thermal cladding is a versatile and valuable process used to enhance substrate properties in various industries. AI and ML techniques play a crucial role in improving thermal cladding performance, optimising processes, reducing defects, and increasing productivity. The integration of genetic algorithms and BP neural networks through GABP further enhances the accuracy of cladding layer size predictions, contributing to more efficient and cost-effective manufacturing processes.

REFERENCES

[1] A. Mehta, H. Vasudev, N. Jeyaprakash, Role of sustainable manufacturing approach: microwave processing of materials, *International Journal on Interactive Design Manufacturing* (2023) 1–17.

[2] H.A. Abdullah, R.A. Anaee, Characteristics and Morphological Studies of Nd Doped Titanium Thin Film Coating on SS 316L by DC Sputtering, *Diyala Journal of Engineering Sciences* 15(3) (2022).

[3] M. Abdulwahab, V. Aigbodion, M. Enechukwu, Anti-corrosion of isothermally treated Ti-6Al-4V alloy as dental biomedical implant using non-toxic bitter leaf extract, *Chemical Data Collections* 24 (2019) 100271.

[4] N. Abhijith, D. Kumar, D. Kalyansundaram, Development of single-stage TiNbMoMnFe high-entropy alloy coating on 304L stainless steel using HVOF thermal spray, *Journal of Thermal Spray Technology* 31(4) (2022) 1032–1044.

[5] D. Castro, P. Jaeger, A.C. Baptista, J.P. Oliveira, An overview of high-entropy alloys as biomaterials, *Metals* 11(4) (2021) 648.

[6] A. Mehta, H. Vasudev, S. Singh, Sustainable manufacturing approach with novel thermal barrier coatings in lowering CO2 emissions: Performance analysis with probable solutions, *International Journal on Interactive Design Manufacturing* (2023) 1–13.

[7] A.R. Ahmady, A. Ekhlasi, A. Nouri, M.H. Nazarpak, P. Gong, A. Solouk, High entropy alloy coatings for biomedical applications: A review, *Smart Materials in Manufacturing 1 (2023) 100009.*

[8] Q. An, L. Huang, S. Jiang, Y. Bao, M. Ji, R. Zhang, L. Geng, Two-scale TiB/Ti64 composite coating fabricated by two-step process, *Journal of Alloys Compounds 755 (2018) 29–40.*

[9] S. Bajda, Y. Liu, R. Tosi, K. Cholewa-Kowalska, M. Krzyzanowski, M. Dziadek, M. Kopyscianski, S. Dymek, A.V. Polyakov, I.P. Semenova, Laser cladding of bioactive glass coating on pure titanium substrate with highly refined grain structure, *Journal of the Mechanical Behavior of Biomedical Materials 119 (2021) 104519.*

[10] S. Bose, D. Ke, H. Sahasrabudhe, A. Bandyopadhyay, Additive manufacturing of biomaterials, *Progress in Materials Science 93 (2018) 45–111.*

[11] Y. Sharma, A. Mehta, H. Vasudev, N. Jeyaprakash, G. Prashar, C. Prakash, Analysis of friction stir welds using numerical modelling approach: a comprehensive review, *International Journal on Interactive Design Manufacturing (2023) 1–14.*

[12] R.R. Behera, A. Hasan, M.R. Sankar, L.M. Pandey, Laser cladding with HA and functionally graded TiO2-HA precursors on Ti–6Al–4V alloy for enhancing bioactivity and cyto-compatibility, *Surface Coatings Technology 352 (2018) 420–436.*

[13] V. Bhojak, J.K. Jain, T.S. Singhal, K.K. Saxena, C. Prakash, M.K. Agrawal, V. Malik, Friction stir processing and cladding: an innovative surface engineering technique to tailor magnesium-based alloys for biomedical implants, *Surface Review Letters (2023) 2340007.*

[14] T. Chen, W. Li, D. Liu, Y. Xiong, X. Zhu, Effects of heat treatment on microstructure and mechanical properties of TiC/TiB composite bioinert ceramic coatings in-situ synthesized by laser cladding on Ti6Al4V, *Ceramics International 47(1) (2021) 755–768.*

[15] Y. Cheng, H. Yang, Y. Yang, J. Huang, K. Wu, Z. Chen, X. Wang, C. Lin, Y. Lai, Progress in TiO 2 nanotube coatings for biomedical applications: a review, *Journal of Materials Chemistry B 6(13) (2018) 1862–1886.*

[16] R. Comesaña, Special Issue on Surface Treatment by Laser-Assisted Techniques, MDPI Coatings, 2020, p. 580.

[17] G. Singh, A. Mehta, A. Bansal, Electrochemical behaviour and biocompatibility of claddings developed using microwave route, *Journal of Electrochemical Science Engineering failure analysis 13(1) (2023) 173–192.*

[18] O.N. Çelik, Microstructure and wear properties of WC particle reinforced composite coating on Ti6Al4V alloy produced by the plasma transferred arc method, *Applied Surface Science 274 (2013) 334–340.*

[19] Q. Chen, G.A. Thouas, Metallic implant biomaterials, *Materials Science Engineering: R: Reports 87 (2015) 1–57.*

[20] H. Singh, A. Mehta, Y. Sharma, H. Vasudev, Role of expert systems to optimize the friction stir welding process parameters using numerical modelling: a review, *International Journal on Interactive Design Manufacturing (2023) 1–17.*

[21] Z. Deng, D. Liu, G. Liu, Y. Xiong, S. Xin, S. Li, C. Li, T. Chen, Study on the corrosion resistance of SiC particle-reinforced hydroxyapatitesilver gradient bioactive ceramic coatings prepared by laser cladding, *Surface Coatings Technology (2023) 129734.*

[22] C. Domínguez-Trujillo, F. Ternero, J.A. Rodríguez-Ortiz, J.J. Pavón, I. Montealegre-Meléndez, C. Arévalo, F. García-Moreno, Y. Torres, Improvement of the balance between a reduced stress shielding and bone ingrowth by bioactive coatings onto porous titanium substrates, *Surface Coatings Technology* 338 (2018) 32–37.

[23] A.B. Edathazhe, Investigation of properties, corrosion and bioactivity of novel BaO added phosphate glasses and glass-ceramic coating on bio-medical metallic implant materials, National Institute of Technology Karnataka, Surathkal, 2018.

[24] N. Singh, A. Mehta, H. Vasudev, P.S. Samra, A review on the design and analysis for the application of Wear and corrosion resistance coatings, *International Journal on Interactive Design Manufacturing* (2023) 1–25.

[25] G. Ertugrul, A. Hälsig, J. Hensel, J. Buhl, S. Härtel, Efficient multi-material and high deposition coating including additive manufacturing by tandem plasma transferred arc welding for functionally graded structures, *Metals* 12(8) (2022) 1336.

[26] L. Fan, H. CHEN, Y. Dong, X. LI, L. DONG, Y. YIN, Corrosion behavior of Fe-based laser cladding coating in hydrochloric acid solutions, *Acta Metall Sin* 54(7) (2018) 1019–1030.

[27] P.K. Verma, A. Mehta, H. Vasudev, V.J.S.R. Kumar, Performance of thermal spray coated metallic materials for bio-implant applications, *Surface Review Letters* (2023).

[28] A. Farazin, C. Zhang, A. Gheisizadeh, A. Shahbazi, 3D bio-printing for use as bone replacement tissues: A review of biomedical application, *Biomedical Engineering Advances* (2023) 100075.

[29] R. Fathi, H. Wei, B. Saleh, N. Radhika, J. Jiang, A. Ma, M.H. Ahmed, Q. Li, K.K.J.A.M.T. Ostrikov, Past and present of functionally graded coatings: Advancements and future challenges, 26 (2022) 101373.

[30] O.S. Fatoba, O.S. Adesina, A. Popoola, Evaluation of microstructure, microhardness, and electrochemical properties of laser-deposited Ti-Co coatings on Ti-6Al-4V Alloy, *The International Journal of Advanced Manufacturing Technology* 97 (2018) 2341–2350.

[31] J. Feng, Y. Tang, J. Liu, P. Zhang, C. Liu, L. Wang, Bio-high entropy alloys: Progress, challenges, and opportunities, Frontiers in Bioengineering Biotechnology 10 (2022) 977282.

[32] F.J.S.E. Findik, Innovation, Laser cladding and applications, *Sustainable Engineering Innovation* 5(1) (2023) 1–14.

[33] V.S. Yedida, A. Mehta, H. Vasudev, S. Singh, Role of numerical modeling in predicting the oxidation behavior of thermal barrier coatings, *International Journal on Interactive Design Manufacturing* (2023) 1–10.

[34] Q. Fu, W. Liang, J. Huang, W. Jin, B. Guo, P. Li, S. Xu, P.K. Chu, Z. Yu, Research perspective and prospective of additive manufacturing of bio-degradable magnesium-based materials, *Journal of Magnesium Alloys* 6 (2023) 158.

[35] R. Fu, W. Luo, R. Nazempour, D. Tan, H. Ding, K. Zhang, L. Yin, J. Guan, X. Sheng, Implantable and biodegradable poly (l-lactic acid) fibers for optical neural interfaces, *Advanced Optical Materials* 6(3) (2018) 1700941.

[36] M. Ganjali, M. Ganjali, S. Sadrnezhaad, Y. Pakzad, Laser cladding of Ti alloys for biomedical applications, *Laser Cladding of Metals* (2021) 265–292.

[37] M. Ganjali, A. Yazdanpanah, M. Mozafari, Laser deposition of nano coatings on biomedical implants, Emerging Applications of Nanoparticles and Architecture Nanostructures, Elsevier 2018, pp. 235–254.

[38] J. Garcia-Herrera, J. Henao, D. Espinosa-Arbelaez, J. Gonzalez-Carmona, C. Felix-Martinez, R. Santos-Fernandez, J. Corona-Castuera, C. Poblano-Salas, J. Alvarado-Orozco, Laser cladding deposition of a Fe-based metallic glass on 304 stainless steel substrates, *Journal of Thermal Spray Technology* 31(4) (2022) 968–979.

[39] S. Gaytan, L. Murr, E. Martinez, J. Martinez, B. Machado, D. Ramirez, F. Medina, S. Collins, R. Wicker, Comparison of microstructures and mechanical properties for solid and mesh cobalt-base alloy prototypes fabricated by electron beam melting, *Metallurgical Materials Transactions A* 41 (2010) 3216–3227.

[40] F. Ghadami, A.S.R. Aghdam, Improvement of high velocity oxy-fuel spray coatings by thermal post-treatments: A critical review, *Thin Solid Films* 678 (2019) 42–52.

[41] Y. Gomez Taborda, M. Gómez Botero, J.G. Castaño-González, A. Bermúdez-Castañeda, Assessment of physical, chemical, and tribochemical properties of biomedical alloys used in explanted modular hip prostheses: A review, Proceedings of the Institution of Mechanical Engineers, *Part H: Journal of Engineering in Medicine* 236(4) (2022) 457–468.

[42] S. Fetni, Q.D.T. Pham, V.X. Tran, L. Duchêne, H.S. Tran, A.M. Habraken, Thermal field prediction in DED manufacturing process using Artificial Neural Network, *Soft Computing* 24 (2021) 127–137.

[43] B. Mohajernia, S.E. Mirazimzadeh, A. Pasha, R.J. Urbanic, Machine learning approaches for predicting geometric and mechanical characteristics for single P420 laser beads clad onto an AISI 1018 substrate, *The International Journal of Advanced Manufacturing Technology* (2022) 1–20.

[44] W. Sallehhudin, A. Diab, Using machine learning to predict the fuel peak cladding temperature for a large break loss of coolant accident, *Frontiers in Energy Research* 9 (2021) 755638.

[45] U. Chadha, S.K. Selvaraj, A.S. Lamsal, Y. Maddini, A.K. Ravinuthala, B. Choudhary, A. Mishra, D. Padala, V. Lahoti, A. Adefris, Directed energy deposition via artificial intelligence-enabled approaches, *Complexity* 22 (2022) 1–7.

[46] S. Fetni, Q.D.T. Pham, L.D. Van Xuan Tran, H.S. Tran, A.M. Habraken, Q.D.T.P.T. Dau, Thermal field pr Thermal field prediction in DED manuf ediction in DED manufacturing pr acturing process using Artificial Neur ocess using Artificial Neural Network, *Progress in Neuromorphic Photonics* (2022) 73–78.

[47] A. Amer, A. Al-Shehri, V. Cunningham, H. Saiari, A. Alshamrany, A. Meshaikhis, Artificial Intelligence to Enhance Corrosion Under Insulation Inspection, Abu Dhabi International Petroleum Exhibition and Conference, SPE, 2020, p. D012S116R121.

[48]   J. Singh, S. Singh, A.J.J.o.E.S. Verma, Engineering, Artificial intelligence in use of ZrO2 material in biomedical science, 13(1) (2023) 83–97.

[49]   J. Sousa, R. Darabi, A. Reis, M. Parente, L.P. Reis, J.C. de Sa, An Adaptive Thermal Finite Element Simulation of Direct Energy Deposition With Reinforcement Learning: A Conceptual Framework, ASME International Mechanical Engineering Congress and Exposition, American Society of Mechanical Engineers, 2022, p. V02BT02A037.

[50]   U.M.R. Paturi, D.G. Vanga, S. Cheruku, S.T. Palakurthy, N.K.J.M.T.P. Jha, Estimation of abrasive wear of nanostructured WC-10Co-4Cr TIG weld cladding using neural network and fuzzy logic approach, 78 (2023) 449–457.

[51]   G. Chryssolouris, K. Alexopoulos, Z. Arkouli, Artificial Intelligence in Manufacturing Processes, A Perspective on Artificial Intelligence in Manufacturing, Springer 2023, pp. 15–39.

[52]   J. Wang, C. Ai, F. Guo, X. Yun, X. Zhu, Research of on-line monitoring technology based on laser triangulation for surface morphology of extreme high-speed laser cladding coating, *Surface and Coatings Technology* 13(3) (2023) 625.

[53]   B. Heer, A. Bandyopadhyay, Silica coated titanium using Laser Engineered Net Shaping for enhanced wear resistance, *Additive Manufacturing* 23 (2018) 303–311.

[54]   Y. Hu, W. Cong, A review on laser deposition-additive manufacturing of ceramics and ceramic reinforced metal matrix composites, *Ceramics International* 44(17) (2018) 20599–20612.

[55]   N. Hua, Z. Qian, B. Lin, Z. Liao, Q. Wang, P. Dai, H. Fang, P.K. Liaw, Formation of a protective oxide layer with enhanced wear and corrosion resistance by heating the TiZrHfNbFe0. 5 refractory multi-principal element alloy at 1,000° C, *Scripta Materialia* 225 (2023) 115165.

[56]   C. Han, Y. Li, Q. Wang, D. Cai, Q. Wei, L. Yang, S. Wen, J. Liu, Y. Shi, Titanium/hydroxyapatite (Ti/HA) gradient materials with quasi-continuous ratios fabricated by SLM: material interface and fracture toughness, *Materials Design* 141 (2018) 256–266.

[57]   T. Hanawa, Reconstruction and regeneration of surface oxide film on metallic materials in biological environments, *Corrosion Reviews* 21(2–3) (2003) 161–182.

[58]   W. Harun, R. Asri, J. Alias, F. Zulkifli, K. Kadirgama, S. Ghani, J. Shariffuddin, A comprehensive review of hydroxyapatite-based coatings adhesion on metallic biomaterials, *Ceramics International* 44(2) (2018) 1250–1268.

[59]   M.Z. Ibrahim, A. Halilu, A.A. Sarhan, T. Kuo, F. Yusuf, M. Shaikh, M. Hamdi, In-vitro viability of laser cladded Fe-based metallic glass as a promising bioactive material for improved osseointegration of orthopedic implants, *Medical Engineering Physics* 102 (2022) 103782.

[60]   M.Z. Ibrahim, A.A. Sarhan, T. Kuo, M. Hamdi, F. Yusof, C. Chien, C. Chang, T. Lee, Advancement of the artificial amorphous-crystalline structure of laser cladded FeCrMoCB on nickel-free stainless-steel for bone-implants, *Materials Chemistry Physics* 227 (2019) 358–367.

[61] M.Z. Ibrahim, A.A. Sarhan, T. Kuo, F. Yusof, M. Hamdi, Characterization and hardness enhancement of amorphous Fe-based metallic glass laser cladded on nickel-free stainless steel for biomedical implant application, *Materials Chemistry Physics* 235 (2019) 121745.

[62] M.Z. Ibrahim, A.A. Sarhan, M. Shaikh, T. Kuo, F. Yusuf, M. Hamdi, Investigate the effects of the laser cladding parameters on the microstructure, phases formation, mechanical and corrosion properties of metallic glasses coatings for biomedical implant application, *Additive Manufacturing of Emerging Materials* (2019) 299–323.

[63] M.N. Jahangir, M.A.H. Mamun, M.P. Sealy, A review of additive manufacturing of magnesium alloys, AIP conference proceedings, AIP Publishing, 2018.

[64] N. Jeyaprakash, S.S. Karuppasamy, C.-H. Yang, Application of Wear-Resistant Laser Claddings, Handbook of Laser-Based Sustainable Surface Modification and Manufacturing Techniques, CRC Press 2023, pp. 1–26.

[65] X. Ji, C. Luo, J. Jin, Y. Zhang, Y. Sun, L. Fu, Tribocorrosion performance of 316L stainless steel enhanced by laser clad 2-layer coating using Fe-based amorphous powder, *Journal of Materials Research Technology* 17 (2022) 612–621.

[66] W. JIA, X. LIN, Numerical microstructure simulation of laser rapid forming 316L stainless steel, *Acta Metall Sin* 46(2) (2010) 135–140.

[67] N.O. Joy-anne, Y. Su, X. Lu, P.-H. Kuo, J. Du, D. Zhu, Bioactive glass coatings on metallic implants for biomedical applications, *Bioactive Materials* 4 (2019) 261–270.

[68] K. Kanishka, B.J.J.o.M.P. Acherjee, A systematic review of additive manufacturing-based remanufacturing techniques for component repair and restoration, *Journal of Manufacturing Processes* 89 (2023) 220–283.

[69] M. Khanzadeh, W. Tian, A. Yadollahi, H.R. Doude, M.A. Tschopp, L.J.A.M. Bian, Dual process monitoring of metal-based additive manufacturing using tensor decomposition of thermal image streams, *Additive Manufacturing* 23 (2018) 443–456.

[70] N. Kaushik, A. Meena, H.S. Mali, High entropy alloy synthesis, characterisation, manufacturing & potential applications: a review, *Materials Manufacturing Processes* 37(10) (2022) 1085–1109.

[71] A.M. Khorasani, M. Goldberg, E.H. Doeven, G. Littlefair, Titanium in biomedical applications—properties and fabrication: a review, *Journal of biomaterials tissue engineering* 5(8) (2015) 593–619.

[72] V. Koshuro, M. Fomina, A. Voyko, I. Rodionov, A. Zakharevich, A. Skaptsov, A. Fomin, Surface morphology of zirconium after treatment with high-frequency currents, *Composite Structures* 202 (2018) 210–215.

[73] M. Krzyzanowski, D. Svyetlichnyy, S. Bajda, Additive manufacturing of multi layered bioactive materials with improved mechanical properties: modelling aspects, Materials Science Forum, *Trans Tech Publ*, 2021, pp. 888–893.

[74] P. Kumar, N.K. Jain, A. Tiwari, Sustainable Polishing of Directed Energy Deposition–Based Cladding Using Micro-Plasma Transferred Arc, Advances in Sustainable Machining and Manufacturing Processes, CRC Press 2022, pp. 289–302.

[75] S. Kumar, P. Katyal, Factors affecting biocompatibility and biodegradation of magnesium based alloys, *Materials Today: Proceedings* 52 (2022) 1092–1107.

[76] S.S. Kumar, V. Tripathi, R. Sharma, G. Puthilibai, M. Sudhakar, K. Negash, Study on developments in protection coating techniques for steel, *Advances in Materials Science Engineering* 22 (2022) 1–17.

[77] W. Li, P. Xu, Y. Wang, Y. Zou, H. Gong, F. Lu, Laser synthesis and microstructure of micro-and nano-structured WC reinforced Co-based cladding layers on titanium alloy, *Journal of Alloys Compounds* 749 (2018) 10–22.

[78] J. Liu, D. Liu, S. Li, Z. Deng, Z. Pan, C. Li, T. Chen, The effects of graphene oxide doping on the friction and wear properties of TiN bioinert ceramic coatings prepared using wide-band laser cladding, *Surface Coatings Technology* 458 (2023) 129354.

[79] W. Liu, S. Liu, L. Wang, Surface modification of biomedical titanium alloy: micromorphology, microstructure evolution and biomedical applications, *Coatings* 9(4) (2019) 249.

[80] Z. Liu, K.C. Chan, L. Liu, S. Guo, Bioactive calcium titanate coatings on a Zr-based bulk metallic glass by laser cladding, *Materials Letters* 82 (2012) 67–70.

[81] M. Lu, P. McCormick, Y. Zhao, Z. Fan, H. Huang, Laser deposition of compositionally graded titanium oxide on Ti6Al4V alloy, *Ceramics International* 44(17) (2018) 20851–20861.

[82] N. Ma, S. Liu, W. Liu, L. Xie, D. Wei, L. Wang, L. Li, B. Zhao, Y. Wang, Research progress of titanium-based high entropy alloy: methods, properties, and applications, *Frontiers in Bioengineering Biotechnology* 8 (2020) 603522.

[83] A. Mehta, G. Singh, Consequences of hydroxyapatite doping using plasma spray to implant biomaterials, *Journal of Electrochemical Science Engineering failure analysis* 13(1) (2023) 5–23.

[84] A. Mahajan, S. Devgan, D.J.M. Kalyanasundaram, M. Processes, Surface alteration of Cobalt-Chromium and duplex stainless steel alloys for biomedical applications: a concise review, *Materials Manufacturing Processes* 38(3) (2023) 260–270.

[85] M.A. Mahmood, A.C. Popescu, I.N. Mihailescu, Metal matrix composites synthesized by laser-melting deposition: a review, *Materials* 13(11) (2020) 2593.

[86] A.I. Mahmoud Zakaria Alsayed, Laser cladding of FeCrMoCB metallic glass on nickel-free stainless-steel to develop durable and cost-effective biomedical implants/Mahmoud Zakaria Alsayed Abdalfattah Ibrahim, PhD diss, Universiti Malaya, 2019.

[87] J.D. Majumdar, A. Kumar, S. Pityana, I. Manna, Laser surface melting of AISI 316L stainless steel for bio-implant application, Proceedings of the National Academy of Sciences, *India Section A: Physical Sciences* 88 (2018) 387–403.

[88] A.S.H. Makhlouf, A. Barhoum, Emerging applications of nanoparticles and architectural nanostructures: current prospects and future trends, *Metals* 37 (2018).

[89] J. Mesquita-Guimarães, B. Henriques, F. Silva, Bioactive glass coatings, Bioactive glasses, Elsevier 2018, pp. 103–118.

[90] M.H. Miah, D. Singh Chand, G.S. Malhi, S. Khan, Influence of laser scanning power on microstructure and tribological behavior of NI-composite claddings fabricated on TC4 titanium alloy, *Aircraft Engineering Aerospace Technology* (2023) 103–118.

[91] A.-C. Mocanu, F. Miculescu, G.E. Stan, I. Pasuk, T. Tite, A. Pascu, T.M. Butte, L.-T. Ciocan, Modulated Laser Cladding of Implant-Type Coatings by Bovine-Bone-Derived Hydroxyapatite Powder Injection on Ti6Al4V Substrates—Part I: Fabrication and Physico-Chemical Characterization, *Materials* 15(22) (2022) 7971.

[92] A.-C. Mocanu, F. Miculescu, G.E. Stan, T. Tite, M. Miculescu, M.H. Țierean, A. Pascu, R.-C. Ciocoiu, T.M. Butte, L.-T. Ciocan, Development of ceramic coatings on titanium alloy substrate by laser cladding with pre-placed natural derived-slurry: Influence of hydroxyapatite ratio and beam power, *Ceramics International* 49(7) (2023) 10445–10454.

[93] N.P. Msweli, S.O. Akinwamide, P.A. Olubambi, B.A. Obadele, Microstructure and biocorrosion studies of spark plasma sintered yttria stabilized zirconia reinforced Ti6Al7Nb alloy in Hanks' solution, *Materials Chemistry Physics* 293 (2023) 126940.

[94] A. Mthisi, A. Popoola, D. Adebiyi, O. Popoola, Laser Cladding of Ti-6Al-4V Alloy with Ti-Al2O3 Coating for Biomedical Applications, IOP Conference Series: Materials Science and Engineering, IOP Publishing, 2018, p. 012005.

[95] S. Mukherjee, S. Dhara, P. Saha, Laser surface remelting of Ti and its alloys for improving surface biocompatibility of orthopaedic implants, *Materials Technology* 33(2) (2018) 106–118.

[96] K. Munir, A. Biesiekierski, C. Wen, Y. Li, Surface modifications of metallic biomaterials, Metallic Biomaterials Processing and Medical Device Manufacturing, Elsevier 2020, pp. 387–424.

[97] L. Murr, Metallurgy principles applied to powder bed fusion 3D printing/ additive manufacturing of personalized and optimized metal and alloy bio-medical implants: An overview, *Journal of Materials Research Technology* 9(1) (2020) 1087–1103.

[98] R. Nazempour, Q. Zhang, R. Fu, X. Sheng, Biocompatible and implantable optical fibers and waveguides for biomedicine, *Materials* 11(8) (2018) 1283.

[99] T.D. Ngo, A. Kashani, G. Imbalzano, K.T. Nguyen, D.J.C.P.B.E. Hui, Additive manufacturing (3D printing): A review of materials, methods, applications and challenges, *Composites Part B: Engineering* 143 (2018) 172–196.

[100] M. Niinomi, Recent metallic materials for biomedical applications, *Metallurgical Materials Transactions A* 33 (2002) 477–486.

[101] M.P. Nikolova, M.D. Apostolova, Advances in Multifunctional Bioactive Coatings for Metallic Bone Implants, *Materials* 16(1) (2022) 183.

[102] S. Omarov, N. Nauryz, D. Talamona, A. Perveen, Surface Modification Techniques for Metallic Biomedical Alloys: A Concise Review, *Metals* 13(1) (2022) 82.

[103] A.A. Oudah, M.A. Hassan, N. Almuramady, Materials manufacturing processes: Feature and trends, AIP Conference Proceedings, AIP Publishing, 2023.

[104] G. Padmanabham, R. Bathe, Laser materials processing for industrial applications, Proceedings of the National Academy of Sciences, *India Section A: Physical Sciences* 88 (2018) 359–374.

[105] Y. Qu, T. Nguyen-Dang, A.G. Page, W. Yan, T. Das Gupta, G.M. Rotaru, R.M. Rossi, V.D. Favrod, N. Bartolomei, F. Sorin, Superelastic multimaterial electronic and photonic fibers and devices via thermal drawing, *Advanced Materials* 30(27) (2018) 1707251.

[106] Y. Qu, T. Nguyen-Dang, A.G. Page, W. Yan, T.D. Gupta, G.M. Rotaru, R.M. Rossi, V.D. Favrod, N. Bartolomei, F. Sorin, Stretchable optical and electronic fibers via thermal drawing, International Flexible Electronics Technology Conference (IFETC), IEEE, 2018, pp. 1–1.

[107] R. Ranjan, A.K. Das, Improving the Resistance to Wear and Mechanical Characteristics of Cladding Layers on Titanium and its Alloys: A Review, *Tribology in Industry* 44(1) (2023) 136.

[108] J. Reddy, M. Chamanzar, Parylene photonic waveguide arrays: a platform for implantable optical neural implants, CLEO: Applications and Technology, Optica Publishing Group, 2018, p. AM3P. 6.

[109] J.W. Reddy, M. Chamanzar, Low-loss flexible Parylene photonic waveguides for optical implants, *Optics Letters* 43(17) (2018) 4112–4115.

[110] R. Rojo, M. Prados-Privado, A.J. Reinoso, J.C. Prados-Frutos, Evaluation of fatigue behavior in dental implants from in vitro clinical tests: a systematic review, *Metals* 8(5) (2018) 313.

[111] D. Romanov, K. Sosnin, S.Y. Pronin, Y.F. Ivanov, V. Gromov, Structure and Properties of Electroexplosion Molybdenum Coating Deposited on Titanium Alloy VT6, *Metal Science Heat Treatment* 64(11) (2023) 639–647.

[112] A. Santos, J. Teixeira, C. Fonzar, E. Rangel, N. Cruz, P.N. Lisboa-Filho, A Tribological Investigation of the Titanium Oxide and Calcium Phosphate Coating Electrochemical Deposited on Titanium, *Metals* 13(2) (2023) 410.

[113] A. Sharma, A. Singh, V. Chawla, J. Grewal, A. Bansal, Microwave processing and characterization of alumina reinforced HA cladding for biomedical applications, *Materials Today: Proceedings* 57 (2022) 650–656.

[114] R.K. Sharma, G.P.S. Sodhi, V. Bhakar, R. Kaur, S. Pallakonda, P. Sarkar, H. Singh, Sustainability in manufacturing processes: Finding the environmental impacts of friction stir processing of pure magnesium, *CIRP Journal of Manufacturing Science Technology* 30 (2020) 25–35.

[115] A. Shearer, M. Montazerian, J.J. Sly, R.G. Hill, J.C. Mauro, Trends and perspectives on the commercialization of bioactive glasses, *Acta Biomaterialia* (2023) 2246–2257.

[116] J. Singh, J.P. Singh, S. Kumar, H.S. Gill, Short review on hydroxyapatite powder coating for SS 316L, *Journal of Electrochemical Science Engineering* 13(1) (2023) 25–39.

[117] K. Singh, S. Mohan, S. Konovalov, M. Graf, Effect of nano-hydroxyapatite and post heat treatment on biomedical implants by sol-gel and HVOF spraying, Nanomaterials for Sustainable Tribology, CRC Press 2023, pp. 257–285.

[118] S. Tayebati, K.T. Cho, A hybrid machine learning framework for clad characteristics prediction in metal additive manufacturing, arXiv preprint arXiv:.01872 (2023).

[119] P. Sun, N. Yan, S. Wei, D. Wang, W. Song, C. Tang, J. Yang, Z. Xu, Q. Hu, X. Zeng, Microstructural evolution and strengthening mechanisms of Inconel 718 alloy with different W addition fabricated by laser cladding, *Materials Science Engineering: A* 868 (2023) 144535.

[120] M.A. Rahman, T. Saleh, M.P. Jahan, C. McGarry, A. Chaudhari, R. Huang, M. Tauhiduzzaman, A. Ahmed, A.A. Mahmud, M.S. Bhuiyan, Review of Intelligence for Additive and Subtractive Manufacturing: Current Status and Future Prospects, *Micromachines* 14(3) (2023) 508.

[121] M. Salonvaara, A. Desjarlais, A.J. Aldykiewicz Jr, E. Iffa, P. Boudreaux, J. Dong, B. Liu, G. Accawi, D. Hun, E. Werling, Application of Machine Learning to Assist a Moisture Durability Tool, *Energies* 16(4) (2023) 2033.

[122] W.Y.S. Lim, J. Cao, A. Suwardi, T.L. Meng, C.K.I. Tan, H. Liu, Recent advances in laser-cladding of metal alloys for protective coating and additive manufacturing, *Journal of Adhesion Science Technology* 36(23–24) (2022) 2482–2504.

[123] M. Gao, S. Li, W. Guan, H. Xie, X. Wang, J. Liu, H. Wang, Excellent thermal shock resistance of NiCrAlY coatings on copper substrate via laser cladding, *Journal of Materials Science Technology* 130 (2022) 93–102.

[124] J. Zheng, D. Zheng, L. Qiao, Y. Ying, Y. Tang, W. Cai, W. Li, J. Yu, J. Li, S. Che, High permeability and low core loss Fe-based soft magnetic composites with Co-Ba composite ferrite insulation layer obtained by sol-gel method, *Journal of Alloys Compounds* 893 (2022) 162107.

[125] M. Dehghan Manshadi, N. Alafchi, A. Tat, M. Mousavi, A. Mosavi, Comparative analysis of machine learning and numerical modeling for combined heat transfer in Polymethylmethacrylate, *Polymers* 14(10) (2022) 1996.

[126] M. Kumar, S. Kumar, K. Jha, A. Mandal, Composite coating by TIG cladding with different rare earth oxides, *Surface Engineering* 38(3) (2022) 271–287.

[127] L. Wang, S. Chen, X. Sun, J. Chen, J. Liang, M. Wang, Effects of Y2O3 on the microstructure evolution and electromagnetic interference shielding mechanism of soft magnetic FeCoSiMoNiBCu alloys by laser cladding, *Additive Manufacturing* 55 (2022) 102811.

[128] S.F.S. Ferreiro, J. Larreina, M. Tena, J. Leunda, I. Garmendia, A. Arnaiz, Artificial intelligence methodology for smart and sustainable manufacturing industry, *IFAC-PapersOnLine* 54(1) (2021) 1041–1046.

[129] I.d.V. Tomaz, F.H.G. Colaço, S. Sarfraz, D.Y. Pimenov, M.K. Gupta, G. Pintaude, Investigations on quality characteristics in gas tungsten arc welding process using artificial neural network integrated with

genetic algorithm, *The International Journal of Advanced Manufacturing Technology* 113(11–12) (2021) 3569–3583.

[130] M.A. Mahmood, A.C. Popescu, M. Oane, A. Channa, S. Mihai, C. Ristoscu, I.N. Mihailescu, Bridging the analytical and artificial neural network models for keyhole formation with experimental verification in laser melting deposition: A novel approach, *Results in Physics* 26 (2021) 104440.

[131] H.Z. Imam, Y. Zheng, P. Martinez, R. Ahmad, Vision-based damage localization method for an autonomous robotic laser cladding process, *Procedia CIRP* 104 (2021) 827–832.

[132] M. Braik, H. Al-Zoubi, H. Al-Hiary, Artificial neural networks training via bio-inspired optimisation algorithms: modelling industrial winding process, case study, *Soft Computing* 25 (2021) 4545–4569.

[133] S. Chatterjee, S.S. Mahapatra, V. Bharadwaj, B.N. Upadhyay, K.S. Bindra, Prediction of quality characteristics of laser drilled holes using artificial intelligence techniques, *Engineering with Computers* 37 (2021) 1181–1204.

[134] Y. Zhang, Y. Xu, Y. Sun, W. Cheng, Surface quality optimization of laser cladding based on surface response and genetic neural network model, *Surface Topography: Metrology Properties* 10(4) (2022) 044007.

[135] G. He, Y. Du, Q. Liang, Z. Zhou, L. Shu, Modeling and Optimization Method of Laser Cladding Based on GA-ACO-RFR and GNSGA-II, *International Journal of Precision Engineering Manufacturing-Green Technology* (2022) 1–16.

[136] A.R. Dhar, D. Gupta, S.S. Roy, A.K. Lohar, Forward and backward modeling of direct metal deposition using metaheuristic algorithms tuned artificial neural network and extreme gradient boost, *Progress in Additive Manufacturing* (2022) 1–15.

[137] F. Mumali, Artificial neural network-based decision support systems in manufacturing processes: A systematic literature review, *Computers Industrial Engineering* 165 (2022) 107964.

[138] W.Y.S. Lim, J. Cao, A. Suwardi, T.L. Meng, C.K.I. Tan, H.J.J.o.A.S. Liu, Technology, Recent advances in laser-cladding of metal alloys for protective coating and additive manufacturing, 36(23–24) (2022) 2482–2504.

[139] Y. Chen, Y. Hu, S. Zhang, X. Mei, Q. Shi, Optimized erosion prediction with MAGA algorithm based on BP neural network for submerged low-pressure water jet, *Applied Sciences* 10(8) (2020) 2926.

[140] N. Hua, H. Huang, X. Zhang, Investigating the Working Efficiency of Typical Work in High-Altitude Alpine Metal Mining Areas Based on a SeqGAN-GABP Mixed Algorithm, *Advances in Civil Engineering* 2021 (2021) 1–12.

[141] M. Feng, W. Yao, J. An, Z. Huang, Y. Yuan, Y. Yao, Prediction Method of Graphene Defect Modification Based on Neural Network, *Mobile Information Systems* 2022 (2022).

[142] X.a. Fan, X. Gao, G. Liu, N. Ma, Y. Zhang, Research and prospect of welding monitoring technology based on machine vision, *The International Journal of Advanced Manufacturing Technology* 115 (2021) 3365–3391.

[143] W. Hu, J. Fang, Z. Liu, J. Tan, Intelligent design and optimization of wind turbines, Wind Energy Engineering, Elsevier 2023, pp. 315–325.

[144] V. Eyupoglu, E. Polat, B. Eren, R.A. Kumbasar, Two-dimensional assessment of cobalt transport and separation through ionic polymer inclusion membrane: experimental optimization and artificial neural network modeling, *Journal of Dispersion Science Technology* 44(5) (2023) 763–778.

Chapter 8

Evolution of nickel-based superalloy claddings for against high-temperature oxidation and wear

Jashanpreet Singh and Harjot Singh Gill

8.1 INTRODUCTION TO CRITICAL RAW MATERIALS (CRMS)

Though it is not on the EU Critical Raw Materials (CRMs) list for the year 2020, nickel (Ni) is nonetheless subject to strict oversight by the EU Commission [1]. This is due to the fact that its demand is anticipated to increase in the coming years, primarily as a result of the expanding market for batteries used in electric vehicles. Because of this, it is not possible to rule out the possibility that nickel may be added to the CRMs summary in the coming years. Ni is the most common feedstock for the production of alloys, such as steel and non-ferrous alloys utilized in aerospace industry. The list of CRMs that will be implemented by the EU in 2020 already includes titanium, cobalt, and tantalum. This chapter is focused Ni-superalloys used for the fabrication of aerospace engine parts and developed to withstand in high-temperature conditions. Carbon, silicon, nickel, titanium, chromium, cobalt, molybdenum, tungsten, tantalum, niobium, hafnium, radium, and ruthenium are only some of the critical and near-critical elements (CREs) found in these alloys, and many of the coatings and claddings often used to protect aviation engines also include CRMs.

A turbojet engine consists of the following components at its most fundamental level: an air intake, a compressor or fan, a combustor, a gas turbine, and a smoke nozzle (Figure 8.1) [2]. In order for the engine to function, air must be drawn in, compressed into as tiny of a volume as is humanly feasible, mixed with fuel, and then set ablaze inside the engine's combustion chamber. The force required to push the airplane forward is generated by the hot gases that are expelled from the combination of hot air and fuel. To acquire engines with additional power, highly fuel-efficient, lowered CO_2 and NO_x exhaust, and decreased noise frequency, it is necessary to focus on the aerodynamic features for combustor and gas turbines for enhanced cooling system for blades of turbine, and advancement in alloys as well as manufacturing techniques [3,4]. This is essential to increase the engine's efficiency,

DOI: 10.1201/9781032713830-8

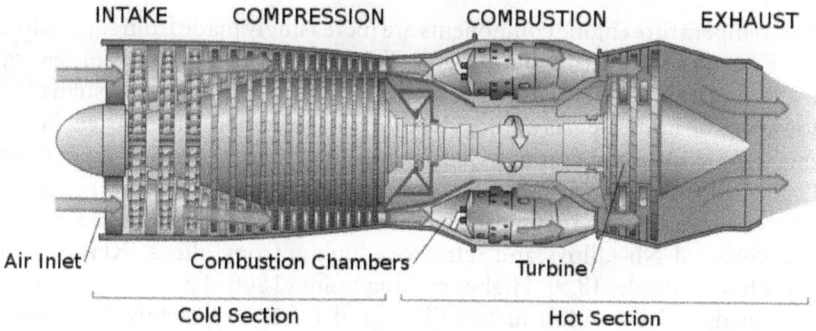

Figure 8.1 Illustration of a turbojet engine [2] (Permission under CC BY 4.0 licence).

which can be accomplished by paying attention to aerodynamic designs. According to the laws of thermodynamics, the high value of temperature inside the combustion chamber results in greater performance. Combustion chamber materials have evolved from high-temperature resistant steels to nickel- or cobalt-based alloys as engine temperatures have risen [5].

The air that is drawn in the engine is compressed in the portion of the compressor known as the axial or radial depending to the flow vein form. The materials that are often used in this section are heat-resistant alloys based on the elements iron, nickel, and titanium. The air pressure may be increased by up to 30 times, and the compressor can achieve temperatures of up to 1000°C. There is a portion of the engine called the combustor where compressed air is mixed with fuel to create a high-intensity airflow that propels the aircraft. These gases are then routed into the turbine engine, where they create thrust and then exit the engine through the exhaust. Internally, the combustor may reach temperatures of 1500–1600°C. Combustion chamber exhaust is sent to the turbine section, which has a high-pressure (HP) turbine and a low-pressure (LP) turbine. A shaft or spool that runs the length of the combustion chamber connects the turbine to the compressor. The blades of the turbine are what really pull the energy out of the high-pressure gases. It follows that the blades and aerofoil of the high-pressure turbine section experience the largest stress in airplane jet engines. High temperatures, hostile conditions, and heavy loads cause massive thermal gradients (700–1200°C) and significant pressures on these components (100–800 MPa). Turbine blade and, by extension, turbine engine performance might be impeded by the high temperatures encountered during the operation. This increases the risk of failure of the turbine blades from creep, corrosion, or fatigue [6], and failure from creep or corrosion may be caused by vibrations in the turbine engine.

8.2 TURBOJET ENGINE AND ITS COMPONENTS

High-temperature engine components are increasingly made from superalloys based on nickel and titanium. Figure 8.2 displays the location of several nickel- and titanium-based alloys used in airplane propulsion systems [7]. However, the hottest parts of today's cutting-edge aeroengines may get hotter than the critical temperature, that is, melting of nickel-based superalloys (1200–1300°C). Numerous studies are being conducted to discover possible replacement materials that may be used to get over this limitation. Silicide-, Co-, Mo- and Nb- alloys and refractory high-entropy alloys (RHEAs) are only a few examples [8,9]. Higher melting point (1800–1900°C) aluminides and silicides of transition metals (TM) and refractory metals (RM) were proposed during the "far beyond Ni-superalloys era" [10]. Suitably alternative materials should be able to satisfy many requirements simultaneously. For example, high heat tolerance, corrosion and oxidation resistance, fracture toughness, low weight, and low mass are all desirable qualities.

To guarantee the combustor section of the engine can withstand the severe heat loads that are imposed on it during flight, actively cooled panels are utilized. The fuel is cooled by passing through the combustor's heat exchanger plates before being injected. Keeping the material and the coolant below their critical temperatures is achieved in this manner. Materials like

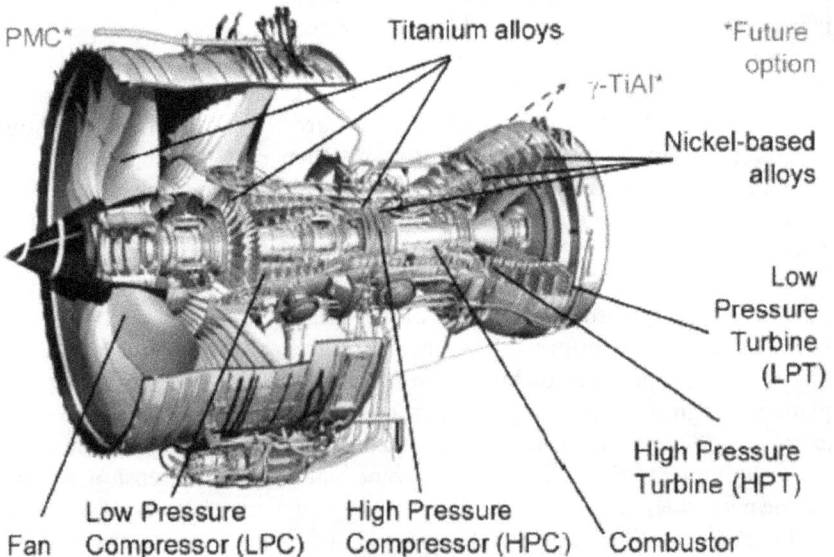

Figure 8.2 Turbojet engine with nickel and titanium alloys [7] (Permission under CC BY 4.0 licence).

niobium alloy x750 (Inconel X-750) and niobium alloy 752 (Nb 752) are only two examples of what may be utilized for the engine's active panels. Potentially useful alternative materials with fuel/coolant heat gain, Inconel X-750 was demonstrated to be able to sustain high heat transfer rates far below the fuel coking temperature [11]. Furthermore, the weight-to-area ratio of this material is only mild. In comparison with Inconel X-750, Vermaak and his colleagues [12] investigated the practicability of using a variety of alternative metal alloys for active cooling applications. Both the low softening temperature and the pressure-induced stresses of Ti-21S alloy severely restrict its usefulness. Although GRCop-84's conductivity exceeds that of Inconel X-750 by nearly an order of magnitude, it requires a substantially larger structure to endure the same mechanochemical environment. C-103, a refractory alloy based on niobium, has some very close chemical and physical properties to Inconel X-750, in contrast.

In addition to cooling, the development of appropriate thermal barrier coatings (TBCs) and thermal barrier claddings (TBCLs) is an additional key step that is required to maintain the temperature of the parts below the critical value. Thermal spray coatings and cladding are most widely used in the aeronautical, aerospace, and mechanical industry [13–32]. Coatings and claddings are widely used in the industry for the protection of machinery from erosion, corrosion, fatigue, oxidation, cavitation, and wet conditions [33–44]. This step is important regardless of the base material that is used for the hottest regions of the engine. As a direct result of this, continued progress is made by conducting an in-depth study on the total system consisting of superalloy, bond coat, and thermal barrier [45]. Recent developments in turbojet engines design have focused on minimizing specific fuel consumption (SFC) while simultaneously improving thrust-to-weight ratio, heat resistance, and lightweight construction. Titanium aluminide, which has greater characteristics, is thus being employed in place of nickel alloys, which were formerly utilized in the low-pressure portion of the turbines [5,46]. The advantages of TiAl-based alloys include low density, high value of mechanical strength (stiffness, yield strength, and resistance to creep) at service extreme heat to 900°C, although, at operating temperatures greater than 750°C, oxidative corrosion may set in [47,48]. In addition, in this scenario, the whole system consisting of a bond coat, a thermal barrier, and an alloy based on titanium must be tuned in order to ensure that titanium-based alloys can function well under demanding service circumstances. It is important to take note of the fact that there is a clear upward tendency in the worldwide market for the aerospace sector. In 2019, Airbus projected an annual increase of 4.3%, which would result in a need for around 39,200 new passenger and dedicated cargo aircraft over the following 20 years. Even though the COVID-19 pandemic is projected to cause a significant drop in the volume of passenger travel in the year 2020, analysts anticipate both a long-term recovery and a return to strong expansion in the aviation sector [49]. Because of this, there is an increasing need for brand new commercial aircraft that are more environmentally friendly and have better fuel economy [50]. The optimization of aircraft loads, and the subsequent reduction in fuel consumption and emissions per passenger, as well as the improvement of aircraft dependability, will have a very significant influence on prices, as well as on the environment and on public safety. Because engines are the most important parts of aircraft, it is imperative that they continue to advance in terms of safety, dependability, and fuel efficiency. To this end, machine learning is proving useful in predicting how long an aircraft engine has left to run before it has to be replaced or repaired [51–53]. A study may be found in a study [54] that discusses the role that machine learning has played in transforming and reshaping the current scientific, technical, and industrial environments for aeronautical engineering.

In relation to this, the worldwide additive manufacturing (AM) industry is valued at $12 billion, and it is expected that this would increase to around $30 billion in 2028. The AM procedures of alloys for aerospace are anticipated to become less expensive and more efficient thanks to machine learning [55]. The use of AM methods enables a reduction in the amount of material waste and energy consumption, as well as the facilitation of prototyping and the enhancement of component performance. The metallic components that make up the substrate might come in the form of a wrought part, a forging, a casting, or a damaged or flawed component [56]. AM may be used in the aerospace sector for the purposes of manufacturing components, repairing components, and manufacturing and repairing tooling. When it comes to blades and vanes, the potential of additive manufacturing appears to be larger in mending than on manufacturing of the components. This is because of the small decrease in buy-to-fly ratio that it may give [56]. The wingtip fences for the A320ceo were made via additive manufacturing and were supplied by Satair. Nevertheless, there are a few issues that need to be resolved before the commercialization of the AM technique, and more work is still essential for its optimization. Since additive manufacturing may be conceptualized as a multilayer or rapid-run welding process, the part that is produced using this method is predisposed to the development of weld-cracking flaws [57]. In addition, there is still a need for solutions to the issues of large residual stresses that are brought on by the fast cooling rates associated with additive manufacturing processes.

8.3 EVOLUTION OF NI-BASED SUPERALLOYS

About half of the mass of a modern aviation engine is made up of Ni-based superalloys [58], which is why they are the material of choice for turbine engines. Ni superalloys have excellent strength and fatigue resistance in service because they are resistant to oxidation, creep, and stress rupture at high temperatures. Creep is a problem for many alloys, even those used in lower-temperature zones (where temperatures are typically between 150 and 350°C), but nickel superalloys are resistant to temperatures several hundred degrees higher [59]. Ni superalloys are used in higher-temperature zones. Because Ni-based superalloys have a complicated composition and micro-structure, the fabrication technique that is used has a significant impact on the characteristics of the final component, which may be a turbine blade or another component of the aeroengines. As a general rule, nickel superalloys are separated into two categories: wrought superalloys and cast superalloys. However, casting techniques have undergone significant development since the early 1960s onward (Figure 8.3), which has made it possible to create alloys with improved performance characteristics [59]. The starting step in the normal process for the manufacturing of a superalloy part is melting the basic raw materials. This is just a quick rundown of the whole operation. Once this is complete, the resultant alloy may either be sent straight to the casting and wroughting stage, or it can be exposed to further remelting operations, such as arc remelting, to improve the grain size and decrease the degree of particle segregation. Traditional casting techniques provide components with a polycrystalline structure that are only resistant to low temperatures (Figure 8.3). It was not until the development of more refined casting methods like directional solidification (DS) that HP turbine blade alloys were sufficiently strong to withstand creep and fatigue in operation.

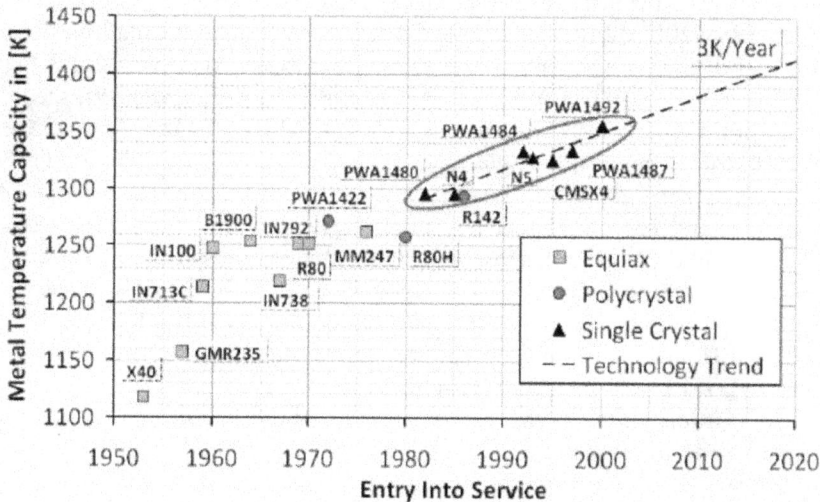

Figure 8.3 Operational temperature of superalloys in gas turbine [59] (Permissions under CC BY 3.0 licence).

The production of single or monocrystal turbine blades can also be accomplished using powder metallurgy, which is an alternate method to the casting process that involves the molten alloy. The superalloy powders that are produced from the first melting step are subjected to further processing and are consolidated by the use of hot isostatic pressing (HIP) in this method [60,61]. According to what has been described, this technique makes it possible to alloy a greater number of elements while maintaining a consistent composition, grain size distribution, and phase composition, as well as reducing segregation and porosity. However, additional negative effects, such as contamination or the formation of previous particle boundaries, might be produced (PPB). To enhance the mechanical characteristics of the cast or HIP alloy, further processing processes, including heat treatment and ageing, must be carried out before the fabrication of the final component.

Advancement of AM methods for the fabrication of engine parts has been bolstered by the trend towards the fabrication of metals in powder form. These AM methods allow for the quick conversion of 3D designs to the parts that are eventually made, and the creation of geometrically complicated items in near net shape form. After the AM process is complete, it may be subjected to post-processing heat treatments including HIP and ageing to repair any imperfections. Although the microstructure of AM-made Ni-superalloys components is distinct from that of cast and wrought components, this is still achievable. This is because the alloy has a dendritic structure, similar to that of cast metals. The alloy's isotropic characteristics are modified by this structure.

Apart from the alloys mentioned in Figure 8.3, other most widely used Ni-superalloys for gas turbine engines are conventionally cast alloys (Rene 80, MAR-M200/M246, and Inconel 713LC), directionally cast alloys (Rene 80DS and CM247 LC DS), first generation (Gen) mono-crystal alloys (Rene N4, PWA 1480, CMSX-2, CMSX-3, RR2000, and SRR99), second generation single-crystal alloys (Rene N5, PWA 1484, CMSX-4, CMSX-6, SC180), third generation single-crystal alloys (Rene N6, CMSX-10, TMS-75/113/121), fourth generation single-crystal alloys (MC-NG, PWA1497, TMS 138, and TMS 138A), fifth generation single-crystal alloys (TMS-162, TMS-173, and TMS-196), sixth generation single-crystal alloys (TMS-238), wrought superalloys (Inconel 600, Inconel 718, ATI-718 Plus, Rene 41, Nimonic 80A, Nimonic 105, Waspaloy, Hastelloy X, Hastelloy S, Udimet 500, and Udimet 700), and powder-processed superalloys (N18, Rene 95, Rene 88DT, Inconel 100, Astroloy) [62–64].

The extraordinary mechanical properties of Ni-based superalloys are linked to the microstructure of these alloys. They are characterized by a unique two-phase microstructure made up of a FCC-matrix phase and a cuboidal γ' precipitate with an FCC-based L12-ordered structure, as seen in Figure 8.4 [65]. This microstructure is characterized by having an FCC structure. It is possible for Ni-superalloys to have about 15 different elements, the vast majority of which are CRMs. Some of the elements that make

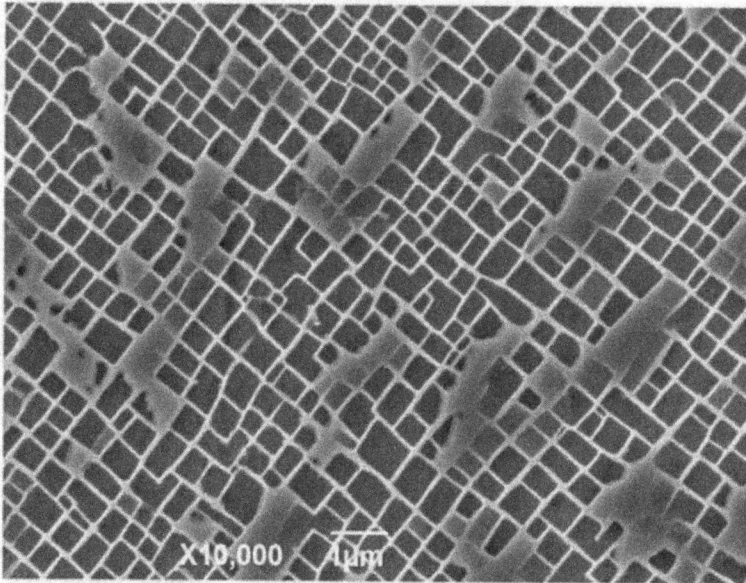

Figure 8.4 The phase forms the matrix channels and the γ' phase forms the cuboids in these transmission electron microscopy images of a typical Ni-based superalloy [65] (Permissions under CC BY 4.0 licence).

up CRMs, such as tungsten, cobalt, rhenium, chromium, molybdenum, iron, and yttrium may facilitate as alloy elements to stabilize the phase and increase the strength of the solution [66,67].

However, the γ' phase that is accountable for the amplification of precipitation may be induced by the introduction of additional substances. Primary solutes in Ni-based superalloys are generally less than 10 at.% and include Al, Ti, and Ta. This is the critical concentration for the production of the signature γ' phase, an intermetallic complex with the combination $Ni_3(A, Ti, Ta)$ [28]. The phase's presence is mainly accountable for the material's strength and low susceptibility to creep deformation at extreme temperatures. Molybdenum is added to Ni-based alloys' γ' phase as a room-temperature and elevated-temperature strengthener [68]. To prevent the alloys from being weakened by Cr_7C_3 sensitization, Nb, Ti, and carbide formers are added. As a room-temperature and high-temperature strengthener, molybdenum is included in Ni-based alloys' γ' phase. A Cr_2O_3 scale, which is enhanced by the addition of a trace quantity of chromium, makes the material more resistant to oxidation and corrosion at high temperatures [6]. Stable chemical oxides, such as Cr_2O_3 and Al_2O_3 scales, are reported to grow on the surface of a Ni-based alloy, protecting the underlying alloy element from further increased oxidation and contributing to the alloy's high temperature resistance [69]. One of the explanations why alloys based on nickel can withstand high temperatures indefinitely is because of this.

The successful implementation of fourth and fifth generation Ni-based superalloys is hindered by their reduced oxidation resistance brought on by the growing proportion of refractory elements like ruthenium, molybdenum, and rhenium. Due to their higher vapour pressures, these refractory oxide species may contribute to the decomposition of Al_2O_3 that forms on the surface during heating [70]. The sixth generation mono-crystal Ni-superalloy TMS-238 was recently developed for usage in turbojet engine blades [70]. Due to its elevated-temperature creep strength and improved oxidation resistance, this alloy may achieve a creep life of 1000 h at 137 MPa and 1120°C [71]. The raw ingredients for TMS-238 alloy, of which rhenium makes up 60% and ruthenium the remaining 30%, may be purchased for about $2400/ L [49]. The sixth generation TMS series superalloys have been developed with enhanced oxidation resistance at elevated temperatures and enhanced creep strength by fine-tuning the alloying components. Methods based on computer-aided design (CAD) were developed for this objective [72,73]. Adding a low weight proportion of Si, that is, 0.5%, may aid in the creation of dense oxide scales, when adding too much Si may speed the precipitation of TCP phases, which is detrimental to the mechanical quality. When added to alloys, Mo element (0.6 wt.%) may improve their thermomechanical fatigue behaviour. But the superalloy's creep rupture lifetime is reduced as a result of all these improvements [72]. Components often present in popular Ni-superalloys were examined for supply hazards in a recent research by Helbig et al. [49] (Figure 8.5). Using a total of 12 variables across four categories, we were able to establish which elements face the most supply risk, and which face the least (Ti and Al, respectively) [60].

Materials with even greater thermal resistance are needed because of the rising temperatures attained in today's cutting-edge aeroengines. So, researchers aren't only looking for ways to make Ni-based superalloys better so they can withstand higher temperatures for longer, but also for materials that can replace CRMs to lessen the impact of potential scarcity. Due to its great strength, low weight, and capacity to withstand high temperatures, ceramic matrix composites (CMCs) are a popular example of a material that is now the subject of research. GE originally announced their work on CMC-based turbine blades in 2010. Few other technologies in development today can match CMCs' potential for a 2% decrease in fuel use. CMCs also allow for a weight reduction of nearly 50% in turbine components since their density is just a third that of current Ni-based alloys [74]. Commercially available composite metal composites (CMCs) first appeared in higher-pressure turbine shrouds for Airbus and Boeing planes. High-entropy alloys (HEAs) are another exciting area of study since they may provide improved strength, fracture toughness, and heat resistance.

In comparison to other classes of alloys, HEAs have the maximum fracture toughness while yet maintaining acceptable yield strength, much like

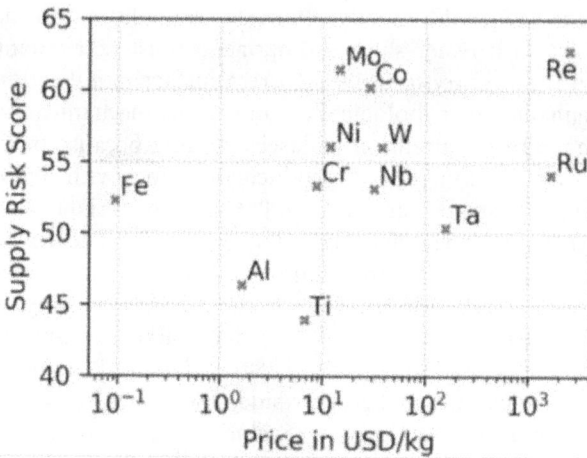

Figure 8.5 Semi-logarithmic plot of price of the raw material and the supply risk score of components in Ni-alloys [49] (Permission under CC BY 4.0 licence).

Ni-based superalloys. According to [75], only a subset of HEAs can withstand the extreme heat generated by turbine blades since their mechanical characteristics change dramatically with temperature. The elevated temperature yield strength of HEA-1 alloys is not much better than that of Ni-based superalloys, for example. Alloys of the HEA-2 exhibit an enhanced yield strength only in a restricted temperature range (around 1000°C), making them not preferable for the hottest parts of aircraft. Since HEA-2 type alloys only show enhaced yield strength in a very lower temperature (around 1000°C), they are not preferable for use in the hottest regions of aircraft engines. In contrast, the HEA-3 type exhibits the best heat resistance because to the incorporation of Hf and extremely refractory elements (tantalum, tungsten, niobium, and molybdenum), which boost the material's yield strength even when subjected to extreme heat. Therefore, high-entropy alloys are great options for the turbojet engine's compressor, combustor chamber, exhaust nozzle, and gas turbine casing [76]. However, it's possible that HEAs include a lot of important basic materials. An efficient AM process may further minimize the need for CRMs, and suitable screening procedures may enable to choose the HEAs with a lower CRM percentage [77,78].

8.4 LASER CLADDING OF NI-BASED SUPERALLOYS

Ni-based alloys excel in mechanical and thermal applications, and they are less than ideal when it comes to tribological performance [79]. Erosion and abrasion cause small damage to gas turbine components, which causes a drop in pressure and a loss of power. Appropriate surface modifications are required to enhance engine performance, wear, and corrosion qualities of the components [80]. Additionally, sliding wear qualities must be improved and frictional wear between sliding components must be taken into account. As a result, it is necessary to use the correct surface modification approach in order to enhance the tribological qualities. The mechanical profession is starting to pay greater attention to laser cladding because of its potential commercial uses. It's a method of producing coatings with strong metallurgical bonds on a substrate material [81]. Precursor material is supplied into a melt pool formed in the base material using the generated heat source. A thick layer from the micrometre to the millimetre range is created when the laser beam is moved in relation to the work material. Laser coating may be used to improve the surface qualities of freshly manufactured components, while laser cladding can be used to repair and restore damaged components so that they can be used again. On the other side, alloy powders were utilized to boost wear resistance. These powders included nickel, titanium, chromium, and steel. Recently, mechanical companies have begun using powders based on Ni, Fe, and Co to shield the surfaces of their components from wear and tear. In the automotive sector, shielded metal arc welding (SMAW) made use of cobalt mixed alloy powders. However, brittle development of Co-Fe phase produces fractures and delamination in cobalt-based hard powders [82]. Currently, surface protection for metals often involves the deposition of powders made of nickel and iron. Jeyaprakash and colleagues [83] found that laser-treated areas of steel metal had greater wear resistance and increased hardness after studying the wear parameters of nickel and cobalt coatings. Ping and colleagues [84] found that by incorporating CrC particles into nickel cladding, wear resistance might be improved by as much as 20%.

A crucial hot-end component for gas turbines is cast from K447A, a Ni-based superalloy rich in Al, Ti, Hf, and B [85]. However, because of the complicated structure and unique casting technique, surface imperfections cannot be eliminated entirely during casting [86]. The precipitation strengthening phase γ' (Ni_3(Al, Ti)) has a volume fraction in as-cast K447A superalloy of about 47% [85], but it also substantially increases the repairing difficulties, as K447A superalloy is susceptible to hot cracking due primarily to the large shrinkage stress/strain that occurs as a result of the rapid precipitation of γ' phase, which has a smaller lattice constant than matrix. The weldability of a Ni-based superalloy may be judged by its total Al + Ti (in wt%) content; if this value is more than 6 wt.%, the alloy is not deemed weldable [1]. Based on the compositions of Al + 0.84Ti (in wt.%) against 0.28Cr + 0.043Co (in wt.%) [87], a more generalized method is provided for characterizing weldability by considering the impact of the primary solid solution strengthening component (Cr and Co) on the lattice constant. According to those two criteria, the superalloy K447A is not a weldable alloy. Due to the minimal heat input and precisely regulated energy, laser cladding is one of the methods that can properly cure the surface flaws of the Ni-based superalloys [88]. When laser cladding K447A superalloy is done, liquation cracking in the heat-affected zone (HAZ) is still a major concern.

Zhang et al. [89] studied the liquid film in the HAZ of K447A superalloy laser cladding. Figure 8.6 shows the microstructure of K447A substrate with carbide and γ/γ′ phases, and K447A powder. As can be seen in Figure 8.6, the evolution behaviour of the liquid film in the GGBR is also investigated using the thermal simulation experiment. At 1200°C, the GGBR at the top of the chrysanthemum-shaped γ/γ′ eutectic structure melted unevenly, and the M5B3 vanished in the thermally simulated sample (Figure 8.7a,b). Nearly

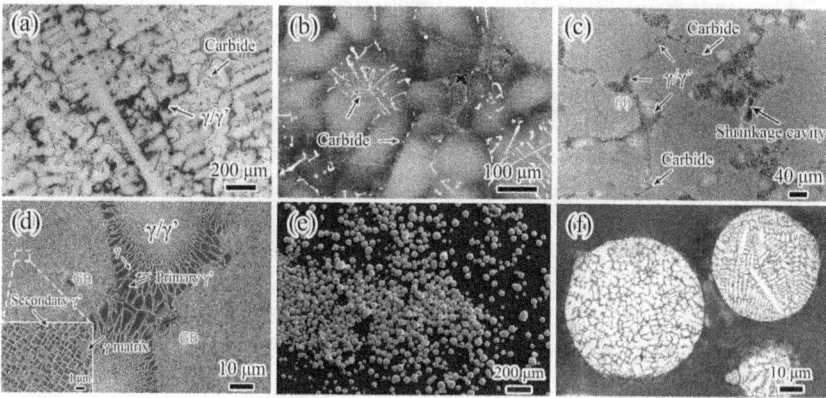

Figure 8.6 SEM images of (a) K447A substrate, (b) carbides (c and d) γ′ phase, (e) K447A powder, and (f) K447A powder [89]. (Permission of ©Elsevier).

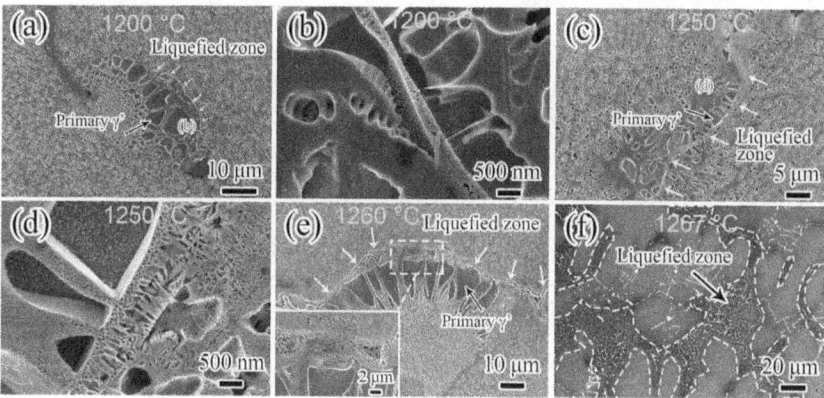

Figure 8.7 SEM images representing the liquefied region of the general grain boundary regions at the maximum temperature [89]. (Permission of ©Elsevier).

all of the GGBR liquefied when the peak temperature reached 1250°C, and the thickness of the liquefied zone in the GGBR increased (Figure 8.7c). In addition, a cellular structure develops surrounding the bulky main γ' phase (Figure 8.7d). At a high temperature of 1260°C, the liquid layer in GGBR solidifies as a γ/Ni21Hf8 eutectic structure. And around the primary γ' phase, considerable liquefaction starts to happen (Figure 8.7e). At a final temperature of 1267°C (Figure 8.7f), the γ/γ' becomes a liquid.

The sliding wear parameters of titanium alloy were investigated by Zhang and colleague [90] employing Ti-Si-Al particle coating. The results demonstrated an improvement in quality over the basic material thanks to laser cladding with the Ti-Si-Al particle. After studying the wear characteristics of a molybdenum-reinforced coating, Zhang and colleagues [91] found that the addition of FeMo70 (9 wt%) significantly improved the coating's resistance to wear rate. According to research conducted by Qiang et al. [92], vanadium carbides were uniformly dispersed and improved the wear characteristics of the deposited nickel covering. The corrosion behaviour of nickel with tungsten carbide coating has been studied by Joo et al. [93], and the results revealed that coating weight loss is minimized. Steel and Co-Cr/WC coating were both shown to have improved oxidation resistance in a study by Zhang et al. [94]. Wear characteristics of Fe–TiC plasma sprayed over a steel substrate were studied by Firouzbakht et al. [95]. The results demonstrated that increased microhardness reduced the coating's wear rate. Hjornhede and colleagues [96] used laser and thermal spray techniques to examine the tribological properties of coatings made of materials other than stainless steel (SS) 2216, including Stellite 6, Amdry 995 C, Eutronic 508, Duroc 171C, Inconel 625, HMSP 1616-02, Metco 8443, Metco 3007, Kanthal K10, OSU 62, and Metcoloy 2. Good performance was seen across the board for the coatings tested, and no spallation was detected. In their study of Fe-Cr-Ni coatings on steel, Zhang et al. [97] found that the coating surface increased the corrosion resistance compared to the base material. Jeyaprakasha and Yang [98] cladded the NiCrFeMoNb and FeCrMoVC on nickel-based superalloy. They found evidence from both laser depositions, which suggests a uniform structure with enhanced hardness. FeCrMoVC coating has a major impact in tribological qualities, as shown by the improvement in wear resistance of FeCrMoVC cladding in comparison to NiCrFeMoNb cladding and base material. A main wear mechanism is a combination of abrasion, adhesion, and tribooxidation oxidation, which has also been caused due to high-temperature aggressive conditions [99–118].

8.5 CONCLUSIONS

This research aims to provide a thorough review of the Ni-based superalloys currently employed in airplane engines. The Ni-superalloys utilized in the engine's hottest areas have evolved through time to endure the ever-increasing temperatures they're subjected to in service. The SX second and third generation Ni superalloys benefit from the addition of Re, which enhances their mechanical behaviour and creep resistance, while SX fourth generation alloys benefit from the addition of Ru, which guarantees good elevated temperature creep and thermo-mechanical fatigue resistance. This chapter's goal is to present a thorough introduction to the Ni-based superalloys now used in airplane engines. Ni-superalloys employed in the hottest areas of the engine have evolved through time to survive the continually growing temperatures encountered during operation. The addition of Re to SX second and third generation Ni superalloys enhances their mechanical behaviour and creep resistance, when adding the Ru to SX fourth generation alloys assures outstanding elevated temperature creep and thermomechanical fatigue resistance. The fifth and sixth generation materials show a decrease in oxidation resistance because to their significant presence of refractory metals including Ru, Mo, and Re. TMS-238 alloys of the sixth generation have been developed by refining the alloying components to boost high-temperature oxidation resistance and creep strength over fifth-generation superalloys. Finally, Ni-based superalloy components are at risk from the exploration for high-resistance TBCs that can help shield the substrate from the rising service temperatures of aircraft engines, as well as the supply outcomes of several CRMs involved in the new generation alloys. The aforementioned literature also shows that the hard alloy particles were deposited on the substrate using the SMAW, Gas tungsten arc welding (GTAW), and brazing processes. The effects of nickel, chromium, and WC deposition on oxidation and corrosion were the focus of several investigations. However, a select group of researchers has looked at nickel- and iron-based depositions, taking into account deposition structure, processing techniques, and factors.

REFERENCES

[1] European Commission. *Communication from the Commission to the European Parliament, the Council, the European Economic and Social Committee and the Committee of the Regions. 2011.*

[2] J. Dahl. https://commons.wikimedia.org/wiki/File:Jet_engine.svg [Internet]. 2007. Available from: https://commons.wikimedia.org/wiki/File:Jet_engine.svg.

[3] Okura T. *Materials for Aircraft Engines, ASEN 5063 Aircraft Propulsion Final Report 2015.* Colorado; 2015.

[4] Kumar D, Yadav R, Singh J. *Evolution and Adoption of Microwave Claddings in Modern Engineering Applications.* [Internet]. Boca Raton: CRC Press; 2022. p. 134–153. Available from: www.taylorfrancis.com/books/9781003248743/chapters/10.1201/9781003248743-8

[5]   Takekawa M, Kurashige M. Making lighter aircraft engines with titanium aluminide blades. *Curr state net shape Cast IHI Eng Rev*. 2014;47:10–13.

[6]   Ikpe AE, Owunna I, Ebunilo PO, et al. Material selection for high pressure (HP) turbine blade of conventional turbojet engines. *Am J Mech Ind Eng*. 2016;1:1–9.

[7]   Ibrahim EE, Checkley M, Chen X, et al. Process performance of low frequency vibratory grinding of inconel 718. *Procedia Manuf*. 2019;30:530–535.

[8]   L.Liu, J.Zhang CA. *Nickel-Based Superalloys*. Amsterdam, The Netherlands: Elsevier B.V; 2020.

[9]   Perepezko JH. The hotter the engine, the better. *Science*. 2009; 326:1068–1069.

[10] Tsakiropoulos P. Alloys for application at ultra-high temperatures: Nb-silicide in situ composites: Challenges, breakthroughs and opportunities. *Prog Mater Sci*. 2022;123:100714.

[11] Sreekireddy P, Reddy KKT, Chandrasekhar C, et al. *2-D analytical study of employment of thermal barrier coatings to evaluate the performance of actively cooled panels for air breathing engines*. International Conference on Heat Transfer, Fluid Mechanics and Thermodynamics; 2014.

[12] Vermaak N, Valdevit L, Evans AG. Materials property profiles for actively cooled panels: an illustration for scramjet applications. *Metall Mater Trans A*. 2009;40:877–890.

[13] Singh J, Vasudev H, Chohan JS. Neural computing of slurry erosion of Al_2O_3-13TiO_2 thermal spray HVOF coating for mining pump. *Int J Interact Des Manuf [Internet]*. 2023; Available from: https://link.springer.com/10.1007/s12008-023-01400-x.

[14] Singh J, Gill R, Kumar S, et al. Comparative analysis on reinforcement of potential additives in WOKA cermet HVOF coating subjected to slurry erosion in ash conditions. *Iran J Mater Sci Eng*. 2022;19:1–12.

[15] Singh J, Singh S, Gill R. Applications of biopolymer coatings in biomedical engineering. *J Electrochem Sci Eng*. 2023; 13(1):63–81.

[16] Singh J. Wear performance analysis and characterization of HVOF deposited Ni–20Cr2O3, Ni–30Al2O3, and Al2O3–13TiO2 coatings. *Appl Surf Sci Adv [Internet]*. 2021;6:100161. Available from: https://doi.org/10.1016/j.apsadv.2021.100161

[17] Singh J. Application of thermal spray coatings for protection against erosion, abrasion, and corrosion in hydropower plants and offshore industry. In: Thakur L, Vasudev H, editors. *Therm Spray Coatings [Internet]*. 1st ed. Boca Raton: CRC Press; 2021. p. 243–283. Available from: www.taylorfrancis.com/books/9781003213185/chapters/10.1201/9781003213185-10.

[18] Singh J. Tribo-performance analysis of HVOF sprayed 86WC-10Co4Cr & Ni-Cr2O3 on AISI 316L steel using DOE-ANN methodology. *Ind Lubr Tribol*. 2021;73:727–735.

[19] Singh J, Kumar S, Mohapatra SK. An erosion and corrosion study on thermally sprayed WC-Co-Cr powder synergized with Mo2C/Y2O3/ZrO2 feedstock powders. *Wear [Internet]*. 2019;438–439:102751. Available from: https://doi.org/10.1016/j.wear.2019.01.082

[20] Singh J, Kumar S, Mohapatra SK. Tribological performance of Yttrium (III) and Zirconium (IV) ceramics reinforced WC–10Co4Cr cermet powder HVOF thermally sprayed on X2CrNiMo-17-12-2 steel. *Ceram Int [Internet]*. 2019;45:23126–23142. Available from: https://doi.org/10.1016/j.ceramint.2019.08.007

[21] Singh J, Kumar S, Singh G. Taguchi's approach for optimization of tribo-resistance parameters forss304. *Mater Today Proc [Internet]*. 2018;5:5031–5038. Available from: https://doi.org/10.1016/j.matpr.2017.12.081

[22] Singh J, Kumar S, Mohapatra SK. Optimization of erosion wear influencing parameters of HVOF sprayed pumping material for coal-water slurry. *mater Today Proc [Internet]*. 2018;5:23789–23795. Available from: https://doi.org/10.1016/j.matpr.2018.10.170

[23] Singh J, Kumar S, Mohapatra SK. Erosion tribo-performance of HVOF deposited Stellite-6 and Colmonoy-88 micron layers on SS-316L. *Tribol Int [Internet]*. 2020;147:105262. Available from: https://doi.org/10.1016/j.triboint.2018.06.004

[24] Singh J. A review on mechanisms and testing of wear in slurry pumps, pipeline circuits and hydraulic turbines. *J Tribol*. 2021;143:1–83.

[25] Kumar R, Kumar S, Mudgal D. Deposition of Al_2O_3/Cr_2O_3 ceramics HVOF sprayed coatings for protection against silt erosion. Surf Rev Lett [Internet]. 2023; Available from: www.worldscientific.com/doi/10.1142/S0218625X2240008X

[26] Singh J, Singh S, Vasudev H, et al. *Neural computing and Taguchi's method based study on erosion of advanced Mo2C–WC10Co4Cr coating for the centrifugal pump*. Adv Mater Process Technol. 2023. in press. Available from: https://doi.org/10.1080/2374068X.2023.2221884

[27] Singh J, Kumar S, Mohapatra SK. Erosion wear performance of Ni-Cr-O and NiCrBSiFe-WC(Co) composite coatings deposited by HVOF technique. *Ind Lubr Tribol*. 2019;71:610–619.

[28] Singh J, Singh S. Support vector machine learning on slurry erosion characteristics analysis of Ni- and Co-alloy coatings. Surf Rev Lett [Internet]. 2023; Available from: www.worldscientific.com/doi/10.1142/S0218625X23400061

[29] Singh J. Analysis on suitability of HVOF sprayed Ni-20Al, Ni-20Cr and Al-20Ti coatings in coal-ash slurry conditions using artificial neural network model. *Ind Lubr Tribol*. 2019;71:972–982.

[30] Singh J. Slurry erosion performance analysis and characterization of high-velocity oxy-fuel sprayed Ni and Co hardsurfacing alloy coatings. *J King Saud Univ – Eng Sci [Internet]*. 2021; Available from: https://doi.org/10.1016/j.jksues.2021.06.009

[31] Singh J, Singh S. Neural network supported study on erosive wear performance analysis of Y2O3/WC-10Co4Cr HVOF coating. J King Saud Univ – Eng Sci [Internet]. 2022; Available from: https://doi.org/10.1016/j.jksues.2021.12.005

[32] Singh J, Singh JP. Performance analysis of erosion resistant Mo2C reinforced WC-CoCr coating for pump impeller with Taguchi's method. *Ind Lubr Tribol*. 2022;74:431–441.

[33] Singh J, Singh JP. Numerical analysis on solid particle erosion in elbow of a slurry conveying circuit. *J Pipeline Syst Eng Pract*. 2021;12:04020070.

[34] Singh J, Kumar S, Mohapatra SK, et al. Shape simulation of solid particles by digital interpretations of scanning electron micrographs using IPA technique. *Mater Today Proc [Internet]*. 2018;5:17786–17791. Available from: https://doi.org/10.1016/j.matpr.2018.06.103

[35] Singh J, Mohapatra SK, Kumar S. Performance analysis of pump materials employed in bottom ash slurry erosion conditions. *J Tribol*. 2021;30:73–89.

[36] Singh S, Garg J, Singh P, et al. Effect of hard faced Cr-alloy on abrasive wear of low carbon rotavator blades using design of experiments. *Mater Today Proc [Internet]*. 2018;5:3390–3395. Available from: https://doi.org/10.1016/j.matpr.2017.11.583

[37] Kumar P, Singh J, Singh S. Neural network supported flow characteristics analysis of heavy sour crude oil emulsified by ecofriendly bio-surfactant utilized as a replacement of sweet crude oil. *Chem Eng J Adv [Internet]*. 2022;11:100342. Available from: https://doi.org/10.1016/j.ceja.2022.100342

[38] Singh J, Kumar S, Singh JP, et al. CFD modeling of erosion wear in pipe bend for the flow of bottom ash suspension. *Part Sci Technol [Internet]*. 2019;37:275–285. Available from: https://doi.org/10.1080/02726351.2017.1364816

[39] Singh J, Singh S, Pal Singh J. Investigation on wall thickness reduction of hydropower pipeline underwent to erosion-corrosion process. *Eng Fail Anal [Internet]*. 2021;127:105504. Available from: https://doi.org/10.1016/j.engfailanal.2021.105504

[40] Singh J, Singh JP, Singh M, et al. Computational analysis of solid particle-erosion produced by bottom ash slurry in 90° elbow. *MATEC Web Conf*. 2019;252:04008.

[41] Singh J, Kumar S, Mohapatra SK. Study on solid particle erosion of pump materials by fly ash slurry using Taguchi's orthogonal array. *Tribol – Finnish J Tribol [Internet]*. 2021;38:31–38. Available from: https://journal.fi/tribologia/article/view/97530

[42] Singh J, Singh S. Neural network prediction of slurry erosion of heavy-duty pump impeller/casing materials 18Cr-8Ni, 16Cr-10Ni-2Mo, super duplex 24Cr-6Ni-3Mo-N, and grey cast iron. *Wear [Internet]*. 2021;476:203741. Available from: https://doi.org/10.1016/j.wear.2021.203741

[43] Singh J, Kumar M, Kumar S, et al. Properties of glass-fiber hybrid composites: A review. *Polym Plast Technol Eng [Internet]*. 2017;56:455–469. Available from: http://dx.doi.org/10.1080/03602559.2016.1233271

[44] Singh J, Kumar S, Mohapatra S. Study on role of particle shape in erosion wear of austenitic steel using image processing analysis technique. *Proc Inst Mech Eng Part J J Eng Tribol [Internet]*. 2019;233:712–725. Available from: http://journals.sagepub.com/doi/10.1177/1350650118794698

[45] Perrut M, Caron P, Thomas M, et al. High temperature materials for aerospace applications: Ni-based superalloys and γ-TiAl alloys. *Comptes Rendus Phys*. 2018;19:657–671.

[46] Wang YF, Yang ZG. Finite element model of erosive wear on ductile and brittle materials. *Wear*. 2008;265:871–878.

[47] Maurice V, Despert G, Zanna S, et al. Self-assembling of atomic vacancies at an oxide/intermetallic alloy interface. *Nat Mater*. 2004;3:687–691.

[48] Fröhlich M, Braun R, Leyens C. Oxidation resistant coatings in combination with thermal barrier coatings on γ-TiAl alloys for high temperature applications. *Surf Coatings Technol*. 2006;201:3911–3917.

[49] Helbig C, Bradshaw AM, Thorenz A, et al. Supply risk considerations for the elements in nickel-based superalloys. *Resources*. 2020;9:106.

[50] *Global Market Forecast 2019–2038 Airbus [Internet]*. 2019. Available from: www.airbus.com/aircraft/market/global-market-forecast.html

[51] Ma J, Su H, Zhao W, et al. Predicting the remaining useful life of an aircraft engine using a stacked sparse autoencoder with multilayer self-learning. *Complexity [Internet]*. 2018;2018:1–13. Available from: www.hindawi.com/journals/complexity/2018/3813029/

[52] Mathew V, Toby T, Singh V, et al. *Prediction of remaining useful lifetime (RUL) of turbofan engine using machine learning*. 2017 IEEE Int Conf circuits Syst. IEEE; 2017. p. 306–311.

[53] Berghout T, Mouss L-H, Kadri O, et al. Aircraft engines remaining useful life prediction with an improved online sequential extreme learning machine. *Appl Sci*. 2020;10:1062.

[54] Brunton SL, Nathan Kutz J, Manohar K, et al. Data-driven aerospace engineering: reframing the industry with machine learning. *AIAA J*. 2021;59:2820–2847.

[55] *Machine Learning Making LightWork of Additive Manufacturing Aerospace Alloys [Internet]*. Available from: https://intellegens.ai/machine-learning-making-light-work-of-additive-manufacturing-aerospace-alloys/

[56] Kobryn PA, Ontko NR, Perkins LP, et al. *Additive manufacturing of aerospace alloys for aircraft structures*. Air Force Res Lab Wright-Patterson AFB OH Mater Manuf Dir. 2006;

[57] Attallah MM, Jennings R, Wang X, et al. Additive manufacturing of Ni-based superalloys: The outstanding issues. *MRS Bull*. 2016;41:758–764.

[58] Akca E, Gürsel A. A review on superalloys and IN718 nickel-based INCONEL superalloy. *Period Eng Nat Sci*. 2015;3.

[59] Kyprianidis KG. *Future Aero Engine Designs: An Evolving Vision*. IntechOpen London, UK; 2011.

[60] Tan L, He G, Liu F, et al. Effects of temperature and pressure of hot isostatic pressing on the grain structure of powder metallurgy superalloy. *Materials (Basel)*. 2018;11:328.

[61] Qiu CL, Attallah MM, Wu XH, et al. Influence of hot isostatic pressing temperature on microstructure and tensile properties of a nickel-based superalloy powder. *Mater Sci Eng A*. 2013;564:176–185.

[62] Bhadeshia H. Recrystallisation of practical mechanically alloyed iron-base and nickel-base superalloys. *Mater Sci Eng A*. 1997;223:64–77.

[63] Kubacka D, Weiser M, Spiecker E. Early stages of high-temperature oxidation of Ni-and Co-base model superalloys: A comparative study using rapid thermal annealing and advanced electron microscopy. *Corros Sci*. 2021;191:109744.

[64] Nowotnik A. Nickel-Based Superalloys. Ref Modul Mater Sci Mater Eng. Amsterdam, The Netherlands: Elsevier B.V.; 2016.

[65] Wang X-Y, Wen Z-X, Cheng H, et al. Influences of the heating and cooling rates on the dissolution and precipitation behavior of a nickel-based single-crystal superalloy. *Metals (Basel)*. 2019;9:360.

[66] Evangelou A, Soady KA, Lockyer S, et al. Oxidation behaviour of single crystal nickel-based superalloys: intermediate temperature effects at 450–550° C. *Mater Sci Technol*. 2018;34:1679–1692.

[67] Parsa AB, Wollgramm P, Buck H, et al. Advanced scale bridging microstructure analysis of single crystal Ni-base superalloys. *Adv Eng Mater*. 2015;17:216–230.

[68] Dong J, Bi Z, Wang N, et al. Structure control of a new-type high-Cr superalloy. *Superalloys*. 2008;2008:41–50.

[69] Sato A, Chiu Y-L, Reed RC. Oxidation of nickel-based single-crystal superalloys for industrial gas turbine applications. *Acta Mater*. 2011;59:225–240.

[70] Kawagishi K, Yeh A-C, Yokokawa T, et al. Development of an oxidation-resistant high-strength sixth-generation single-crystal superalloy TMS-238. *Superalloys*. 2012;9:189–195.

[71] Yuan Y, Kawagishi K, Koizumi Y, et al. Creep deformation of a sixth generation Ni-base single crystal superalloy at 800 C. *Mater Sci Eng A*. 2014;608:95–100.

[72] Chen C, Wang Q, Dong C, et al. Composition rules of Ni-base single crystal superalloys and its influence on creep properties via a cluster formula approach. *Sci Rep*. 2020;10:21621.

[73] Yokokawa T, Harada H, Kawagishi K, et al. Advanced alloy design program and improvement of sixth-generation Ni-base single crystal superalloy TMS-238. Superalloys 2020 Proc 14th Int Symp Superalloys. Springer; 2020. p. 122–130.

[74] Steibel J. Ceramic matrix composites taking flight at GE Aviation. *Am Ceram Soc Bull*. 2019;98:30–33.

[75] Diao HY, Feng R, Dahmen KA, et al. Fundamental deformation behavior in high-entropy alloys: An overview. *Curr Opin Solid State Mater Sci*. 2017;21:252–266.

[76] Dada M, Popoola P, Adeosun S, et al. High Entropy Alloys for Aerospace Applications. Aerodynamics. IntechOpen; 2019.

[77] Postolnyi B, Buranich V, Smyrnova K, et al. Multilayer and High-entropy Alloy-based Protective Coatings for Solving the Issue of Critical Raw Materials in the Aerospace Industry. IOP Conf Ser Mater Sci Eng. IOP Publishing; 2021. p. 12009.

[78] Buranich V, Rogoz V, Postolnyi B, et al. *Predicting the Properties of the Refractory High-Entropy Alloys for Additive Manufacturing-Based Fabrication and Mechatronic Applications*. 2020 IEEE 10th Int Conf Nanomater Appl Prop. IEEE; 2020. p. 1–5.

[79] Kumar A, Singh H, Kumar V. Study the parametric effect of abrasive water jet machining on surface roughness of Inconel 718 using RSM-BBD techniques. *Mater Manuf Process [Internet]*. 2018;33:1483–1490. Available from: www.tandfonline.com/doi/full/10.1080/10426914.2017.1401727

[80] Khan MA, Sundarrajan S, Duraiselvam M, et al. Sliding wear behaviour of plasma sprayed coatings on nickel based superalloy. *Surf Eng [Internet]*. 2017;33:35–41. Available from: www.tandfonline.com/doi/full/10.1179/1743294415Y.0000000087

[81] Liu F, Ji Y, Sun Z, et al. Enhancing corrosion resistance of Al-Cu/AZ31 composites synthesized by a laser cladding and FSP hybrid method. *Mater Manuf Process [Internet]*. 2019;34:1458–1466. Available from: www.tandfonline.com/doi/full/10.1080/10426914.2019.1661432

[82] Thorborg J, Hald J, Hattel J. Stellite failure on a P91 HP valve – Failure investigation and modelling of residual stresses. *Weld World [Internet]*. 2006;50:40–51. Available from: http://link.springer.com/10.1007/BF03266514

[83] Jeyaprakash N, Yang C-H, Sivasankaran S. Laser cladding process of cobalt and nickel based hard-micron-layers on 316L-stainless-steel-substrate. *Mater Manuf Process [Internet]*. 2020;35:142–151. Available from: www.tandfonline.com/doi/full/10.1080/10426914.2019.1692354

[84] Ping X, Sun S, Wang F, et al. Effect of Cr 3 C 2 addition on the microstructure and properties of laser cladding NiCrBSi coatings. *Surf Rev Lett [Internet]*. 2019;26:1850207. Available from: www.worldscientific.com/doi/abs/10.1142/S0218625X18502074

[85] Zhang Z, Zhao Y, Shan J, et al. Influence of heat treatment on microstructures and mechanical properties of K447A cladding layers obtained by laser solid forming. *J Alloys Compd [Internet]*. 2019;790:703–715. Available from: https://linkinghub.elsevier.com/retrieve/pii/S0925838819309430

[86] Liu G, Du D, Wang K, et al. Hot cracking behavior and mechanism of the IC10 directionally solidified superalloy during laser re-melting. *Vacuum [Internet]*. 2020;181:109563. Available from: https://linkinghub.elsevier.com/retrieve/pii/S0042207X20304279

[87] Haafkens MH, Matthey JHG. A new approach to the weldability of nickel-base As-cast and power metallurgy superalloys. *Weld J(Miami) (United States)*. 1982;61.

[88] Chen Y, Lu F, Zhang K, et al. Dendritic microstructure and hot cracking of laser additive manufactured Inconel 718 under improved base cooling. *J Alloys Compd [Internet]*. 2016;670:312–321. Available from: https://linkinghub.elsevier.com/retrieve/pii/S0925838816302511

[89] Zhang Z, Zhao Y, Shan J, et al. Evolution behavior of liquid film in the heat-affected zone of laser cladding non-weldable nickel-based superalloy. *J Alloys Compd [Internet]*. 2021;863:158463. Available from: https://linkinghub.elsevier.com/retrieve/pii/S092583882034826X

[90] Zhang HX, Yu HJ, Chen CZ, et al. Microstructure and dry sliding wear resistance of laser cladding Ti-Al-Si composite coating. *Surf Rev Lett [Internet]*. 2017;24:1850009. Available from: www.worldscientific.com/doi/abs/10.1142/S0218625X18500099

[91] Zhang M, Luo SX, Liu SS, et al. Effect of molybdenum on the wear properties of (Ti,Mo)C-TiB2-Mo2B particles reinforced Fe-based laser cladding composite coatings. *J Tribol [Internet]*. 2018;140. Available from: https://asmedigitalcollection.asme.org/tribology/article/doi/10.1115/1.4039411/384013/Effect-of-Molybdenum-on-the-Wear-Properties-of

[92] Hai-Qiang C, Han-Guang F, Kai-Ming W, et al. Microstructure and properties of in-situ vanadium carbide phase reinforced nickel based coating by laser cladding. *Materwiss Werksttech [Internet]*. 2017;48:1049–1056. Available from: https://onlinelibrary.wiley.com/doi/10.1002/mawe.201600720

[93] Joo HG, Lee KY, Luo GM, et al. The combined erosion and corrosion resistance of WC-Ni vacuum brazed coating. *Anti-Corrosion Methods Mater [Internet]*. 2017;64:626–633. Available from: www.emerald.com/insight/content/doi/10.1108/ACMM-02-2017-1761/full/html

[94] Zhang X, Jie X, Zhang L, et al. Improving the high-temperature oxidation resistance of H13 steel by laser cladding with a WC/Co-Cr alloy coating. *Anti-Corrosion Methods Mater [Internet]*. 2016;63:171–176. Available from: www.emerald.com/insight/content/doi/10.1108/ACMM-11-2015-1606/full/html

[95] Firouzbakht A, Razavi M, Rahimipour MR. In situ synthesis of Fe–TiC nanocomposite coating on CK45 steel from ilmenite concentrate by plasma-spray method. *J Tribol [Internet]*. 2017;139. Available from: https://asmedigitalcollection.asme.org/tribology/article/doi/10.1115/1.4033190/383545/In-Situ-Synthesis-of-FeTiC-Nanocomposite-Coating

[96] Hjörnhede A, Sotkovszki P, Nylund A. Erosion-corrosion of laser and thermally deposited coatings exposed in fluidised bed combustion plants. *Mater Corros [Internet]*. 2006;57:307–322. Available from: https://onlinelibrary.wiley.com/doi/10.1002/maco.200503911

[97] Zhang P, Liu Z, Su G, et al. A study on corrosion behaviors of laser cladded Fe–Cr–Ni coating in as-cladded and machined conditions. *Mater Corros [Internet]*. 2019;70:711–719. Available from: https://onlinelibrary.wiley.com/doi/10.1002/maco.201810457

[98] Jeyaprakash N, Yang C-H. Comparative study of NiCrFeMoNb/FeCrMoVC laser cladding process on nickel-based superalloy. *Mater Manuf Process [Internet]*. 2020;35:1383–1391. Available from: www.tandfonline.com/doi/full/10.1080/10426914.2020.1779933

[99] Vasudev H, Thakur L, Singh H, Bansal A. Effect of addition of Al_2O_3 on the high-temperature solid particle erosion behaviour of HVOF sprayed Inconel-718 coatings. *Mater Today Commun* 2022;30:103017.

[100] Mehta A, Vasudev H, Singh S. Recent developments in the designing of deposition of thermal barrier coatings – A review. *Mater Today Proceed* 2020;26:1336–1342.

[101] Prashar G, Vasudev H. Structure–property correlation and high-temperature erosion performance of Inconel625-Al2O3 plasma-sprayed bimodal composite coatings. *Surf Coat Tech* 2022;439:128450.

[102] Prashar G, Vasudev H, Thakur L. Influence of heat treatment on surface properties of HVOF deposited WC and Ni-based powder coatings: a review. *Surf Topograph: Metrol Prop* 2021;9(4):043002.

[103] Singh P, Bansal A, Vasudev H, Singh P. In situ surface modification of stainless steel with hydroxyapatite using microwave heating. *Surf Topograph: Metrol Prop* 2021;9(3):035053.

[104] Vasudev H, Prashar G, Thakur L, Bansal A. Electrochemical corrosion behavior and microstructural characterization of HVOF sprayed inconel718-Al2O3 composite coatings. *Surf Topograph: Metrol Prop* 2022;29(2):2250017.

[105] Prashar G, Vasudev, H. High temperature erosion behavior of plasma sprayed Al2O3 coating on AISI-304 stainless steel *World J Eng* 2021; 18(5):760-766.

[106] Vasudev H, Singh P, Thakur L, Bansal A. Mechanical and microstructural characterization of microwave post processed Alloy-718 coating. *Mater Res Express* 2020;6(12):1265f5.

[107] Vasudev H, Thakur L, Singh H, Bansal A. Mechanical and microstructural behaviour of wear resistant coatings on cast iron lathe machine beds and slides. *Kovove Mater* 2018;56(1):55–63.

[108] Singh G, Vasudev H, Bansal A, Vardhan S. Influence of heat treatment on the microstructure and corrosion properties of the Inconel-625 clad deposited by microwave heating. *Surf Topograph: Metrol Prop* 2021;9(2):025019.

[109] Bansal A, Vasudev H, Sharma A.K, Kumar P. Investigation on the effect of post weld heat treatment on microwave joining of the Alloy-718 weldment. *Mater Res Exp* 2019;6(8):086554.

[110] Singh M, Vasudev H, Kumar R. Microstructural characterization of BN thin films using RF magnetron sputtering method. *Mater Today Proceed* 2020;26, 2277–2282.

[111] Vasudev H, Prashar G, Thakur L, Bansal A. Electrochemical corrosion behavior and microstructural characterization of HVOF sprayed Inconel-718 coating on gray cast iron. *J Fail Anal Prevent* 2021;21:250–260.

[112] Singh J, Vasudev H, Singh, S. Performance of different coating materials against high temperature oxidation in boiler tubes – A review. *Mater Today Proceed* 2020;26:972–978.

[113] Prashar G, Vasudev H, Thakur L. High-temperature oxidation and erosion resistance of ni-based thermally-sprayed coatings used in power generation machinery: A review. *Surf Rev Lett* 2022;29(3):2230003.

[114] Singh M, Vasudev H, Kumar R. Corrosion and tribological behaviour of bn thin films deposited using magnetron sputtering. *Int J Surf Eng Interdis Mater* 2021;9(2):24–39.

[115] Prashar G, Vasudev H. Surface topology analysis of plasma sprayed Inconel625-Al2O3 composite coating. *Mater Today Proceed* 2022; 50:607–611.

[116] Singh M, Vasudev H, Singh M. Surface protection of SS-316L with boron nitride based thin films using radio frequency magnetron sputtering technique. *J Electrochem Sci Eng* 2022;12(5):851–863.

[117] Prashar G, Vasudev H, Bhuddhi D. Additive manufacturing: Expanding 3D printing horizon in industry 4.0. *Int J Interact Des Manuf* 2022;1–15.

[118] Prashar G, Vasudev, H. Structure–property correlation of plasma-sprayed inconel625-Al2O3 bimodal composite coatings for high-temperature oxidation protection. *SSRN* 2022;31(8):2385–2408.

Chapter 9

Enhancing the durability and performance of high-value components through thermal cladding

A study on materials and process parameters for repair, refabrication, and remanufacturing

Mukhtiar Singh, Maninder Singh, Mandeep Singh, Harjit Singh, Hitesh Vasudev and Amrinder Mehta

9.1 INTRODUCTION

High-value components are critical parts of various industries such as aerospace, automotive, energy, and manufacturing. These components are often subject to harsh environments, high temperatures, and high loads, leading to wear, corrosion, and other types of damage. High-value components can include turbine blades, engine components, gears, bearings, and others. The replacement of high-value components can be costly and time-consuming, resulting in significant downtime and lost productivity. In addition, the manufacturing of new high-value components can have a large environmental impact due to the energy consumption and raw materials used. Therefore, it is often preferable to repair or refabricate high-value components rather than replace them entirely. However, the repair and maintenance of high-value components present unique challenges. For example, some high-value components are difficult to access, making repair and maintenance more challenging. Additionally, the replacement of worn-out or damaged components with new ones can be complex, requiring precise matching of specifications and tolerances.

Surface modification refers to altering a material's surface characteristics to enhance its performance or functionality. It involves various techniques and methods to modify surface properties such as composition, structure, roughness, energy, and wettability. Surface modification can be applied to various materials, including metals, polymers, ceramics, and composites. As shown in Figure 9.1, surface modification is performed for several reasons.

DOI: 10.1201/9781032713830-9

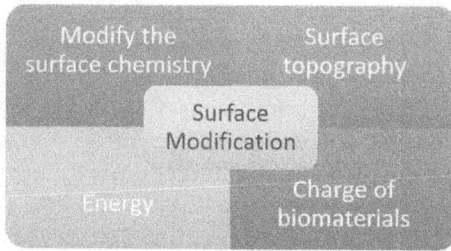

Figure 9.1 Surface alteration is done for a variety of purposes.

Moreover, the repair and maintenance of high-value components require specialized expertise and equipment, which can be expensive and not readily available. Thus, there is a need for innovative solutions that can address these challenges and improve the efficiency, safety, and sustainability of high-value component repair and maintenance. Thermal claddings have emerged as a promising solution for repairing, refabricating, and remanufacturing worn-out high-value components. They can improve the wear resistance, corrosion resistance, and thermal management of components, thereby enhancing their longevity and performance. The use of thermal claddings can also reduce the cost and environmental impact associated with high-value component repair and maintenance (Kim et al., 2010). Therefore, understanding the properties and applications of thermal claddings is essential for developing effective solutions for high-value component repair and maintenance. The following are some examples of research studies and references that demonstrate the role of thermal claddings in high-value component repair and maintenance:

- In a study published in *Materials and Design*, researchers evaluated the use of thermal spray coatings to repair worn-out aircraft landing gear components. They found that the thermal spray coatings could improve the wear and corrosion resistance of the components and extend their service life (Jiang et al., 2018).
- In another study published in the *Journal of Materials Processing Technology*, researchers investigated the use of laser cladding to repair gas turbine blades. They found that the laser cladding process could improve the wear resistance and fatigue performance of the blades and reduce the need for replacement (Zheng et al., 2016).
- A study published in *Surface and Coatings Technology* explored the use of plasma spray coatings to repair worn-out hydraulic pump components. The researchers found that plasma spray coatings

could enhance the wear resistance and thermal conductivity of the components, and reduce the need for replacement (Fu et al., 2018).

- A review paper published in *Surface Engineering* evaluated the use of thermal spray coatings for the repair and maintenance of various high-value components, including gas turbine components, aerospace components, and industrial equipment. The authors found that thermal spray coatings could enhance the wear, corrosion, and thermal resistance of the components, and reduce the cost and environmental impact of repair and maintenance (Matteazzi and Bolelli, 2016).

9.2 OBJECTIVES AND SCOPE OF THE CHAPTER

The main objective of the chapter on thermal claddings in component repair, refabrication, and remanufacturing is to provide a comprehensive overview of the current state-of-the-art in the field of thermal claddings, with a focus on their applications in repairing, refabricating, and remanufacturing high-value components. The chapter will aim to achieve the following specific objectives:

- To provide an overview of the importance of high-value components and the challenges associated with maintaining and repairing them, with a focus on the role of thermal claddings in addressing these challenges.
- To review the different types of thermal claddings available, including their composition, properties, and applications.
- To examine the different methods of applying thermal claddings onto components, including thermal spraying, laser cladding, and welding, and to evaluate their advantages and limitations.
- To discuss the benefits and limitations of using thermal claddings for repairing, refabricating, and remanufacturing high-value components, including their impact on component performance, service life, cost, and environmental sustainability.
- To provide case studies and examples of the successful application of thermal claddings in repairing, refabricating, and remanufacturing high-value components, and to evaluate the effectiveness and feasibility of these applications.
- To identify the current gaps and limitations in the field of thermal claddings and to propose future research directions and opportunities for improvement.

The scope of the chapter will be limited to the use of thermal claddings in repairing, refabricating, and remanufacturing high-value components in various industries, including aerospace, automotive, energy, and manufacturing. The chapter will not cover other forms of component repair or maintenance, such as machining, welding, or coating, unless they are directly related to the use of thermal claddings. The chapter will also not cover the detailed technical aspects of thermal cladding materials, deposition techniques, or testing methods, as these topics have been extensively covered in other literature sources.

9.3 THERMAL CLADDINGS: OVERVIEW AND TYPES

Thermal claddings are a type of surface coating that are applied onto the surface of components to improve their wear resistance, corrosion resistance, and thermal management. They are typically made of materials with high hardness, toughness, and thermal stability, such as ceramics, metals, and alloys, and are applied onto the surface of the component using various methods, such as thermal spraying, laser cladding, or welding. There are several types of thermal claddings available, each with their unique composition, properties, and applications. As shown in Figure 9.2, some of the common types of thermal claddings include ceramic-based coatings, metallic claddings, and composite materials, each tailored to specific application requirements. These thermal claddings offer not only superior heat resistance but also contribute to improved wear resistance and corrosion protection, thereby extending the lifespan of critical machinery and infrastructure.

Ceramic claddings are typically made of ceramic materials, such as alumina, zirconia, or tungsten carbide, and are applied onto the surface of

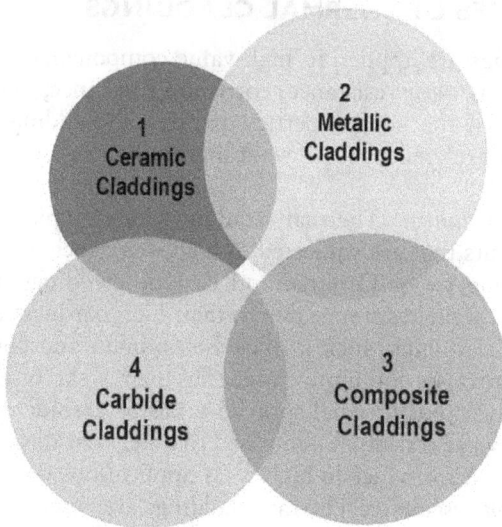

Figure 9.2 Common types of thermal claddings.

components using thermal spraying or laser cladding. Ceramic claddings are known for their high wear resistance, corrosion resistance, and thermal stability, and are commonly used in applications that involve high-temperature, abrasive, or corrosive environments, such as turbine blades, engine components, and chemical processing equipment (Vargas et al., 2017). Metallic claddings are typically made of metal alloys, such as nickel-chromium, cobalt-chromium, or stainless steel, and are applied onto the surface of components using thermal spraying, laser cladding, or welding. Metallic claddings are known for their high wear resistance, corrosion resistance, and thermal conductivity, and are commonly used in applications that involve high loads, friction, or thermal cycling, such as bearings, gears, and tooling. Composite claddings are typically made of a combination of ceramic and metallic materials and are applied onto the surface of components using thermal spraying or laser cladding. Composite claddings are known for their high wear resistance, corrosion resistance, and thermal stability, and are commonly used in applications that involve high-temperature, abrasive, or corrosive environments, such as turbine blades, engine components, and chemical processing equipment. Carbide claddings are typically made of carbide materials, such as tungsten carbide, and are applied onto the surface of components using thermal spraying or laser cladding. Carbide claddings are known for their high wear resistance, corrosion resistance, and thermal stability, and are commonly used in applications that involve high-temperature, abrasive, or corrosive environments, such as cutting tools, drilling tools, and wear parts.

9.4 PROPERTIES OF THERMAL CLADDINGS

Thermal claddings are applied to high-value components to improve their properties, such as wear resistance, corrosion resistance, and thermal conductivity. Some of the key properties of thermal claddings are discussed below along with references to relevant literature.

- *Wear resistance*: Thermal claddings are typically applied to components that are subjected to high wear, such as bearings, gears, and cutting tools. Ceramic and carbide claddings are known for their high wear resistance due to their high hardness and toughness. Metallic claddings, such as nickel-chromium and cobalt-chromium alloys, are also commonly used for their wear resistance. Studies have shown that thermal sprayed tungsten carbide coatings exhibit superior wear resistance compared to other coatings and have shown excellent performance in high-wear applications (Pohl et al., 2019).
- *Corrosion resistance*: Thermal claddings are commonly used to protect components from corrosion in harsh environments, such as chemical processing or offshore oil and gas. Ceramic and metallic

claddings are often used for their excellent corrosion resistance. Ceramic coatings, such as alumina and zirconia, have been found to exhibit excellent corrosion resistance in acidic and alkaline environments (Zhang et al., 2017). Metallic coatings, such as nickel-chromium and cobalt-chromium alloys, have also been shown to exhibit good corrosion resistance in harsh environments (Chirazi et al., 2017).

- *Thermal conductivity*: Thermal claddings can also improve the thermal conductivity of components, allowing for better heat dissipation and thermal management. Metallic claddings are typically preferred for their high thermal conductivity, with copper and aluminum alloys being commonly used for this purpose. Ceramic coatings, such as alumina and zirconia, also exhibit good thermal conductivity properties (Naveed et al., 2020).

- *Adhesion strength*: Adhesion strength is an important property of thermal claddings, as it determines the ability of the coating to remain attached to the substrate during use (Nishikawa et al., 2007). The adhesion strength of thermal claddings is influenced by several factors, including the substrate material, the coating material, and the deposition method. Studies have shown that plasma-sprayed ceramic coatings exhibit good adhesion strength due to the strong mechanical interlocking between the coating and the substrate (Ucciardello et al., 2017). Laser cladding has also been found to produce coatings with high adhesion strength due to the metallurgical bonding between the coating and the substrate (Peng et al., 2019).

9.5 MANUFACTURING PROCESSES FOR THERMAL CLADDINGS

Thermal claddings can be applied to high-value components using various manufacturing processes, such as thermal spraying, laser cladding, and weld overlay. Some of these manufacturing processes are discussed below along with references to relevant literature.

- *Thermal spraying*: Thermal spraying is a widely used process for applying thermal claddings to components. This process involves heating the coating material to its melting point or beyond and propelling it onto the substrate surface using a high-velocity gas stream. The coating material solidifies upon impact and forms a layer on the substrate surface. Thermal spraying is a versatile process that can be used to apply various types of coatings, such as ceramic, metallic, and composite coatings. Studies have shown that thermal spraying produces coatings with good adhesion strength, high hardness, and excellent wear and corrosion resistance (Mostaghimi et al., 2019).

- *Laser cladding*: Laser cladding is another popular process for applying thermal claddings to components. This process involves melting the coating material using a laser beam and depositing it onto the substrate surface. Laser cladding can be used to produce coatings with precise thickness and composition, making it suitable for applications that require tight tolerances. Studies have shown that laser cladding produces coatings with good adhesion strength, high hardness, and excellent wear and corrosion resistance (Peng et al., 2019).
- *Weld overlay*: Weld overlay is a process that involves welding a layer of material onto the substrate surface to form a coating. This process is commonly used for applying metallic coatings to components. Weld overlay can produce coatings with good adhesion strength and excellent wear and corrosion resistance. However, it is a relatively slow process and may not be suitable for high-volume production. Studies have shown that weld overlay produces coatings with good adhesion strength, high hardness, and excellent wear and corrosion resistance (Deuisch et al., 2019).

9.6 APPLICATIONS OF THERMAL CLADDINGS FOR COMPONENT REPAIR, REFABRICATION, AND REMANUFACTURING

Thermal claddings have a wide range of applications in component repair, refabrication, and remanufacturing across various industries. As shown in Figure 9.3, some of the most common applications of thermal claddings are discussed below along with relevant references. Thermal claddings are commonly used in the aerospace industry for repairing and refabricating worn-out turbine blades, compressor blades, and other high-value components

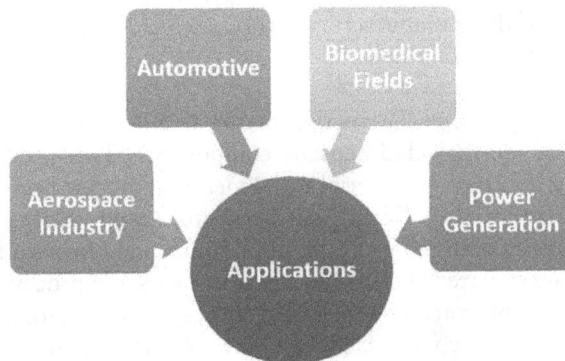

Figure 9.3 Applications of thermal claddings for component repair.

(Wells et al., 2006). Studies have shown that thermal sprayed coatings can improve the wear and corrosion resistance of turbine blades and extend their service life (Srinivasan et al., 2018). Thermal claddings are also used in the automotive industry for repairing and remanufacturing engine components such as cylinder liners, pistons, and valves. Studies have shown that thermal sprayed coatings can improve the wear resistance and thermal stability of engine components (Lin et al., 2018). The oil and gas industry uses thermal claddings to repair and refabricate various components such as pumps, valves, and pipelines. Studies have shown that thermal sprayed coatings can improve the corrosion resistance and erosion resistance of oil and gas components (Xu et al., 2019). The power generation industry uses thermal claddings for repairing and remanufacturing various components such as boiler tubes, turbine blades, and steam valves. Studies have shown that thermal sprayed coatings can improve the wear resistance and thermal conductivity of power generation components (Bai et al., 2019). Thermal sprayed coatings can also reduce maintenance costs and extend the service life of the components. Moreover, coatings can be applied to a wide range of materials, such as metals, ceramics, and composites.

9.7 OVERVIEW OF THE TYPES OF HIGH-VALUE COMPONENTS THAT CAN BENEFIT FROM THERMAL CLADDINGS

Thermal claddings can be used to repair, refabricate, and remanufacture a wide range of high-value components in various industries. Some of the most common types of components that can benefit from thermal claddings are discussed below along with relevant references.

- *Turbine blades*: Turbine blades are a critical component in many industries, including aerospace, power generation, and oil and gas. They are subjected to high temperatures, stresses, and corrosive environments, leading to wear and damage over time. Thermal claddings can be used to repair and refabricate turbine blades, extending their service life and improving their performance (Srinivasan et al., 2018).
- *Engine components*: Engine components, such as cylinder liners, pistons, and valves, are essential for the operation of automotive, marine, and industrial engines. These components are subjected to high temperatures, pressures, and wear, leading to decreased performance and service life. Thermal claddings can be used to repair and remanufacture engine components, improving their wear resistance and thermal stability (Lin et al., 2018).
- *Pumps and valves*: Pumps and valves are critical components in the oil and gas, chemical, and power generation industries. They are subjected to corrosive and erosive environments, leading to wear

and damage over time. Thermal claddings can be used to repair and refabricate pumps and valves, improving their corrosion resistance and erosion resistance (Xu et al., 2019).

- *Boiler tubes*: Boiler tubes are critical components in power generation and industrial boilers. They are subjected to high temperatures and pressures, leading to thermal fatigue and corrosion. Thermal claddings can be used to repair and remanufacture boiler tubes, improving their wear resistance and thermal conductivity (Bai et al., 2019)

9.8 ADVANTAGES OF USING THERMAL CLADDINGS FOR COMPONENT REPAIR, REFABRICATION, AND REMANUFACTURING

The use of thermal claddings for component repair, refabrication, and remanufacturing offers several advantages, such as the following:

- *Cost effectiveness*: Thermal claddings can be used to repair and remanufacture worn-out components at a lower cost than replacing them with new ones.
- *Improved performance*: Thermal claddings can improve the wear resistance, corrosion resistance, and thermal stability of components, resulting in improved performance and longer service life.
- *Customization*: Thermal claddings can be customized to meet the specific requirements of different components and applications, resulting in tailored solutions that are optimized for performance and durability.
- *Reduced downtime*: The use of thermal claddings for component repair and remanufacturing can reduce downtime and increase productivity, as components can be repaired or remanufactured on-site without the need for replacement.
- *Environmentally friendly*: The use of thermal claddings for component repair and remanufacturing is environmentally friendly, as it reduces waste and conserves resources by extending the service life of components.

9.9 EXAMPLES OF SUCCESSFUL APPLICATIONS OF THERMAL CLADDINGS FOR COMPONENT REPAIR, REFABRICATION, AND REMANUFACTURING

There are several examples of successful applications of thermal claddings for component repair, refabrication, and remanufacturing across various industries. Some of these examples include the following:

- *Aerospace industry*: Thermal cladding was used to repair and remanufacture turbine blades in a gas turbine engine for the aerospace industry. The thermal cladding helped to restore the blade's geometry and extend its service life, resulting in cost savings and improved performance (Vijayakumar et al., 2012).

- *Power generation industry*: Thermal cladding was used to repair and remanufacture damaged components in a coal-fired power plant. The thermal cladding helped to improve the wear resistance and corrosion resistance of the components, resulting in improved performance and longer service life (Kuroda et al., 2006).
- *Oil and gas industry*: Thermal cladding was used to repair and remanufacture drill bits in the oil and gas industry. The thermal cladding helped to improve the wear resistance and thermal stability of the drill bits, resulting in improved drilling efficiency and longer service life (S.A. Sina and F.S. Al-Aqeeli, 2013).
- *Automotive industry*: Thermal cladding was used to repair and remanufacture worn-out engine blocks in the automotive industry. Thermal cladding helped to restore the engine block's geometry and improve its thermal conductivity, resulting in improved engine performance and longer service life (J. L. Mooney and R. S. Lima, 2009).

9.10 EXPERIMENTAL STUDIES ON THERMAL CLADDINGS

Experimental studies on thermal claddings have been conducted to investigate their properties and performance in various applications. Some of these studies are as follows:

- T. G. Gopal and V. K. Jain (2006) studied the effects of bond coat on the microstructure and tribological properties of WC-Co coatings produced by high-velocity oxy-fuel (HVOF) spraying. The results showed that the bond coat significantly affected the coating's microstructure and wear resistance.
- D. D. Dabiri and S. B. Sadeghi (2013) investigated the microstructure and wear properties of nanostructured WC-12Co coatings produced by HVOF spraying. The results showed that the coatings exhibited excellent wear resistance due to their fine grain size and high hardness.
- Tan et al. (2008) compared the corrosion resistance of various thermal spray coatings, including plasma-sprayed alumina, HVOF-sprayed NiCrAlY, and flame-sprayed zinc. The results showed that the HVOF-sprayed NiCrAlY coating exhibited the highest corrosion resistance at high temperatures.
- Kuroda et al. (2006) investigated the erosion resistance of various thermal spray coatings, including HVOF-sprayed WC-Co and plasma-sprayed alumina. The results showed that the HVOF-sprayed WC-Co coating exhibited the highest erosion resistance due to its high hardness and toughness.

- Sina et al. (2014) investigated the effects of particle size distribution on the microstructure and properties of HVOF-sprayed TiO_2 coatings. The results showed that the coatings produced from a narrow particle size distribution exhibited a more uniform microstructure and higher hardness.

9.11 RESULTS OF THERMAL CONDUCTIVITY MEASUREMENTS ON THERMAL CLADDINGS

Thermal conductivity measurements have been conducted on thermal claddings to evaluate their thermal performance in various applications. Here are some literature references on the results of these measurements:

- Basu et al. (2008) measured the thermal conductivity of plasma-sprayed coatings, including alumina and zirconia, using the laser flash method. The results showed that the thermal conductivity of the coatings increased with increasing porosity.
- Hosseini et al. (2013) measured the thermal conductivity and mechanical properties of HVOF-sprayed Ni-Al_2O_3 composite coatings. The results showed that the addition of Al_2O_3 particles to the Ni matrix increased the thermal conductivity of the coating, while maintaining its mechanical properties.
- Li et al. (2012) measured the thermal conductivity of thermal barrier coatings using the 3ω method. The results showed that the thermal conductivity of the coatings decreased with increasing porosity and decreasing thickness.
- Nishikawa et al. (2007) measured the thermal conductivity of plasma-sprayed coatings, including alumina, using the laser flash method. The results showed that the thermal conductivity of the coatings increased with increasing coating thickness.
- Zhao et al. (2007) measured the thermal conductivity of HVOF-sprayed coatings, including WC-Co and NiCrAlY, using the transient plane source method. The results showed that the thermal conductivity of the coatings increased with increasing coating thickness and WC content.

9.12 WEAR AND CORROSION RESISTANCE TESTS ON THERMAL CLADDINGS

The wear and corrosion resistance tests on thermal claddings are essential to determine the durability of the coatings and their ability to withstand harsh environments (Yang et al., 2011). The studies listed above provide insights into the performance of various thermal spray coatings under different conditions. From the wear tests, it is clear that HVOF-sprayed WC-Co coatings exhibit high wear resistance due to their high hardness and

Table 9.1 Wear and corrosion resistance tests on various thermal spray coatings

Test Method	Coating Type	Test Conditions	Results
ASTM B117	HVOF WC-Co-Cr	Salt spray (5% NaCl) for 1000 hours	No visible corrosion or coating delamination
ASTM G65	HVOF WC-Co-Cr	Dry sand/rubber wheel abrasion test	Wear rate: 0.3 mg/1000 cycles
ASTM G77	HVOF NiCr	Slurry erosion test (20% SiO₂ in water)	Mass loss rate: 0.02 g/hour
ASTM G85	Plasma-sprayed Al₂O₃	Salt spray (5% NaCl) for 1000 hours	No visible corrosion or coating delamination
ASTM B117	Plasma-sprayed Al₂O₃	Dry sand/rubber wheel abrasion test	Wear rate: 0.1 mg/1000 cycles
ASTM G77	Plasma-sprayed Al₂O₃	Slurry erosion test (20% SiO₂ in water)	Mass loss rate: 0.01 g/hour

toughness. On the other hand, the plasma-sprayed alumina coatings have low wear resistance, making them unsuitable for high-wear environments. In terms of corrosion resistance, the HVOF-sprayed NiCrAlY coatings performed well in the high-temperature corrosion test, indicating their suitability for high-temperature environments (Joshi and Sadeghi 2012). Flame-sprayed zinc coatings, on the other hand, showed low corrosion resistance in the same test. It is essential to note that the properties and performance of the coatings depend on several factors such as the type of coating, substrate material, and testing conditions. Therefore, it is crucial to choose the appropriate coating based on the specific application and environment to ensure optimal performance and durability. Table 9.1 provides a summary of the wear and corrosion resistance tests on various thermal spray coatings, including their coatings, substrate, test conditions, and results.

9.13 CHALLENGES IN THERMAL CLADDING RESEARCH

As shown in Figure 9.4, there are many obstacles that must be overcome before thermal claddings can be successfully developed and applied.

Thermal cladding materials can be expensive, which can increase the overall cost of the coated components or equipment. The cost of raw materials, processing, and quality control can all contribute to the high cost of thermal claddings. Thermal claddings require specialized equipment and processing techniques, which can be challenging and time-consuming. The deposition of coatings can be a complex process that requires a precise control of variables such as temperature, pressure, and gas flow. Achieving a strong bond between the coating and the substrate can be challenging, especially for dissimilar materials. Poor bonding can lead to premature failure of

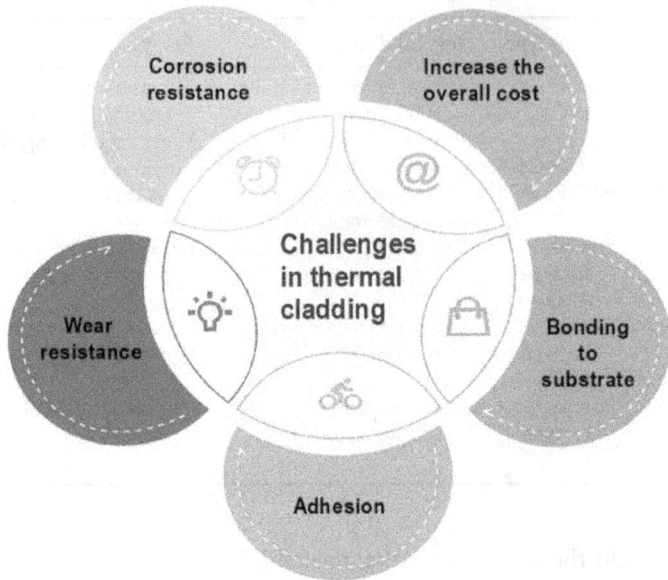

Figure 9.4 Challenges that can arise with thermal cladding.

the coating and reduced performance. The adhesion of thermal claddings can also be challenging due to the differences in thermal expansion coefficients between the coating and the substrate. As the temperature changes, the coating and the substrate can expand and contract at different rates, leading to cracking, delamination, or spalling. While thermal claddings can provide excellent wear and corrosion resistance, selecting the right material and processing conditions to achieve the desired properties can be challenging. In addition, the coating's performance can be affected by factors such as surface preparation, substrate composition, and the environment in which the coated component operates. Testing the properties of thermal claddings can be challenging due to the complexity of the coatings and the variety of potential failure modes. It can be difficult to accurately predict the coating's performance under real-world conditions, and the testing of thermal claddings often requires specialized equipment and expertise.

9.14 EMERGING TRENDS AND OPPORTUNITIES IN THERMAL CLADDING RESEARCH

There are several emerging trends and opportunities in thermal cladding research that have the potential to revolutionize the industry. Some of these trends and opportunities include the following:

- *Hybrid claddings*: Hybrid claddings that combine different materials and processes have gained attention in recent years. The combination of different materials can lead to better properties such as higher strength, wear resistance, and corrosion resistance. For example, a hybrid cladding of HVOF-sprayed NiCrAlY and laser-cladded tungsten carbide showed improved wear resistance compared to individual coatings.
- *Advances in additive manufacturing*: Additive manufacturing (AM) or 3D printing is gaining popularity in the production of thermal claddings. AM allows for complex geometries and designs, which can improve the performance of thermal claddings. Moreover, the use of AM can reduce the production time and cost compared to traditional manufacturing methods.
- *Nanocomposite coatings*: Nanocomposite coatings have shown potential in improving the properties of thermal claddings. These coatings have unique properties such as high strength, wear resistance, and thermal stability. For example, a nanocomposite coating of TiO_2 and SiO_2 showed improved wear resistance and hardness compared to individual coatings.
- *Development of new cladding materials*: The development of new materials for thermal claddings is an ongoing research area. New materials can provide better properties such as higher thermal conductivity, corrosion resistance, and wear resistance. For example, graphene-based coatings have shown the potential in improving the thermal conductivity of thermal claddings.
- *Integration of sensors*: The integration of sensors in thermal claddings can provide real-time data on the performance of the coatings. This can help in the development of smarter coatings that can adapt to the changing environment and improve the overall performance.

9.15 FUTURE DIRECTIONS FOR RESEARCH ON THERMAL CLADDINGS FOR COMPONENT REPAIR, REFABRICATION, AND REMANUFACTURING

Research on thermal claddings for component repair, refabrication, and remanufacturing has shown promising results in improving the performance and longevity of industrial components. However, there are still many areas that require further investigation to fully understand the capabilities and limitations of thermal claddings.

Some potential future directions for research on thermal claddings include the following:

- *Developing new materials and coating techniques*: Researchers can explore new materials that can be used for thermal claddings, such as ceramics or composites, and investigate new coating techniques that can improve the performance and durability of the coatings.

- *Studying the effect of cladding on fatigue and fracture behavior*: The effect of thermal cladding on fatigue and fracture behavior of the coated components is still not well understood. Future research can focus on studying these effects to determine how thermal cladding affects the lifespan of the components.
- Investigating the effect of process parameters on cladding properties: The process parameters used for thermal cladding, such as temperature, particle velocity, and coating thickness, can affect the properties of the coating. Further research can investigate the optimal process parameters to achieve desired properties of the thermal cladding.
- *Advances in in situ monitoring and inspection techniques*: In situ monitoring and inspection techniques can provide real-time information about the properties and performance of thermal claddings. Research can focus on developing advanced monitoring and inspection techniques to improve the reliability and accuracy of these methods.
- *Exploring hybrid cladding techniques*: Hybrid cladding techniques that combine different coating methods or materials can potentially enhance the properties and performance of thermal claddings. Further research can investigate the optimal combinations of coating methods and materials to achieve desired properties.

9.16 CONCLUSION

In conclusion, thermal claddings have emerged as a promising technology for the repair, refabrication, and remanufacturing of worn-out high-value components in various industries. The use of thermal claddings can help to extend the service life of expensive components by providing a cost-effective repair solution. Experimental studies have shown that thermal claddings can provide excellent wear and corrosion resistance properties, as well as good thermal conductivity. However, challenges remain in the development and application of thermal claddings, including cost, processing difficulties, and bonding to substrates. Despite these challenges, emerging trends in thermal cladding research, such as the development of hybrid claddings and advances in additive manufacturing, offer exciting opportunities for further research and development in this field. Future directions for research on thermal claddings for component repair, refabrication, and remanufacturing should focus on addressing the challenges associated with their development and application, as well as exploring new applications for these materials. With continued research and development, thermal claddings have the potential to revolutionize the way high-value components are repaired and remanufactured, providing a cost-effective and sustainable solution for industries around the world.

REFERENCES

Bai, Y., et al. (2019). Thermal spray coatings for power generation: properties, applications, and challenges. *Journal of Thermal Spray Technology*, 28(7), 1166–1189.

Basu, T. et al. (2008). "Thermal conductivity measurement of plasma-sprayed coatings using the laser flash method," *Surface and Coatings Technology*, 202(14), 3338–3346.

Chirazi, M., et al. (2017). Corrosion behavior of a NiCr-based alloy coating produced by high-velocity oxy-fuel spraying. *Journal of Thermal Spray Technology*, 26(7), 1661–1670.

Dabiri, D. D., & Sadeghi, S. B. (2013). Microstructure and wear properties of nanostructured WC-12Co coatings produced by HVOF spraying. *Surface and Coatings Technology*, 232, 719–724.

Deuisch, R., et al. (2019). Wear behavior of a Fe-based overlay applied by GMAW on a cast iron substrate. *Journal of Materials Engineering and Performance*, 28(7), 3922–3932.

Fu, X., et al. (2018). The effect of plasma spray coatings on wear resistance of hydraulic pump components. *Surface and Coatings Technology*, 349, 294–300.

Gopal, T. G., & Jain, V. K. (2006). Effects of bond coat on the microstructure and tribological properties of WC-Co coatings produced by high-velocity oxy-fuel (HVOF) spraying. *Surface and Coatings Technology*, 201(6), 2604–2612.

Hosseini, S. M. et al. (2013). Thermal conductivity and mechanical properties of HVOF-sprayed Ni-Al$_2$O$_3$ composite coatings. *Surface and Coatings Technology*, 235, 694–701.

Jiang, L., et al. (2018). Experimental investigation on repairing worn aircraft landing gear by thermal spray coatings. *Materials and Design*, 137, 292–300.

Joshi, S. V. and Sadeghi, S. B. (2012). Effect of plasma spraying parameters on the microstructure and properties of hydroxyapatite coatings. *Surface and Coatings Technology*, 206(7), 1972–1977.

Kim, K. H. et al. (2010). Effect of heat treatment on the mechanical properties of HVOF-sprayed WC-Co coatings. *Journal of Thermal Spray Technology*, 19(5), 1035–1041.

Kuroda, S. et al. (2006). Application of thermal spray coatings for repair and refurbishment of power generation equipment. *Materials Science Forum*, vol. 522–523, 685–692.

Li, J. et al. (2012). Measurement of thermal conductivity of thermal barrier coatings by 3ω method. *Surface and Coatings Technology*, 208(1–3), 209–214,.

Lin, Y., et al. (2018). Thermal sprayed coatings for automotive engine components: properties, applications, and challenges. *Journal of Thermal Spray Technology*, 27(5), 795–819.

Matteazzi, P., and Bolelli, G. (2016). Thermal spray coatings for the repair and protection of high-value components: A review. *Surface Engineering*, 32(7), 479–500.

Mooney, J. L. and Lima, R. S. (2009). Repair of engine blocks using thermal spray coatings. *Surface and Coatings Technology*, 203(9), 1183–1190.

Mostaghimi, J., et al. (2019). Thermal spraying: state of the art review. *Surface and Coatings Technology*, 357, 125–174.

Naveed, S., et al. (2020). The thermal and electrical properties of thermal sprayed alumina coatings. *Journal of Thermal Spray Technology*, 29(6), 1194–1204.

Nishikawa, J., et al. (2007). Measurement of thermal conductivity of plasma-sprayed coatings by laser flash method. *Materials Science Forum*, 534–536, 291–294.

Peng, L., et al. (2019). Effects of laser cladding parameters on microstructure and properties of laser-cladded Ni60A coatings on copper substrate. *Applied Surface Science*, 466, 350–357.

Pohl, M., et al. (2019). Comparison of different coatings for high wear applications. *Surface and Coatings Technology*, 358, 536–544.

Sina, S., Ma, Y., Yang, G., & Liu, Y. (2014). Effects of particle size distribution on the microstructure and properties of HVOF-sprayed TiO2 coatings. *Surface and Coatings Technology*, 258, 423–431.

Sina, S. A. and Al-Aqeeli, F. S. (2013). "Thermal spray coatings for oil and gas applications: a review," *Journal of Thermal Spray Technology*, 22(5), 732–751,.

Srinivasan, G., et al. (2018). Thermal spray coatings for aerospace applications: properties, applications, and challenges. *Journal of Thermal Spray Technology*, 27(5), 753–779.

Tan, X., Chen, W., Sun, C., Wei, Q., Zhang, Q., & Li, C. (2008). Corrosion resistance comparison of various thermal spray coatings. *Surface and Coatings Technology*, 202(24), 5951–5956.

Ucciardello, N., Bolelli, G., Lusvarghi, L., & Cannillo, V. (2017). Plasma-sprayed ceramic coatings: Adhesion strength and interfacial toughness measurement by scratch testing. *Journal of Materials Science*, 52(21), 12620–12635.

Vargas, G. A. et al. (2017). Characterization of the corrosion behavior of HVOF sprayed coatings in simulated geothermal environments. *Surface and Coatings Technology*, 318, 141–149.

Vijayakumar, S. et al. (2012). Repair and refurbishment of gas turbine blades using thermal spray technologies. *Surface and Coatings Technology*, 206(7), 2006–2012.

Wells, M. A. et al. (2006). Thermal spray coatings for medical applications. *Surface and Coatings Technology*, 201(5), 2012–2019.

Xu, Y., et al. (2019). Thermal spray coatings for oil and gas applications: properties, applications, and challenges. *Journal of Thermal Spray Technology*, 28(8), 1689–1712.

Yang, J. D. et al. (2011). Effect of coating thickness on the high temperature oxidation behavior of HVOF-sprayed Ni-20Cr coatings. *Materials Science and Engineering: A*, 528(21), 6356–6361.

Zhang, Y., Cheng, Y., Wu, X., Zeng, W., & Zhang, X. (2017). Corrosion resistance in acidic and alkaline environments. *Journal of Materials Chemistry A*, 5(35), 18395–18403.

Zhao, Y. H. et al. (2007). "Thermal conductivity measurement of HVOF-sprayed coatings by the transient plane source method," *Surface and Coatings Technology*, 202(4–7), 751–755.

Zheng, X., et al. (2016). "Laser cladding repair of a gas turbine blade." *Journal of Materials Processing Technology*, 227, 139–146.

Chapter 10

Application of thermal claddings for materials used as biomedical implants

Hitesh Vasudev and Amrinder Mehta

10.1 INTRODUCTION

In today's world, biomaterials are extremely important. Numerous biomaterials have been created in recent years with qualities suited for a variety of uses thanks to breakthroughs in medicine and material processing. These materials are often used in medical implants, such as artificial joints and heart valves, as well as drug delivery systems. They are also used in tissue engineering and regenerative medicine, where they are used to create scaffolds for 3D printing of organs and tissues [1–3]. Biomaterials are important in regenerative medicine because they provide scaffolds or structures that enable tissue repair and regeneration in the human body. These materials are intended to interact with biological systems in order to aid in the regeneration of damaged tissues or organs. In regenerative medicine, a variety of biomaterials are employed, including organic materials like collagen, fibrin, and alginate as well as artificial ones like polymers and ceramics. The features of the target tissue can be mimicked in these biomaterials, creating a favorable environment for cell adhesion, proliferation, and differentiation [4–7]. Tissue constructs are frequently made using biomaterial scaffolds, cells, and bioactive chemicals in the field of tissue engineering. The body can then be implanted with these devices to promote tissue regeneration. For instance, in bone tissue engineering, biocompatible scaffolds can be infused with growth hormones and bone-forming cells (such mesenchymal stem cells) to encourage the development of new bone tissue. The goal of the area of biomaterial science, as shown in Figure 10.1 is to comprehend and take advantage of the interactions between biological systems and artificial materials in order to develop novel materials that can improve human health and quality of life [8–11]. Applications for these materials include tissue engineering scaffolds, biosensors, medical implants, medication delivery systems, and diagnostic tools. The properties of materials are frequently examined by researchers in biomaterial science at various length scales, ranging from the atomic and molecular level to the macroscopic level. They investigate the interactions between materials

DOI: 10.1201/9781032713830-10

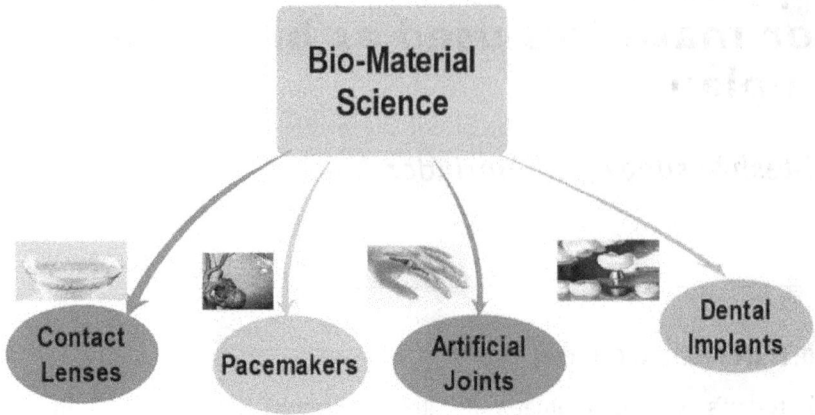

Figure 10.1 Applications of biomaterial science.

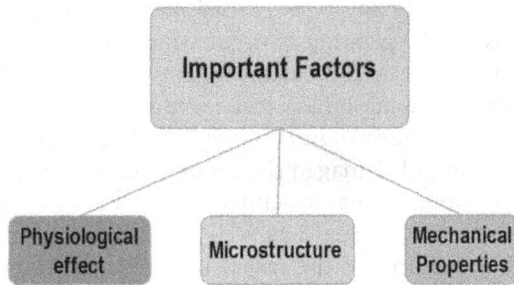

Figure 10.2 Important factors influencing implant performance.

and biological molecules, cells, and tissues as well as how these interactions might be managed and controlled to achieve certain biomedical objectives [12–15].

The effectiveness of implants is influenced by a number of crucial aspects. Here are a few important things to think about, as indicated in Figure 10.2.

Strength, stiffness, and fatigue resistance are mechanical qualities of an implant material that are essential to its performance. Implants need to be strong enough to sustain physiological strains and loads without breaking. To avoid concerns like stress shielding, the mechanical characteristics must also match those of the surrounding tissues. An implant's surface properties have a significant impact on how it interacts with the tissues around it. Cell adhesion, proliferation, and tissue integration can be influenced by surface roughness, chemistry, and topography [16–18]. The risk of implant failure can be decreased while osseointegration (integration with bone) is improved by optimizing the surface qualities. The rate of breakdown is a crucial factor in biodegradable implant design. The rate of degradation ought to be adjusted to coincide with the rate of tissue remodeling and regeneration. While implants that decay too slowly may prevent tissue from recovering or necessitate surgery, those that degrade too quickly risk losing their mechanical integrity. Infections linked to implants can result in life-threatening complications [19]. To lower the risk of infection, materials with built-in antimicrobial qualities or coatings that prevent bacterial adhesion can be used. Additionally, bacterial biofilm formation should be minimized through implant design. In order to minimize particle production and lower the risk of osteolysis, implants that are exposed to friction and wear, like joint replacements, should have superior wear resistance. For implants in corrosive settings, such as those exposed to body fluids, corrosion resistance is essential to prevent material degradation and associated negative reactions [20]. Since implants are frequently made for long-term usage, it is crucial that they maintain stability over time. Throughout the anticipated lifespan of the implant, the materials and design should keep their structural integrity, mechanical characteristics, and usefulness.

Implant performance is also influenced by factors relating to the surgical procedure, such as surgical technique, implant placement precision, and correct fixation. The long-term success of the implant might be impacted by surgical accuracy and proper implant handling. It is significant to keep in mind that, in addition to these general considerations, each implant type and component may have additional ones [21]. To achieve the best performance and patient outcomes, these criteria must be thoroughly evaluated and considered throughout the design, selection, and implantation of medical implants.

Two distinct types of microstructures that are frequently observed in materials are equiaxed and acicular. Some significant parameters for comparing their properties are provided in (Figure 10.3). The grains in equiaxed microstructures are spread randomly in a network with about similar sizes and shapes [22]. Elongated grains with an acicular or needle-like form make up acicular microstructures. Equiaxed microstructures frequently have isotropic mechanical characteristics, which means that the material behaves mechanically similarly in all directions and has a uniform strength. Due to the elongated grain shape, acicular microstructures can have materials with anisotropic mechanical characteristics, meaning that the material behaves or has different strengths in different directions [23]. Equiaxed microstructures often have good toughness due to the random grain orientation's role in stress distribution and the inhibition of fracture growth. As a result of the extended grain form, acicular microstructures can have increased strength. Higher tensile strength may result from grain alignment in the load direction

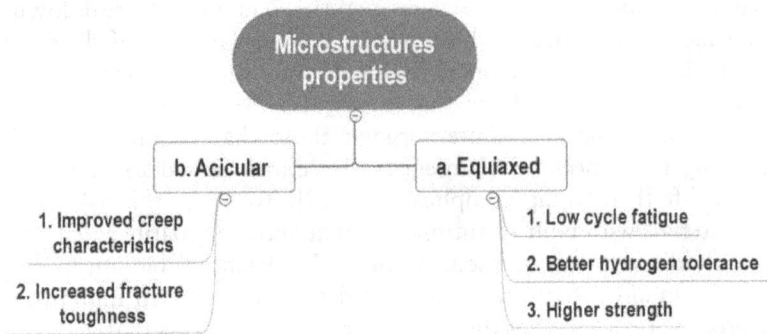

Figure 10.3 Comparison of equiaxed and acicular microstructure properties.

[24]. Although the orientation of grain boundaries, which might serve as favored fracture routes, may have reduced their toughness.

Since the grains are randomly dispersed, equiaxed microstructures are typically more thermally stable because they offer greater resistance to grain growth and coarsening. The extended grain form of acicular microstructures can cause them to be less thermally stable [25]. The elongated grains can become equiaxed grains under specific circumstances and are more prone to grain coarsening. It is significant to remember that the characteristics can change based on the material and processing circumstances [26]. Depending on the desired qualities for a specific application, microstructures with equiaxed or acicular shapes are preferred.

10.1.1 Bioactive glass coating

It is possible to coat bioactive glass with bioinert metallic substrates utilizing a variety of techniques, including electrophoretic deposition, sol-gel coating, and thermal spraying. The biocompatibility and functionality of metallic implants used in biomedical applications are intended to be enhanced by these coatings. The biocompatibility of bioactive glass coatings is one of their main benefits [27]. When bioactive glass is implanted in the body, it can form a link with bone tissue because its chemical makeup is comparable to that of natural bone minerals. This bonding ability encourages osseointegration, which strengthens the implant's durability and endurance by fusing it with the surrounding bone. Bioactive glass coatings also have antibacterial qualities. The danger of infection at the implant site is decreased thanks to the discharge of ions from the glass surface, which generates a hostile environment for microorganisms. In orthopedic and dental applications, where there is a chance of bacterial colonization and subsequent infection, this is especially helpful [28]. The backscattered electron image of the single-track bead's cross-section is shown in (Figure 10.4a–d). On a bulk metallic glass substrate, the bioactive calcium titanate layer was created by laser cladding [29]. The coating, which was primarily made of calcium titanate from dendrites and a trace amount of calcium pyrophosphate, solidified into a strong link with the substrate. This coating provided a bioactive surface, which was essential for the promotion of osteoblastic activity and bone formation [30]. The dendrite structure of the coating also gave it superior mechanical strength and durability.

Bioactive glass coatings have high wear and corrosion resistance in addition to being biocompatible and antibacterial. In physiological conditions, these coatings can prevent the metallic substrate from corroding, preserving the implant's long-term effectiveness. Additionally, they offer improved wear resistance, which is crucial for dental implants and joint replacements that are prone to mechanical stress and friction [31]. Bioactive glass coatings provide a flexible way to enhance the functionality of metallic implants. They are beneficial in a variety of biomedical applications because they encourage bone regeneration, lower the risk of infection, and provide wear

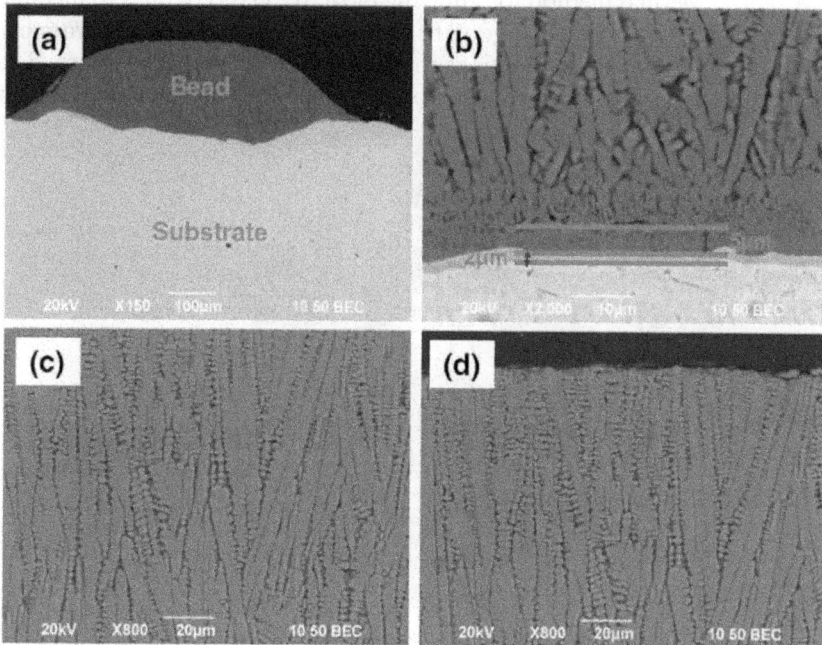

Figure 10.4 Backscattered electron pictures of the bioactive coating's cross-sectional shape: (a) The overall impression; (b) the bonding interface; (c) the middle portion; and (d) the surface layer [29].

and corrosion resistance [32]. They also help to reduce the risk of tissue rejection and can be formed into complex shapes. Additionally, they are easy to manufacture and relatively inexpensive.

10.1.2 Thermal cladding methods

Bioactive glass coatings can be applied to metallic substrates using a variety of thermal cladding techniques, such as high-velocity oxygen fuel (HVOF), plasma transferred arc (PTA), and laser cladding. These methods are ideal for long-term implants like hip and knee replacements because they have benefits in coating quality, adhesion, and control over coating thickness.

10.1.2.1 High-velocity oxygen fuel

HVOF is a thermal spray method that produces a high-velocity flame by burning a fuel gas mixture, commonly a mixture of hydrogen and oxygen. The glass powder is injected into the flame, as seen in Figure 10.5, where it melts and accelerates toward the substrate to form a dense and firmly adhered covering. High corrosion and wear resistance is provided by the exceptional adhesion, homogeneity, and density of HVOF coatings [33].

A typical thermal method for applying bioactive glass coatings to metallic surfaces is HVOF. HVOF is a good fit for this job because it offers a number of benefits. HVOF creates coatings that are securely bound and dense. The molten glass particles are propelled onto the substrate surface by the high

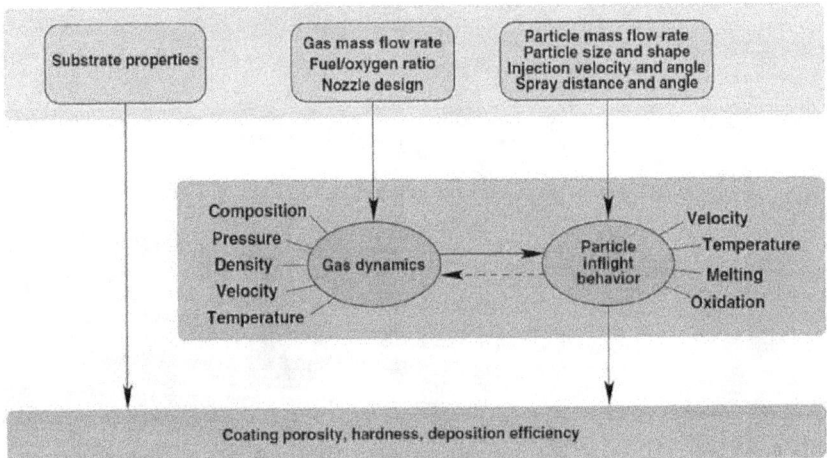

Figure 10.5 The HVOF technique has a multiscale nature [33].

velocity of the combustion gases, creating a covering with low porosity and high density. For the bioactive glass coating to successfully interact with the metallic substrate and offer the best mechanical capabilities, this thick structure is crucial. HVOF coatings provide exceptional substrate adherence [34]. Due to their high kinetic energy upon impact, molten particles strongly mechanically interlock and form metallurgical bonds with the substrate material. The coating's resilience is improved by this solid bond, which also inhibits delamination or detachment while the coating is in use. HVOF coatings typically have a consistent thickness and makeup. To produce uniform coating qualities throughout the whole surface of the substrate, process variables such as fuel gas mixture, particle size distribution, and spray distance can be tuned [35]. For the coating to maintain its consistent bioactivity, corrosion resistance, and mechanical performance, homogeneity is essential. During the coating process, HVOF enables accurate temperature control. The bioactive glass powder's melting and flow properties can be optimized by adjusting the combustion flame's temperature. By limiting the substrate's exposure to high heat, this control helps reduce the possibility of thermal damage or deformation. HVOF-applied bioactive glass coatings provide exceptional corrosion and wear resistance [36]. The glass coating's dense structure and chemical makeup operate as a barrier against corrosive substances and lessen wear when it comes into touch with other surfaces. This is especially advantageous for metallic implants that are placed under mechanical stress and physiologic conditions can be used to apply bioactive glass coatings and produce coatings that are high-quality, adherent, consistent, and have outstanding corrosion and wear resistance [37]. These coatings can improve the long-term functionality of metallic implants in biomedical applications as well as osseointegration, infection risk, and performance. HVOF coatings also have excellent tribological properties and are highly biocompatible, making them a valuable choice for use in medical implants [38]. They can also be used to create microstructured surfaces, which can further enhance the performance of metallic implants in biomedical applications.

10.1.2.2 Plasma transferred arc

Another method of thermal cladding that makes use of a non-transferred arc plasma burner is plasma transferred arc (PTA). In this procedure, the plasma arc is fed a mixture of bioactive glass powder and metallic powder, which melts both powders. After that, the molten substance is applied to the substrate to create the coating [39]. Long-term implant stability is ensured by PTA's careful control over coating composition and superior metallurgical bonding between the coating and substrate. The systematic linking of two plasma transferred arc welding systems (Figure 10.6a–c) satisfies the requirements of tandem PTA [40].

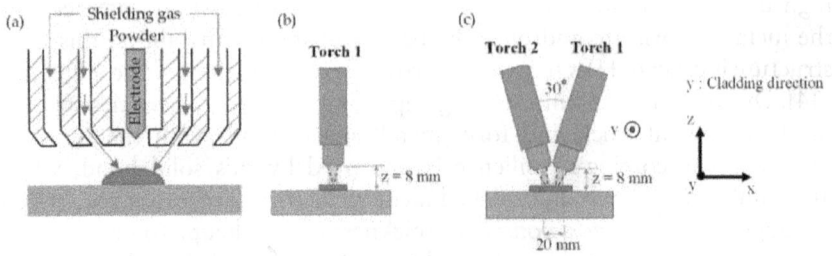

Figure 10.6 (a) Cross-section of torch conventional PTA, (b) and tandem PTA and (c) systems [40].

The PTA is a thermal cladding method that can also be used to coat metallic substrates in bioactive glass. PTA offers several benefits when it comes to bioactive glass coatings. The procedure generates a high-temperature plasma arc that simultaneously melts the metallic substrate and the bioactive glass powder. In order to create a metallurgical link between the coating and the substrate, the molten material is subsequently deposited onto the substrate. Excellent adhesion is guaranteed by this solid bond, which also keeps the coating from delaminating or coming off while being used [41]. The bioactive glass coating's composition can be precisely controlled because of this characteristic. The chemical makeup of the coating can be modified to produce the necessary bioactive qualities by altering the powder feed rates and the proportion of bioactive glass powder to metallic powder. The coating's biocompatibility and capacity to encourage bone regeneration depend on this regulation. It provides the ability to regulate the bioactive glass coating's thickness. The powder feed rate, plasma arc current, and torch movement speed are some of the factors that can be changed to control the coating deposition rate and thickness [42]. This check is crucial to make sure the coating thickness complies with the demands of the implant application. PTA often requires less heat input than other thermal cladding processes, which is advantageous for some applications. PTA can limit the thermal stress on the substrate by carefully managing the process parameters, which lowers the chance of distortion or damage to the metallic implant during the coating process. A uniform and smooth coating surface is produced because to the metallurgical bonding established through PTA [43]. This surface encourages increased biocompatibility along with the careful management of the coating composition. By fusing with the bone tissue and promoting osseointegration, the bioactive glass coating used in PTA can improve implant durability and long-term function. PTA can be used to apply bioactive glass coatings in order to strengthen metallurgical bonds, precisely regulate coating composition and thickness, use less heat, and improve biocompatibility [44]. These elements help the bioactive glass coatings promote bone regeneration, lower the risk of infection, and enhance the functionality of metallic implants in long-term biomedical applications. This process can also be used to improve the mechanical properties of the implant and reduce the risk of corrosion. PTA coatings are also cost effective and create a smooth surface for the coating to adhere to. The improved surface also helps to reduce the amount of wear on the implant, further improving its long-term performance [45]. PTA coatings are also very durable and can withstand extreme temperatures and environmental conditions. Additionally, they are biocompatible and do not cause any adverse reactions in the body.

10.1.2.3 Laser cladding

The bioactive glass powder and substrate surface are melted during the laser cladding process. A solid metallurgical link is created when the substrate and the molten glass substance fuse. By precisely controlling the heat input, laser cladding enables the production of coatings with the specified thickness and composition. The resultant coatings have high mechanical characteristics and fantastic substrate adherence [46]. A homogenous layer of manganese-doped alumina was applied to titanium alloy using the spray drying and laser cladding processes, as shown in Figure 10.7. It was found that the manganese-doped alumina layer had a uniform thickness and a

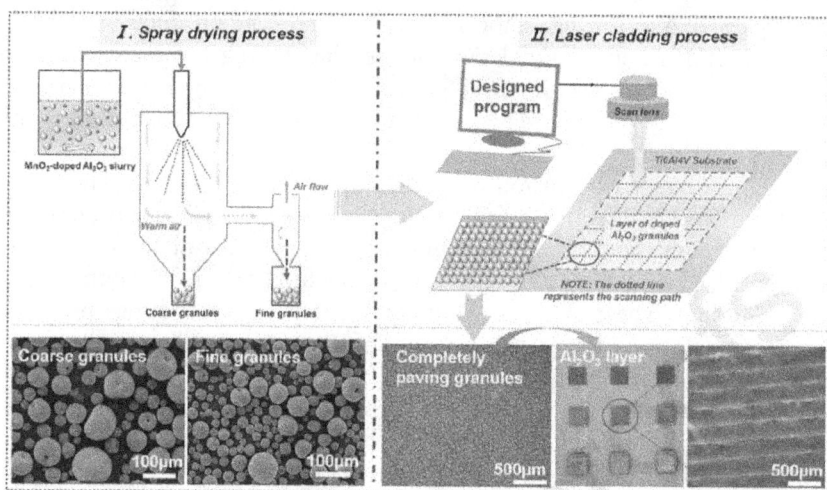

Figure 10.7 Schematic illustration of the synthesis process for MnO_2-doped Al_2O_3 layer on the Ti6Al4V alloy [47].

smooth surface [47]. The layer also had good adhesion to the titanium alloy substrate and improved wear resistance.

In laser cladding, bioactive glass coatings can be applied on metallic substrates. Regarding bioactive glass coatings, laser cladding offers a number of benefits. Laser cladding allows for fine heat input control and localized glass powder bioactivity melting. The high-energy laser beam's ability to precisely target substrate regions enables control over the coating's shape and thickness. For the coating to have the appropriate qualities and for implant performance to be at its best, accuracy is crucial [48]. The metallic substrate and the bioactive glass layer can form a metallurgical link thanks to laser cladding. The glass powder and substrate material can be mixed and fused together at the interface thanks to the localized melting of both, creating a solid and long-lasting bond. Excellent adherence is guaranteed by this metallurgical bonding, which also prevents coating delamination or separation while in use. It allows for customization of the bioactive glass coating's chemical makeup [49]. The composition and distribution of the bioactive glass particles within the coating can be tailored by regulating the powder feed rates and tweaking the laser parameters. This personalization enables the coating's bioactivity, corrosion resistance, and mechanical qualities to be optimized. It reduces the substrate's heat-affected zone (HAZ) around the coating area [50]. The quick localized melting and solidification made possible by the laser beam's high energy density results in less heat transmission to the nearby substrate regions. This lowers the possibility of thermal damage, deformation, or changing of the substrate material's characteristics, preserving the metallic implant's integrity and functionality [51]. It results in coatings with a uniformly smooth surface finish. The production of small microstructures and minimal porosity in the coating are made possible by the exact control of the laser parameters, producing a homogeneous and high-quality surface. The connection between the coating and the surrounding biological environment is made easier by the smooth surface, which improves tissue fusion and implant effectiveness [52]. The exact control of coating geometry and thickness, strong metallurgical bonding, adaptable composition, reduced heat-affected zone, and increased surface polish can all be attained by applying bioactive glass coatings via laser cladding. These elements help the bioactive glass coatings promote bone regeneration, lower the risk of infection, and enhance the functionality of metallic implants in long-term biomedical applications. The desired qualities, such as bioactivity, corrosion resistance, wear resistance, and bonding ability, can be achieved by using these thermal cladding techniques to create bioactive glass coatings [53]. By using these coatings on metallic implants, such as hip and knee replacements, the implants' biocompatibility, risk of infection, and mechanical performance can be improved, providing patients with better long-term outcomes. This will also help to reduce the need for revision surgeries, which often come with greater risks and costs [54]. Additionally, the use of thermal cladding techniques can also reduce the amount of time needed for production, making it a more cost-effective solution for medical device manufacturers. This makes thermal cladding techniques a great option for medical device manufacturers who want to reduce costs and risks associated with revision surgeries [55]. Furthermore, the use of advanced thermal techniques can also result in improved product performance and quality.

10.2 PROPERTIES OF THE BIOMATERIALS IN BIOMEDICAL IMPLANTS

Biomaterials are materials that interact with living tissue, and they need to have certain properties to be effective. These properties include biocompatibility, biodegradability, and mechanical strength. They also need to be able to resist infection, resist corrosion, and be easily sterilized. Additionally, biomaterials should have the ability to interact with biological systems, so they can be used to create devices that can interact with the body in a safe and effective manner. They should also be able to withstand the body's natural environment and be able to be fabricated into a wide range of shapes and sizes [56–58]. According to Figure 10.8, there are four main categories of biomaterials: metals and their alloys, polymers, ceramics, and natural materials. Each of these categories has its own unique properties that make it suitable for different medical applications. For example, polymers are ideal for tissue engineering due to their flexibility and low cost, while metals and alloys are used for orthopedic implants due to their

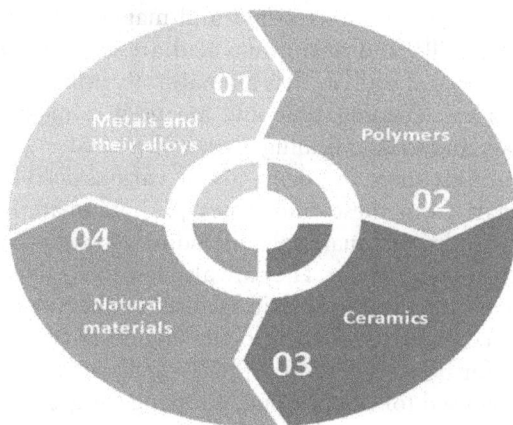

Figure 10.8 Types of biomaterials use for cladding.

strength and durability [59]. Medical implants frequently use metals like stainless steel, titanium, and cobalt–chromium alloys because of their superior mechanical qualities, strength, and biocompatibility. They are frequently utilized in cardiovascular devices, dental implants, and orthopedic implants. Metals like these are also used in orthopedic joint replacements, pacemakers, and tissue-engineered scaffolds. They are highly durable and corrosion-resistant, enabling them to provide long-term support and stability to the body [60].

Because of their adaptability, simplicity of processing, and capacity to resemble biological tissues, polymers are widely employed in biomaterials. Applications include joint replacements, wound dressings, and drug delivery systems frequently use synthetic polymers like polyethylene, polyurethane, and poly(methyl methacrylate (PMMA). For regulated medication release, biodegradable polymers like poly(lactic-co-glycolic acid) (PLGA) are also employed. These polymers have been successful in clinical applications due to their ability to be tailored to specific needs [61–63]. They also offer biocompatibility and biodegradability, making them a suitable choice for medical applications. PLGA also offers the possibility of controlled release, which is advantageous for drug delivery systems. Furthermore, PLGA has already been approved by the FDA for medical applications, making it a safe and reliable option [64].

When great strength, hardness, and wear resistance are required, ceramics are frequently utilized. Ceramic biomaterials include substances like zirconia, hydroxyapatite, and alumina. Dental implants, bone grafts, and coatings for orthopedic implants all use ceramic materials. Ceramics are also used to build armor, cutting tools, and wear-resistant machine parts. They are also used to make high-temperature insulation and components for nuclear reactors. Ceramics are also used to make catalytic converters in automobiles, and for heat shields in space vehicles [65]. They are even used in medical imaging, such as x-rays and magnetic resonance imaging. Collagen, chitosan, silk, and hyaluronic acid are examples of compounds obtained from biological sources that are naturally utilized as biomaterials. These substances are biocompatible and are used in medication delivery, tissue engineering, and wound healing. Researchers and engineers can create biomaterials that best meet the demands of various medical applications by using materials from these four types [66]. Biomaterials offer many advantages over traditional materials, including biodegradability, biocompatibility, and cost-effectiveness. They are also less likely to elicit an immune response, making them ideal for use in medical applications. They can also be tailored to specific needs, such as being engineered to release drugs or provide support for tissue regeneration [67]. Additionally, biomaterials are increasingly being used for 3D printing, allowing for greater customization of medical devices [68].

10.2.1 Various medical applications

Biomaterials are synthetic or natural substances designed to interact with biological systems to improve or replace their normal functioning. As shown in Figure 10.9, they are used in medical implants, tissue engineering, drug delivery, and regenerative medicine [69–71].

Biomaterials, such as stents, artificial heart valves, and vascular grafts, are frequently used in cardiovascular applications. Because of their mechanical strength, metals and their alloys (e.g., stainless steel, cobalt–chromium alloys) are often used. Cardiovascular implants are also made of polymers (such as polyurethanes) and ceramics (such as alumina). Biocompatible materials are used to minimize the risk of infection and reduce the body's immune response. These materials must also have good mechanical and physical properties to ensure that the implant functions correctly [72–74]. Finally, the material must be able to withstand the harsh environment of the body for a long period of time. Biomaterials are utilized to create skin substitutes that can heal wounds and regenerate tissue. Scaffolds made of synthetic polymers such as polyurethanes and hydrogels are used to assist cell development and tissue regeneration. In skin substitute formulations, natural components such as collagen and fibrin are also used. These skin substitutes are used to treat a variety of skin-related diseases and conditions,

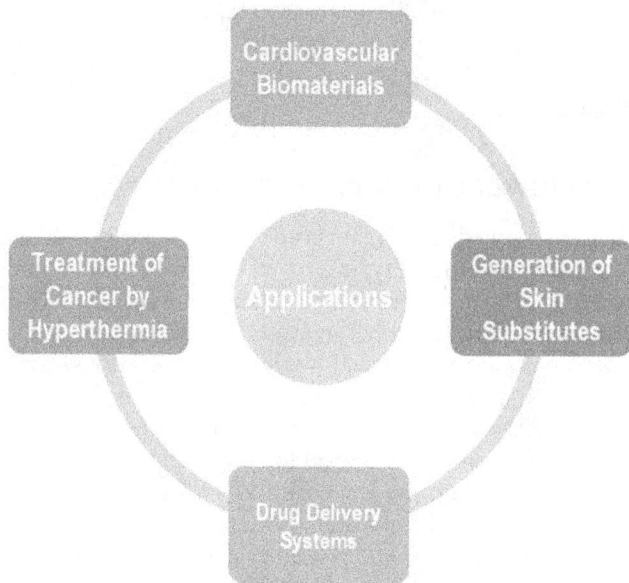

Figure 10.9 Biomaterials utilized in different medical applications.

including burns, chronic wounds, and ulcers [75–77]. They can also be used to improve skin aesthetic appearance, such as reducing wrinkles and blemishes.

In drug delivery systems, biomaterials are used to control the release of therapeutic substances throughout the body. Polymers, particularly bio-degradable polymers such as poly (lactic-co-glycolic acid) (PLGA) are frequently employed to encapsulate medications and allow for their continued release over time. This method improves therapeutic efficacy while reducing the requirement for frequent dosage. Furthermore, the use of biodegradable polymers reduces the risk of side effects associated with long-term exposure to medications [78–80]. Additionally, this method ensures that the therapeutic substance is released in a controlled manner, thereby increasing its efficacy. Biomaterials are used in cancer hyperthermia therapies. Magnetic nanoparticles, which are typically comprised of iron oxide, can be targeted to tumor locations. When exposed to an alternating magnetic field, these nanoparticles generate heat, which can be exploited to destroy cancer cells selectively. This method is less invasive than other treatments and has fewer side effects, making it a promising approach for cancer treatment. Additionally, it has the potential to be combined with other treatments such as chemotherapy and radiation, making it an even more powerful tool for fighting cancer [81–84]. Biomaterials provide a wide range of potential medical uses, advancing several disciplines such as cardiovascular medicine, tissue engineering, drug delivery, and cancer treatment. Biomaterials also offer the potential for personalized treatments, tailored to the specific needs of each individual patient [85]. Furthermore, biomaterials can be used to mimic the body's natural environment, which can help to improve the effectiveness of treatments.

10.3 METALLIC ALLOYS IN BIOMEDICAL IMPLANTS

Metallic alloys in biomedical implants can provide a variety of benefits, such as greater strength and durability, improved biocompatibility, and increased corrosion resistance. These properties make them ideal for use in biomedical implants, which must be able to withstand the body's environment and not cause any adverse reactions. However, metallic alloys can also cause complications such as metal allergy and inflammation. Therefore, careful selection of the alloy is essential for successful implantation. Therefore, careful selection of the alloy is essential for successful implantation [86–88]. It is also important to consider the cost and availability of the alloy when deciding. Numerous metallic elements perform crucial roles in the human body, as depicted in Figure 10.10. These substances, also referred to as trace elements or micronutrients, are needed in minute quantities for a variety of physiological processes. There are some illustrations of metallic components and the roles they play in the human body [89,90].

Magnesium (Mg)
Magnesium is involved in more than 300 enzymatic reactions in the body.

Manganese (Mn)
Manganese acts as a cofactor for various enzymes involved in antioxidant defense.

Chromium (Cr)
Chromium plays a role in insulin metabolism and glucose regulation.

Selenium (Se)
Selenium is an essential component of several enzymes with antioxidant properties.

Metallic Elements

1 2 8 3 7 4 9 5

Iron (Fe)
Hemoglobin and myoglobin heme groups deliver oxygen in the body.

Copper (Cu)
The ferroxidase system enzymes regulate blood iron transit and release from storage.

Zinc (Zn)
FSH and LH play a role in reproductive function.

Cobalt (Co)
Vitamin B12 is present. An overabundance can lead to heart failure.

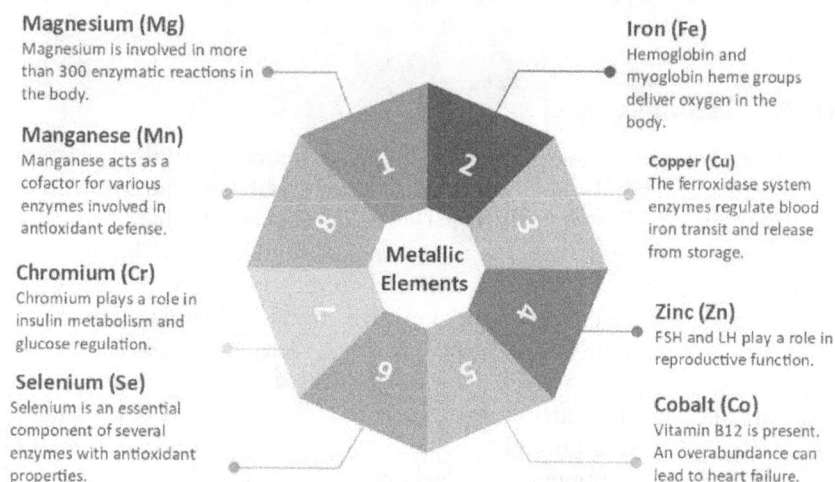

Figure 10.10 Metallic elements and function in human body.

Hemoglobin, the protein in red blood cells that carries oxygen throughout the body, is made up in large part by iron. It also aids the immune system and is involved in energy production. The fundamental function of calcium is to maintain healthy bones and teeth. Additionally, it affects cell activity, blood coagulation, neuron communication, and muscle contraction. Zinc participates in a variety of enzymatic processes and is essential to produce DNA, immunological response, wound healing, and growth and development. Collagen and other connective tissues need copper to produce, and it also helps with energy production, iron metabolism, and antioxidant protection [91–93]. More than 300 enzymatic processes in the body involve magnesium. It is crucial for keeping a healthy heart rhythm, producing energy, synthesizing proteins, and preserving muscle and neuron function. Manganese serves as a cofactor for several enzymes involved in bone growth, cholesterol and amino acid metabolism, antioxidant defense, and metabolism of carbohydrates [94]. Numerous antioxidant enzymes depend on selenium as a key ingredient. It functions in thyroid hormone metabolism, immune system support, and cell damage prevention. Chromium contributes to the regulation of glucose and insulin metabolism. It improves insulin's functionality, which is crucial for controlling blood sugar levels. Vitamin B12 (cobalamin), which contains cobalt as one of its ingredients, is essential to produce red blood cells and the preservation of a healthy nervous system [95]. While these metallic elements are necessary for human health, it's vital to remember that any element can have negative effects if consumed in excess or insufficiently [96]. Maintaining a balanced and diverse diet is always advised to guarantee an adequate supply of essential micronutrients.

10.3.1 Stainless steel

Due to its advantageous properties, including biocompatibility, strength, and corrosion resistance, stainless steel is a popular material for biomedical implants. As shown in Figure 10.11, stainless steel has a number of important properties in relation to biomedical implants. The human body normally tolerates stainless steel well and there is little chance that it may trigger an allergic reaction or other negative effects. Individual biocompatibility can vary, though, and some people may be sensitive to certain alloying components in stainless steel. Its strength and durability are crucial for implants that must tolerate mechanical strains from the body. It has a strong tensile strength and can hold its structural integrity for a long time [97–99].

Chromium, a component of steel, reacts with oxygen to generate a passive oxide layer on the surface. The hostile physiological environment of the body prevents the implant from degrading because of the remarkable corrosion resistance provided by this oxide layer. It comes in a variety of grades and can be altered with various alloying components to achieve particular qualities. Due to its adaptability, stainless steel alloys can be chosen according to their mechanical strength, corrosion resistance, and other properties depending on the particular use and demands of the implant [100]. Compared to other implant materials like titanium or cobalt–chromium alloys, it is very inexpensive.

This makes it an affordable choice for several medicinal applications. Stainless steel has some restrictions despite its benefits. It can be heavier than other materials due to its relatively high density, which may be a

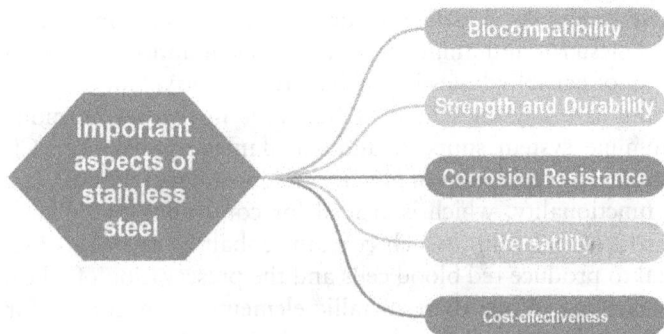

Figure 10.11 Important aspect of stainless steel.

factor in some implant designs. Additionally, extra surface treatments for stainless steel implants, such as polishing or passivation, can be necessary to reduce the danger of particle debris or enhance biocompatibility [101]. This can increase the cost of the implant, making it less desirable for some applications. Furthermore, stainless steel is not as corrosion resistant as other materials, such as titanium alloys, which can be more suitable for certain applications [102]. The interaction of two incongruent hard surfaces in incongruent joints, like the knee and ankle joints, produces highly concentrated heterogeneous stresses (Figure 10.12), which are counteracted by the presence of thick cartilage layers and synovial fluid. These layers of cartilage and synovial fluid act as buffers to ensure that the bones do not rub against each other and cause damage. This allows for the smooth, pain-free movement of the joints [103].

The structures involved in the joint are also designed to absorb and dissipate the forces created by movement. The muscles and ligaments that surround and connect the bones also play a role in this process. They provide stability and support to the joint, allowing it to move without being overstretched or subjected to excessive force. This ensures the joint remains healthy and can move freely [104]. By substituting large concentrations of nitrogen for nickel, new advances were looked into to improve 316L biocompatibility. It was discovered that nitrogen functions as a safe Ni replacement as an austenitic phase stabilizer. The result was the creation of Ni-free, high-concentration nitrogen SS (ASTM F2229) as shown in (Table 10.1). The

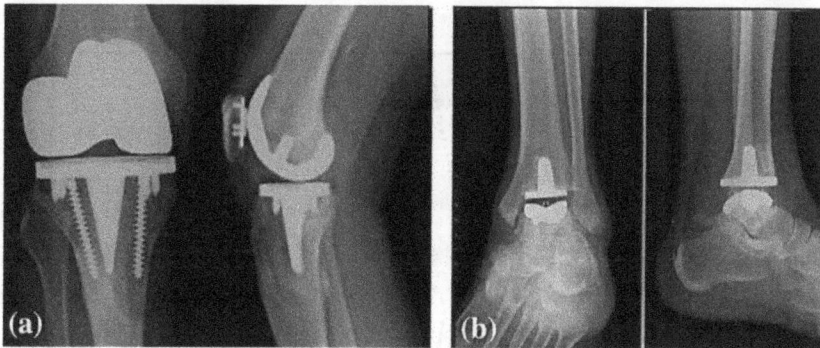

Figure 10.12 Examples of 316L SS implant in (a) knee and (b) ankle [103].

Table 10.1 Mechanical properties of 316L SS [105]

Material	UTS	Yield strength	Modulus of Elasticity,
316L SS	1170 MPa	480 MPa	190 GPa

new material has proven to have excellent corrosion resistance and biocompatibility [105]. It is now widely used in medical implants and instruments. Moreover, its properties can be further improved by heat treatments.

The created Ni-free SS exhibits improved bio-corrosion (Figure 10.13), improved fatigue resistance, and superior biocompatibility close to that of commercially available pure titanium. The Ni-free SS also showed excellent weldability and formability, making it an ideal material for the fabrication of medical implants [106].

Moreover, the Ni-free SS had higher strength than pure titanium, making it a very attractive material for medical applications. It is a substance that is frequently used in biomedical implants, especially for things like orthopedic devices (such bone plates and screws), cardiovascular stents, and surgical equipment. However, a number of variables, such as the particular medical condition, patient characteristics, and the needs of the implant itself, ultimately determine the choice of implant material. Different materials may offer different levels of strength, flexibility, and biocompatibility [107]. It is important to consider all of these factors when selecting implant materials. Ultimately, the material must be suitable for the intended purpose. The material must also be durable and long-lasting, and resist corrosion and wear. Additionally, it should be easy to work with and shape into the desired form.

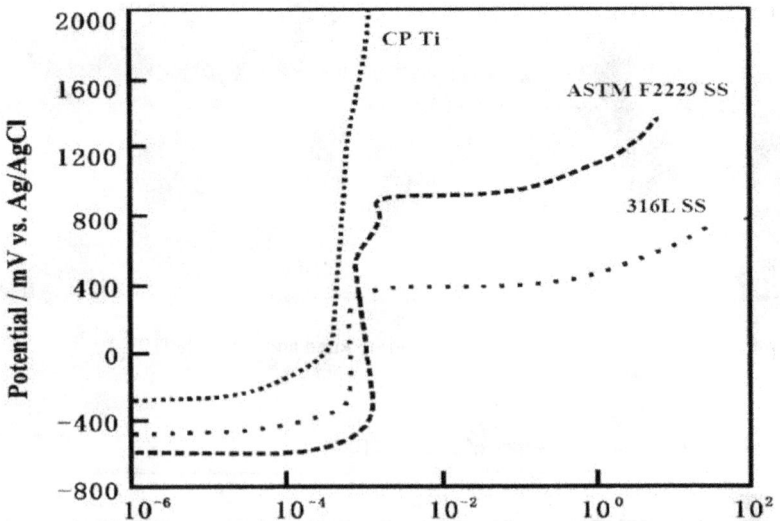

Figure 10.13 Potential/current density curve for 316L SS, ASTM F2229 SS and CP Ti [106].

10.3.2 Cobalt alloys

Due to their superior mechanical properties, corrosion resistance, and bio-compatibility, cobalt alloys, particularly cobalt–chromium (CoCr) alloys, are commonly used in biomedical implants. As depicted in Figure 10.14, cobalt alloys for biomedical implants have a number of important characteristics. The exceptional wear resistance of cobalt alloys reduces the risk of wear debris formation [108].

This is particularly crucial for joint replacements, where the implant surfaces undergo recurrent mechanical movement. Low rates of attrition contribute to the durability and stability of implants. In general, cobalt alloys are considered biocompatible. However, some individuals may develop hypersensitivity or allergic reactions to the implant-emitted cobalt ions. Before using cobalt alloys in biomedical implants, proper patient selection and evaluation of potential allergies are required. Cobalt alloys are relatively radiopaque, which means that they are visible using X-rays and other imaging techniques [109]. This enables simpler postoperative monitoring of the implant's position, stability, and any potential complications. Using techniques such as casting, forging, and additive manufacturing (e.g., selective laser melting), it is simple to produce cobalt alloys, including CoCr alloys. This enables the fabrication of complex implant designs and

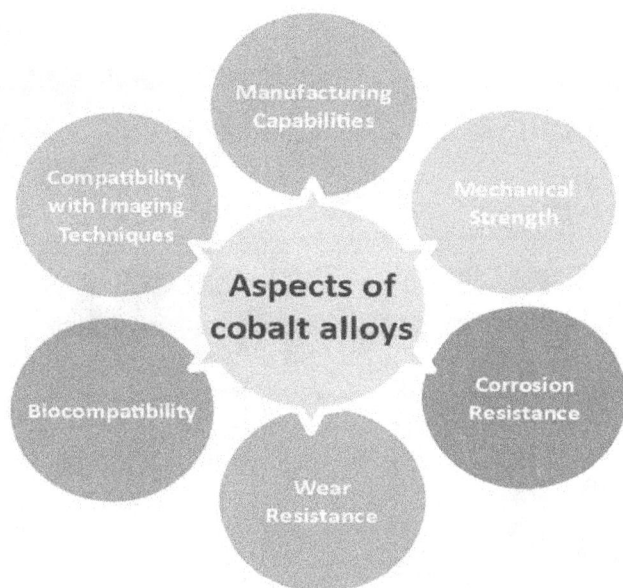

Figure 10.14 Important aspect of cobalt alloys in biomedical implant.

patient-specific customization. Notably, cobalt alloys are not appropriate for all implant varieties or patient populations [110]. Cobalt–chromium alloys, for instance, are typically not recommended for patients with known cobalt allergies. In such instances, titanium or ceramic implants may be considered as alternatives. In addition, cobalt–chromium alloys are generally not recommended for patients with compromised immune systems due to a higher risk of infection [111]. Lastly, it is important to consider the mechanical properties of the implant material when selecting an implant for a patient. For example, cobalt–chromium alloys can be brittle and may not be suitable for large loads or high-impact activities. Titanium and ceramic implants, on the other hand, are more flexible and can better withstand these forces. Subsequently, the metal was created and utilized in orthopedics and joints. Figure 10.15 depicts a knee replacement fabricated from CoCr alloy due to its superior wear resistance and galvanic properties [112]. CoCr alloy is corrosion resistant, and its biocompatible properties make it an ideal material for medical implants. Furthermore, its superior strength to weight ratio makes it an optimal material for medical applications. The metal has also been used to create dental implants, pacemakers, and other medical devices. Additionally, it has been used to create surgical instruments such as bone drills and saws [113].

Figure 10.16 demonstrates that the fatigue behavior of Co-alloys is superior to that of 316L SS [114,115]. The fatigue strength of Co-alloys

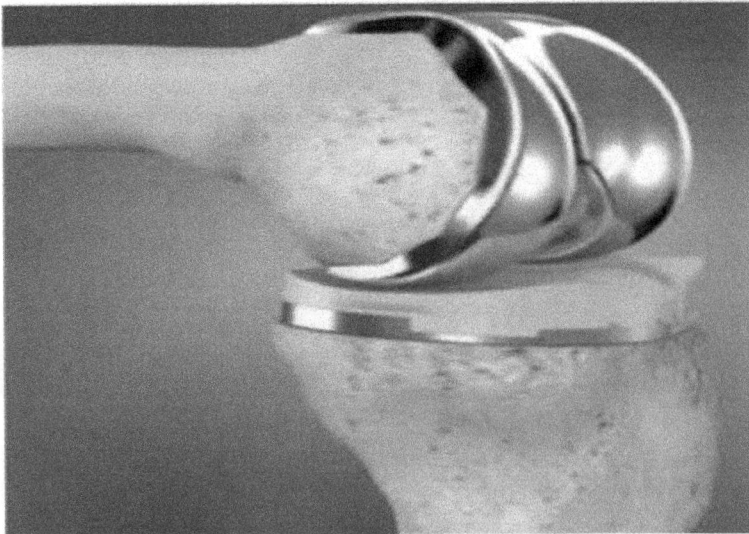

Figure 10.15 Knee replacement made from CoCrMo alloy [114].

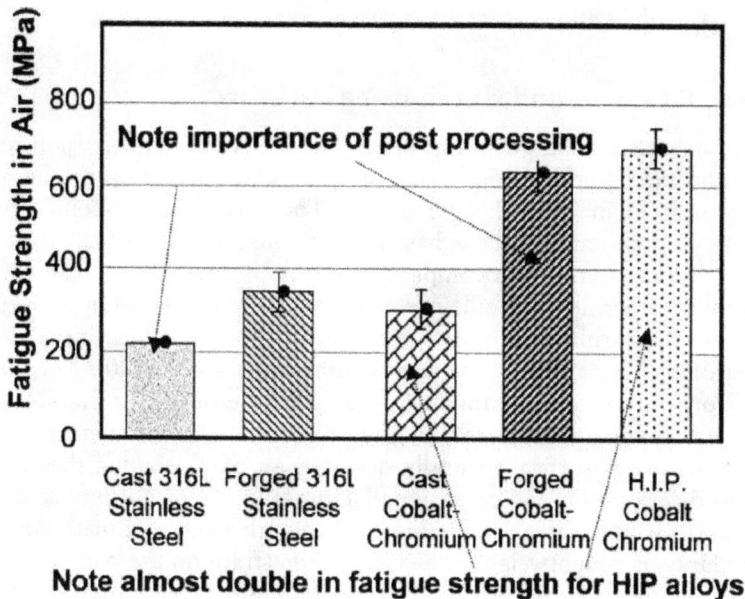

Figure 10.16 Fatigue strength of SS and Co–Cr alloys [115].

is almost twice that of 316L SS. Moreover, the Co-alloy is more resistant to corrosion and has a higher temperature capability than the stainless steel. Co-alloys therefore offer a better solution for applications that require a high level of fatigue resistance and corrosion resistance. Such applications include biomedical implantation and automotive industries [116]. As with any implant material, it is essential to carefully consider the specific medical condition, patient factors, and implant requirements when determining the most appropriate material for biomedical implants. The choice of cobalt alloys should be made in consultation with medical experts and based on the requirements and potential risks of the individual patient [117].

Similarly, the type of cobalt alloy used in automotive manufacturing should be chosen based on the desired properties and performance requirements of the component. Careful consideration should be given to the specific properties of the application and the potential risks and benefits of using cobalt alloys. All factors should be taken into account before selecting the appropriate cobalt alloy. Manufacturers must ensure they use the correct type to avoid failure of the component or risk to the patient [118]. Proper maintenance and inspection of the component should be conducted to ensure it is functioning as expected. Any potential flaws should be addressed quickly and the component replaced if necessary. The safety of the patient should always be the first priority. Regular maintenance should be conducted to ensure the cobalt alloy is functioning correctly [119].

10.3.3 Titanium and titanium-based alloys

Due to its outstanding qualities, such as biocompatibility, corrosion resistance, and mechanical strength, titanium and alloys based on titanium are widely utilized in biomedical implants. The following are some crucial features of titanium and alloys based on titanium for biomedical implants. Due to titanium's strong biocompatibility, there is little chance of an allergic reaction or other negative side effects in people who use it. On its surface, it develops a permanent oxide layer that aids in osseointegration, or the fusion of the implant with the surrounding bone tissue [120]. The strong corrosion resistance of titanium and its alloys ensures the stability and longevity of implants over the long term in the harsh environment of the body. Titanium's surface naturally develops an oxide coating that shields it from deterioration and corrosion. Titanium is known for having a high strength-to-weight ratio. Compared to many other metallic materials used in implants, it is lightweight, which eases the strain on the patient's body. Aluminum and vanadium are added to titanium to create titanium alloys like Ti-6Al-4V, which improve mechanical qualities while remaining lightweight. Since titanium and its alloys have a low elastic modulus, its rigidity is comparable to that of bone [121]. This characteristic reduces stress shielding, a problem that can arise when an implant takes on too much weight and causes bone resorption surrounding the implant. Since titanium is radiolucent, X-rays cannot be blocked by it. Titanium alloys can be modified with radiopaque markers to improve visibility during imaging, although this can make it difficult to see the implant on X-ray images. The mechanical qualities of titanium and alloys based on titanium are adaptable, enabling customization for implant applications [122]. They are simple to process, mold, and construct into intricate shapes, enabling personalization for particular patient requirements. Due to their biocompatibility and capacity to merge with surrounding bone tissue, titanium alloys, such as Ti-6Al-4V, are frequently utilized in dental implants. They offer a solid base for dental prosthesis like bridges and crowns. It's vital to remember that implants made of titanium could still have some restrictions [123]. For instance, they may cost more than other implant materials and necessitate specific fabrication methods or tools. For some dental or cosmetic uses, titanium's dark tint cannot be cosmetically acceptable. There are a huge variety of implant designs accessible today. The initial dental implants included an external connection, and the most typical design is a hexagonal anti-rotational component. Figure 10.17a provides an illustration of this link. A new connection was created since implants are rejected so frequently. Internal connection (Figure 10.17b), in this instance, enabled a better union of the implant and

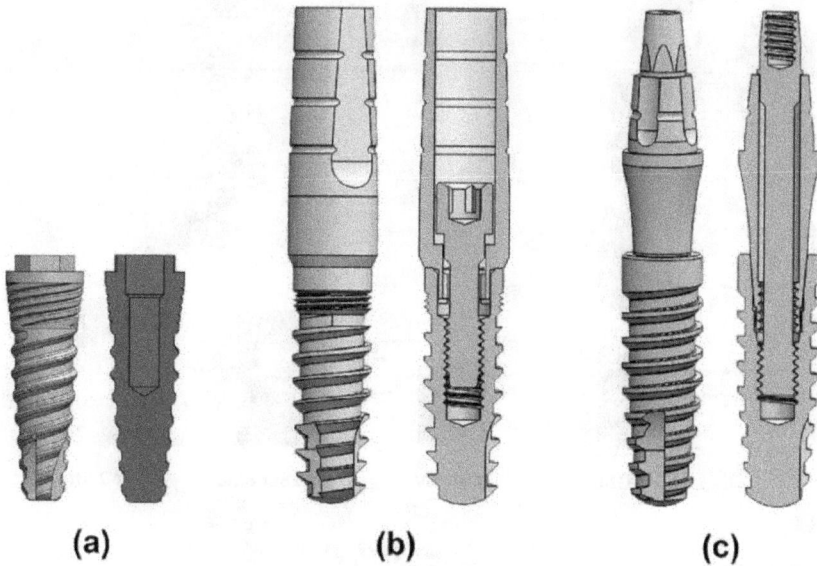

Figure 10.17 Schematic views of dental implant connections: (a) external connection; (b) internal connection and (c) Morse taper connection [127].

abutment. The introduction of the morse taper connection (Figure 10.17c), a different internal connection alternative, came because of its improvement over screw loosening. This connection type was quickly adopted by the dental industry as it was more secure and reliable. It has become the gold standard for dental implants, providing a strong and lasting connection between the implant and abutment. This connection also reduced the risk of bacterial contamination and allowed for better seating of the implant, resulting in improved stability and load-bearing capabilities [124–126]. Furthermore, the morse taper connection is easier to use compared to other connection types, making it the preferred choice for dental implants.

Figure 10.18 illustrates how different materials have varied repassivation times [128]. The figure makes it evident that stainless steel has a longer repassivation period than titanium alloys, meaning that more ions would be liberated from steel than from titanium alloy. This suggests that stainless steel is more corrosion-resistant than titanium alloys and would require less maintenance. Consequently, stainless steel is the better material choice for applications that require corrosion resistance [129]. Titanium and titanium-based alloys have established themselves as dependable and extensively used materials for a variety of biomedical implants, such as orthopedic implants (e.g., joint replacements, bone plates, and screws), dental

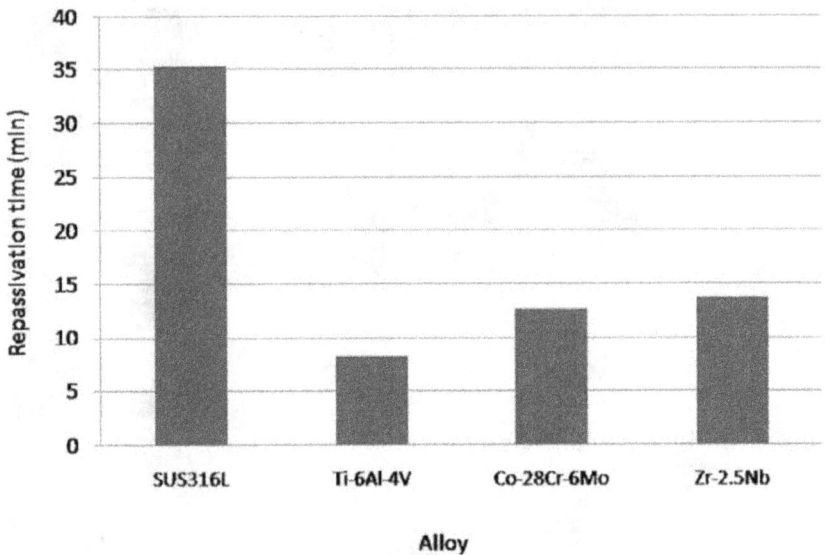

Figure 10.18 A comparison of repassivation time of various metallic alloys [128].

implants, and cardiovascular implants [130–132]. The choice of titanium alloy depends on variables such as the intended use of the implant, its mechanical requirements, and the patient's unique circumstances.

10.4 CONCLUSION

In conclusion, biomaterials serve an essential role in a variety of applications, such as medical implants, tissue engineering, and regenerative medicine. By comprehending and managing the interactions between biological systems and artificial materials, scientists can create novel biomaterials that enhance the health and quality of life of humans. Equiaxed and acicular microstructures are crucial in determining the thermal stability of metallic implants, whereas bioactive glass coatings improve biocompatibility and functionality. Utilizing thermal cladding techniques such as HVOF, PTA, and laser cladding, bioactive glass coatings are applied to metallic substrates, providing long-term stability and enhanced performance. As a result of their biodegradability, biocompatibility, and customizability, biomaterials are ideal for medical applications. Due to their strength, durability, and corrosion resistance, metallic alloys such as stainless steel, titanium, and cobalt–chromium are widely utilized in biomedical implants. The selection of implant material is influenced by variables such as the specific medical condition, the patient's characteristics, and the implant requirements. Due to their superior mechanical properties and biocompatibility, titanium and cobalt–chromium alloys are widely employed. Different connection types, such as external, internal, and morse taper connections, have enhanced the stability and load-bearing capacities of dental implants over time. Stainless steel has a prolonged repassivation time than titanium alloys, indicating superior corrosion resistance. Titanium and titanium alloys have proved to be dependable and widely utilized materials for a variety of biomedical implants. The choice of alloy depends on the intended use of the implant and the patient's needs.

REFERENCES

[1] Y. Hu, W. Cong, A review on laser deposition-additive manufacturing of ceramics and ceramic reinforced metal matrix composites, *Ceramics International* 44(17) (2018) 20599–20612.

[2] R. Fu, W. Luo, R. Nazempour, D. Tan, H. Ding, K. Zhang, L. Yin, J. Guan, X. Sheng, Implantable and biodegradable poly (l-lactic acid) fibers for optical neural interfaces, *Advanced Optical Materials* 6(3) (2018) 1700941.

[3] R. Nazempour, Q. Zhang, R. Fu, X. Sheng, Biocompatible and implantable optical fibers and waveguides for biomedicine, *Materials* 11(8) (2018) 1283.

[4] M. Ganjali, A. Yazdanpanah, M. Mozafari, Laser deposition of nano coatings on biomedical implants, Emerging Applications of Nanoparticles and Architecture Nanostructures, Elsevier2018, pp. 235–254.

[5] Y. Qu, T. Nguyen-Dang, A.G. Page, W. Yan, T. Das Gupta, G.M. Rotaru, R.M. Rossi, V.D. Favrod, N. Bartolomei, F. Sorin, Superelastic multimaterial electronic and photonic fibers and devices via thermal drawing, *Advanced Materials* 30(27) (2018) 1707251.

[6] R. Wandra, C. Prakash, S. Singh, Investigation on surface roughness and hardness of β-Ti alloy by ball burnishing assisted electrical discharge cladding for bio-medical applications, *Materials Today: Proceedings* 50 (2022) 848–854.

[7] Y. Sharma, A. Mehta, H. Vasudev, N. Jeyaprakash, G. Prashar, C. Prakash, Analysis of friction stir welds using numerical modelling approach: a comprehensive review, *International Journal on Interactive Design Manufacturing* (2023) 1–14.

[8] V. Bhojak, J.K. Jain, T.S. Singhal, K.K. Saxena, C. Prakash, M.K. Agrawal, V. Malik, Friction stir processing and cladding: an innovative surface engineering technique to tailor magnesium-based alloys for biomedical implantS, *Surface Review Letters* (2023) 2340007.

[9] S. Bajda, Y. Liu, R. Tosi, K. Cholewa-Kowalska, M. Krzyzanowski, M. Dziadek, M. Kopyscianski, S. Dymek, A.V. Polyakov, I.P. Semenova, Laser cladding of bioactive glass coating on pure titanium substrate with highly refined grain structure, *Journal of the Mechanical Behavior of Biomedical Materials* 119 (2021) 104519.

[10] S. Bose, D. Ke, H. Sahasrabudhe, A. Bandyopadhyay, Additive manufacturing of biomaterials, *Progress in Materials Science* 93 (2018) 45–111.

[11] G. Singh, A. Mehta, A. Bansal, Electrochemical behaviour and biocompatibility of claddings developed using microwave route, *Journal of Electrochemical Science Engineering Failure Analysis* 13(1) (2023) 173–192.

[12] W. Liu, S. Liu, L. Wang, Surface modification of biomedical titanium alloy: micromorphology, microstructure evolution and biomedical applications, *Coatings* 9(4) (2019) 249.

[13] A. Mthisi, A. Popoola, D. Adebiyi, O. Popoola, Laser Cladding of Ti-6Al-4V Alloy with Ti-Al2O3 Coating for Biomedical Applications, IOP Conference Series: Materials Science and Engineering, IOP Publishing, 2018, p. 012005.

[14] M. Lu, P. McCormick, Y. Zhao, Z. Fan, H. Huang, Laser deposition of compositionally graded titanium oxide on Ti6Al4V alloy, *Ceramics International* 44(17) (2018) 20851–20861.

[15] P.K. Verma, A. Mehta, H. Vasudev, V.J.S.R. Kumar, Performance of thermal spray coated metallic materials for bio-implant applications, *Surface Review Letters* (2023) 2340007.

[16] J.D. Majumdar, A. Kumar, S. Pityana, I. Manna, Laser surface melting of AISI 316L stainless steel for bio-implant application, Proceedings of the National Academy of Sciences, *India Section A: Physical Sciences* 88 (2018) 387–403.

[17] T. Chen, W. Li , D. Liu, Y. Xiong, X. Zhu, Effects of heat treatment on microstructure and mechanical properties of TiC/TiB composite bioinert ceramic coatings in-situ synthesized by laser cladding on Ti6Al4V, *Ceramics International* 47(1) (2021) 755–768.

[18] P. Singh, P. Kumar, An overview of biomedical materials and techniques for better functional performance, life, sustainability and biocompatibility of orthopedic implants, *Indian J. Sci. Tech* 11 (2018) 1–7.

[19] S. Mukherjee, S. Dhara, P. Saha, Laser surface remelting of Ti and its alloys for improving surface biocompatibility of orthopaedic implants, *Materials Technology* 33(2) (2018) 106–118.

[20] R.R. Behera, A. Hasan, M.R. Sankar, L.M. Pandey, Laser cladding with HA and functionally graded TiO2-HA precursors on Ti–6Al–4V alloy for enhancing bioactivity and cyto-compatibility, *Surface Coatings Technology* 352 (2018) 420–436.

[21] D.R. Unune, G.R. Brown, G.C. Reilly, Thermal based surface modification techniques for enhancing the corrosion and wear resistance of metallic implants: A review, *Vacuum* (2022) 111298.

[22] N.O. Joy-anne, Y. Su, X. Lu, P.-H. Kuo, J. Du, D. Zhu, Bioactive glass coatings on metallic implants for biomedical applications, *Bioactive Materials* 4 (2019) 261–270.

[23] M.Z. Ibrahim, A.A. Sarhan, M. Shaikh, T. Kuo, F. Yusuf, M. Hamdi, Investigate the effects of the laser cladding parameters on the microstructure, phases formation, mechanical and corrosion properties of metallic glasses coatings for biomedical implant application, *Additive Manufacturing of Emerging Materials* (2019) 299–323.

[24] M. Khanzadeh, W. Tian, A. Yadollahi, H.R. Doude, M.A. Tschopp, L.J.A.M. Bian, Dual process monitoring of metal-based additive manufacturing using tensor decomposition of thermal image streams, *Additive Manufacturing* 23 (2018) 443–456.

[25] G. Padmanabham, R. Bathe, Laser materials processing for industrial applications, Proceedings of the National Academy of Sciences, *India Section A: Physical Sciences* 88 (2018) 359–374.

[26] O.S. Fatoba, O.S. Adesina, A. Popoola, Evaluation of microstructure, microhardness, and electrochemical properties of laser-deposited Ti-Co coatings on Ti-6Al-4V Alloy, *The International Journal of Advanced Manufacturing Technology* 97 (2018) 2341–2350.

[27] R. Wandra, C. Prakash, S. Singh, Experimental investigation and optimization of surface roughness of β-Phase titanium alloy by ball burnishing assisted electrical discharge cladding for implant applications, *Materials Today: Proceedings* 48 (2022) 975–980.

[28] Y. Qu, T. Nguyen-Dang, A.G. Page, W. Yan, T.D. Gupta, G.M. Rotaru, R.M. Rossi, V.D. Favrod, N. Bartolomei, F. Sorin, Stretchable optical and electronic fibers via thermal drawing, International Flexible Electronics Technology Conference (IFETC), IEEE, 2018, p. 1.

[29] Z. Liu, K.C. Chan, L. Liu, S. Guo, Bioactive calcium titanate coatings on a Zr-based bulk metallic glass by laser cladding, *Materials Letters* 82 (2012) 67–70.

[30] Q. An, L. Huang, S. Jiang, Y. Bao, M. Ji, R. Zhang, L. Geng, Two-scale TiB/Ti64 composite coating fabricated by two-step process, *Journal of Alloys Compounds* 755 (2018) 29–40.

[31] A.I. Mahmoud Zakaria Alsayed, Laser cladding of FeCrMoCB metallic glass on nickel-free stainless-steel to develop durable and cost-effective biomedical implants/Mahmoud Zakaria Alsayed Abdalfattah Ibrahim, PhD diss, Universiti Malaya, 2019.

[32] M.N. Jahangir, M.A.H. Mamun, M.P. Sealy, A review of additive manufacturing of magnesium alloys, AIP conference proceedings, AIP Publishing, 2018.

[33] F. Ghadami, A.S.R. Aghdam, Improvement of high velocity oxy-fuel spray coatings by thermal post-treatments: A critical review, *Thin Solid Films* 678 (2019) 42–52.

[34] S.S. Kumar, V. Tripathi, R. Sharma, G. Puthilibai, M. Sudhakar, K. Negash, Study on developments in protection coating techniques for steel, *Advances in Materials Science Engineering* 2022 (2022).

[35] J. Singh, J.P. Singh, S. Kumar, H.S. Gill, Short review on hydroxyapatite powder coating for SS 316L, *Journal of Electrochemical Science Engineering* 13(1) (2023) 25–39.

[36] J. Garcia-Herrera, J. Henao, D. Espinosa-Arbelaez, J. Gonzalez-Carmona, C. Felix-Martinez, R. Santos-Fernandez, J. Corona-Castuera, C. Poblano-Salas, J. Alvarado-Orozco, Laser cladding deposition of a Fe-based metallic glass on 304 stainless steel substrates, *Journal of Thermal Spray Technology* 31(4) (2022) 968–979.

[37] N. Abhijith, D. Kumar, D. Kalyansundaram, Development of single-stage TiNbMoMnFe high-entropy alloy coating on 304L stainless steel using

HVOF thermal spray, *Journal of Thermal Spray Technology* 31(4) (2022) 1032–1044.

[38] R. Fathi, H. Wei, B. Saleh, N. Radhika, J. Jiang, A. Ma, M.H. Ahmed, Q. Li, K.K.J.A.M.T. Ostrikov, Past and present of functionally graded coatings: Advancements and future challenges, *Applied Materials Today* 26 (2022) 101373.

[39] P. Kumar, N.K. Jain, A. Tiwari, Sustainable Polishing of Directed Energy Deposition–Based Cladding Using Micro-Plasma Transferred Arc, Advances in Sustainable Machining and Manufacturing Processes, CRC Press 2022, pp. 289–302.

[40] G. Ertugrul, A. Hälsig, J. Hensel, J. Buhl, S. Härtel, Efficient multi-material and high deposition coating including additive manufacturing by tandem plasma transferred arc welding for functionally graded structures, *Metals* 12(8) (2022) 1336.

[41] O.N. Çelik, Microstructure and wear properties of WC particle reinforced composite coating on Ti6Al4V alloy produced by the plasma transferred arc method, *Applied surface science* 274 (2013) 334–340.

[42] A.A. Oudah, M.A. Hassan, N. Almuramady, Materials manufacturing processes: Feature and trends, AIP Conference Proceedings, AIP Publishing, 2023.

[43] R. Ranjan, A.K. Das, Improving the Resistance to Wear and Mechanical Characteristics of Cladding Layers on Titanium and its Alloys: A Review, *Tribology in Industry* 44(1) (2023) 136.

[44] K. Kanishka, B.J.J.o.M.P. Acherjee, A systematic review of additive manufacturing-based remanufacturing techniques for component repair and restoration, *Journal of Manufacturing Processes* 89 (2023) 220–283.

[45] N. Jeyaprakash, S.S. Karuppasamy, C.-H. Yang, Application of Wear-Resistant Laser Claddings, Handbook of Laser-Based Sustainable Surface Modification and Manufacturing Techniques, CRC Press 2023, pp. 1–26.

[46] F.J.S.E. Findik, Innovation, Laser cladding and applications, *Sustainable Engineering Innovation* 5(1) (2023) 1–14.

[47] X. Zhang, S. Pfeiffer, P. Rutkowski, M. Makowska, D. Kata, J. Yang, T.J.A.S.S. Graule, Laser cladding of manganese oxide doped aluminum oxide granules on titanium alloy for biomedical applications, 520 (2020) 146304.

[48] A.-C. Mocanu, F. Miculescu, G.E. Stan, T. Tite, M. Miculescu, M.H. Țierean, A. Pascu, R.-C. Ciocoiu, T.M. Butte, L.-T. Ciocan, Development of ceramic coatings on titanium alloy substrate by laser cladding with pre-placed natural derived-slurry: Influence of hydroxyapatite ratio and beam power, *Ceramics International* 49(7) (2023) 10445–10454.

[49] Z. Deng, D. Liu, G. Liu, Y. Xiong, S. Xin, S. Li, C. Li, T. Chen, Study on the corrosion resistance of SiC particle-reinforced hydroxyapatitesilver gradient bioactive ceramic coatings prepared by laser cladding, *Surface Coatings Technology* (2023) 129734.

[50] K. Wang, W. Liu, Y. Hong, H.S. Sohan, Y. Tong, Y. Hu, M. Zhang, J. Zhang, D. Xiang, H.J.C. Fu, An overview of technological parameter optimization in the case of laser cladding, *Surface Coatings Technology* 13(3) (2023) 496.

[51] H. Wang, Y. Cheng, R. Geng, B. Wang, Y. Chen, X. Liang, Comparative study on microstructure and properties of Fe-based amorphous coatings prepared by conventional and high-speed laser cladding, *Journal of Alloys Compounds* 952 (2023) 169842.

[52] Y. Zheng, P. Xu, Effect of Nb Content on Phase Transformation and Comprehensive Properties of TiNb Alloy Coating, *Coatings* 13(7) (2023) 1186.

[53] A.R. Ahmady, A. Ekhlasi, A. Nouri, M.H. Nazarpak, P. Gong, A. Solouk, High entropy alloy coatings for biomedical applications: A review, *Smart Materials in Manufacturing* 1 (2023) 100009.

[54] J. Liu, D. Liu, S. Li, Z. Deng, Z. Pan, C. Li, T. Chen, The effects of graphene oxide doping on the friction and wear properties of TiN bioinert ceramic coatings prepared using wide-band laser cladding, *Surface Coatings Technology* 458 (2023) 129354.

[55] M.Z. Ibrahim, A.A. Sarhan, T. Kuo, F. Yusof, M. Hamdi, Characterization and hardness enhancement of amorphous Fe-based metallic glass laser cladded on nickel-free stainless steel for biomedical implant application, *Materials Chemistry Physics* 235 (2019) 121745.

[56] C. Yang, X. Cheng, H. Tang, X. Tian, D. Liu, Influence of microstructures and wear behaviors of the microalloyed coatings on TC11 alloy surface using laser cladding technique, *Surface coatings technology* 337 (2018) 97–103.

[57] T.D. Ngo, A. Kashani, G. Imbalzano, K.T. Nguyen, D.J.C.P.B.E. Hui, Additive manufacturing (3D printing): A review of materials, methods, applications and challenges, *Composites Part B: Engineering* 143 (2018) 172–196.

[58] J. Reddy, M. Chamanzar, Parylene photonic waveguide arrays: a platform for implantable optical neural implants, CLEO: Applications and Technology, Optica Publishing Group, 2018, p. AM3P. 6.

[59] A.B. Edathazhe, Investigation of properties, corrosion and bioactivity of novel BaO added phosphate glasses and glass-ceramic coating on biomedical metallic implant materials, National Institute of Technology Karnataka, Surathkal, 2018.

[60] M. Ganjali, M. Ganjali, S. Sadrnezhaad, Y. Pakzad, Laser cladding of Ti alloys for biomedical applications, *Laser Cladding of Metals* (2021) 265–292.

[61] B. Heer, A. Bandyopadhyay, Silica coated titanium using Laser Engineered Net Shaping for enhanced wear resistance, *Additive Manufacturing* 23 (2018) 303–311.

[62] M.Z. Ibrahim, A.A. Sarhan, T. Kuo, M. Hamdi, F. Yusof, C. Chien, C. Chang, T. Lee, Advancement of the artificial amorphous-crystalline structure of laser cladded FeCrMoCB on nickel-free stainless-steel for bone-implants, *Materials Chemistry Physics* 227 (2019) 358–367.

[63] P. Singh, H. Vasudev, A. Bansal, Effect of post-heat treatment on the microstructural, mechanical, and bioactivity behavior of the microwave-assisted alumina-reinforced hydroxyapatite cladding, Proceedings of the Institution of Mechanical Engineers, *Part E: Journal of Process Mechanical Engineering* (2022) 09544089221116168.

[64] N. Ma, S. Liu, W. Liu, L. Xie, D. Wei, L. Wang, L. Li, B. Zhao, Y. Wang, Research progress of titanium-based high entropy alloy: methods, properties, and applications, *Frontiers in Bioengineering Biotechnology* 8 (2020) 603522.

[65] C. Han, Y. Li, Q. Wang, D. Cai, Q. Wei, L. Yang, S. Wen, J. Liu, Y. Shi, Titanium/hydroxyapatite (Ti/HA) gradient materials with quasi-continuous ratios fabricated by SLM: material interface and fracture toughness, *Materials Design* 141 (2018) 256–266.

[66] W. Harun, R. Asri, J. Alias, F. Zulkifli, K. Kadirgama, S. Ghani, J. Shariffuddin, A comprehensive review of hydroxyapatite-based coatings adhesion on metallic biomaterials, *Ceramics International* 44(2) (2018) 1250–1268.

[67] W. Li, P. Xu, Y. Wang, Y. Zou, H. Gong, F. Lu, Laser synthesis and microstructure of micro-and nano-structured WC reinforced Co-based cladding layers on titanium alloy, *Journal of Alloys Compounds* 749 (2018) 10–22.

[68] A. Farazin, C. Zhang, A. Gheisizadeh, A. Shahbazi, 3D bio-printing for use as bone replacement tissues: A review of biomedical application, *Biomedical Engineering Advances* (2023) 100075.

[69] V. Koshuro, M. Fomina, A. Voyko, I. Rodionov, A. Zakharevich, A. Skaptsov, A. Fomin, Surface morphology of zirconium after treatment with high-frequency currents, *Composite Structures* 202 (2018) 210–215.

[70] T.P. Singh, H. Singh, H. Singh, Characterization of thermal sprayed hydroxyapatite coatings on some biomedical implant materials, *Journal of Applied Biomaterials Functional Materials* 12(1) (2014) 48–56.

[71] J. Mesquita-Guimarães, B. Henriques, F. Silva, Bioactive glass coatings, Bioactive glasses, Elsevier 2018, pp. 103–118.

[72] L. Murr, Metallurgy principles applied to powder bed fusion 3D printing/ additive manufacturing of personalized and optimized metal and alloy biomedical implants: An overview, *Journal of Materials Research Technology* 9(1) (2020) 1087–1103.

[73] D. Zhang, W. Feng, X. Wang, S. Yang, Fabrication of Mg65Zn30Ca5 amorphous coating by laser remelting, *Journal of Non-Crystalline Solids* 500 (2018) 205–209.

[74] M. Krzyzanowski, D. Svyetlichnyy, S. Bajda, Additive manufacturing of multi layered bioactive materials with improved mechanical properties: modelling aspects, Materials Science Forum, Trans Tech Publ, 2021, pp. 888–893.

[75] C. Domínguez-Trujillo, F. Ternero, J.A. Rodríguez-Ortiz, J.J. Pavón, I. Montealegre-Meléndez, C. Arévalo, F. García-Moreno, Y. Torres, Improvement of the balance between a reduced stress shielding and bone ingrowth by bioactive coatings onto porous titanium substrates, *Surface Coatings Technology* 338 (2018) 32–37.

[76] D.T. Waghmare, C.K. Padhee, R. Prasad, M. Masanta, NiTi coating on Ti-6Al-4V alloy by TIG cladding process for improvement of wear resistance: Microstructure evolution and mechanical performances, *Journal of Materials Processing Technology* 262 (2018) 551–561.

[77] L. Fan, H. CHEN, Y. Dong, X. LI, L. DONG, Y. YIN, Corrosion behavior of Fe-based laser cladding coating in hydrochloric acid solutions, Acta Metall Sin 54(7) (2018) 1019–1030.

[78] Y. Cheng, H. Yang, Y. Yang, J. Huang, K. Wu, Z. Chen, X. Wang, C. Lin, Y. Lai, Progress in TiO 2 nanotube coatings for biomedical applications: a review, *Journal of Materials Chemistry B* 6(13) (2018) 1862–1886.

[79] K. Vanmeensel, K. Lietaert, B. Vrancken, S. Dadbakhsh, X. Li, J.-P. Kruth, P. Krakhmalev, I. Yadroitsev, J. Van Humbeeck, Additively manufactured metals for medical applications, *Additive Manufacturing* (2018) 261–309.

[80] A.S.H. Makhlouf, A. Barhoum, Emerging applications of nanoparticles and architectural nanostructures: current prospects and future trends, *Metals* 37 (2018).

[81] R. Soni, S. Pande, Laser Cladded Ti-Alloys in Biomedical Applications: A Review, *Trends in Biomaterials Artificial Organs* 36(2) (2022) 43–48.

[82] H.A. Zaharin, A.M. Abdul Rani, F.I. Azam, T.L. Ginta, N. Sallih, A. Ahmad, N.A. Yunus, T.Z.A. Zulkifli, Effect of unit cell type and pore size on porosity and mechanical behavior of additively manufactured Ti6Al4V scaffolds, *Materials* 11(12) (2018) 2402.

[83] J.W. Reddy, M. Chamanzar, Low-loss flexible Parylene photonic waveguides for optical implants, *Optics Letters* 43(17) (2018) 4112–4115.

[84] A. Shearer, M. Montazerian, J.J. Sly, R.G. Hill, J.C. Mauro, Trends and perspectives on the commercialization of bioactive glasses, *Acta Biomaterialia* 30 (2023) 72–77.

[85] J. Wang, J. Dong, Optical waveguides and integrated optical devices for medical diagnosis, health monitoring and light therapies, *Sensors* 20(14) (2020) 3981.

[86] R. Soni, S. Pande, S. Kumar, S. Salunkhe, H. Natu, H.M.A.M. Hussein, Wear Characterization of Laser Cladded Ti-Nb-Ta Alloy for Biomedical Applications, *Crystals* 12(12) (2022) 1716.

[87] M.A. Mahmood, A.C. Popescu, I.N. Mihailescu, Metal matrix composites synthesized by laser-melting deposition: a review, *Materials* 13(11) (2020) 2593.

[88] R.K. Sharma, G.P.S. Sodhi, V. Bhakar, R. Kaur, S. Pallakonda, P. Sarkar, H. Singh, Sustainability in manufacturing processes: Finding the environmental impacts of friction stir processing of pure magnesium, *CIRP Journal of Manufacturing Science Technology* 30 (2020) 25–35.

[89] W. JIA, X. LIN, Numerical microstructure simulation of laser rapid forming 316L stainless steel, *Acta Metall Sin* 46(2) (2010) 135–140.

[90] R. Comesaña, Special Issue on Surface Treatment by Laser-Assisted Techniques, MDPI Coatings, 2020, p. 580.

[91] K. Singh, S. Mohan, S. Konovalov, M. Graf, Effect of nano-hydroxyapatite and post heat treatment on biomedical implants by sol-gel and HVOF spraying, Nanomaterials for Sustainable Tribology, CRC Press 2023, pp. 257–285.

[92] A. Mahajan, S. Devgan, D.J.M. Kalyanasundaram, M. Processes, Surface alteration of Cobalt-Chromium and duplex stainless steel alloys for biomedical applications: a concise review, *Materials Manufacturing Processes* 38(3) (2023) 260–270.

[93] B.R. Sunil, A.S.K. Kiran, S. Ramakrishna, Surface functionalized titanium with enhanced bioactivity and antimicrobial properties through surface engineering strategies for bone implant applications, *Current Opinion in Biomedical Engineering* 23 (2022) 100398.

[94] D. Castro, P. Jaeger, A.C. Baptista, J.P. Oliveira, An overview of high-entropy alloys as biomaterials, *Metals* 11(4) (2021) 648.

[95] K. Munir, A. Biesiekierski, C. Wen, Y. Li, Surface modifications of metallic biomaterials, Metallic Biomaterials Processing and Medical Device Manufacturing, Elsevier 2020, pp. 387–424.

[96] A.-C. Mocanu, F. Miculescu, G.E. Stan, I. Pasuk, T. Tite, A. Pascu, T.M. Butte, L.-T. Ciocan, Modulated Laser Cladding of Implant-Type Coatings by Bovine-Bone-Derived Hydroxyapatite Powder Injection on Ti6Al4V Substrates—Part I: Fabrication and Physico-Chemical Characterization, *Materials* 15(22) (2022) 7971.

[97] P. Singh, A. Bansal, V.K. Verma, Hydroxyapatite reinforced surface modification of SS-316L by microwave processing, *Surfaces Interfaces* 28 (2022) 101701.

[98] D. Wang, J. Huang, C. Tan, W. Ma, Y. Zou, Y. Yang, Mechanical and corrosion properties of additively manufactured SiC-reinforced stainless steel, *Materials Science Engineering: A* 841 (2022) 143018.

[99] M.P. Nikolova, M.D. Apostolova, Advances in Multifunctional Bioactive Coatings for Metallic Bone Implants, *Materials* 16(1) (2022) 183.

[100] D. Svetlizky, M. Das, B. Zheng, A.L. Vyatskikh, S. Bose, A. Bandyopadhyay, J.M. Schoenung, E.J. Lavernia, N. Eliaz, Directed energy deposition (DED) additive manufacturing: Physical characteristics, defects, challenges and applications, *Materials Today* 49 (2021) 271–295.

[101] Y. Zhang, Dissimilar Bimetallic Structures via Directed Energy Deposition-Based Additive Manufacturing, Washington State University 2021.

[102] A. Mehta, H. Vasudev, N.J.I.J.o.I.D. Jeyaprakash, Manufacturing, role of sustainable manufacturing approach: microwave processing of materials, *International Journal on Interactive Design Manufacturing* (2023) 1–17.

[103] Q. Chen, G.A. Thouas, Metallic implant biomaterials, *Materials Science Engineering: R: Reports* 87 (2015) 1–57.

[104] A. Kansal, A. Dvivedi, P. Kumar, Development and performance study of biomedical porous zinc scaffold manufactured by using additive manufacturing and microwave sintering, *Materials Manufacturing Processes* 38(8) (2023) 1020–1032.

[105] A. Yazdipour, A. Heidarzadeh, Effect of friction stir welding on microstructure and mechanical properties of dissimilar Al 5083-H321 and 316L stainless steel alloy joints, *Journal of Alloys Compounds* 680 (2016) 595–603.

[106] M. Niinomi, Recent metallic materials for biomedical applications, *Metallurgical materials transactions* A 33 (2002) 477–486.

[107] H.A. Abdullah, R.A. Anaee, Characteristics and Morphological Studies of Nd Doped Titanium Thin Film Coating on SS 316L by DC Sputtering, *Diyala Journal of Engineering Sciences* 15(3) (2022).

[108] V. Kaushik, N. Kumar, M. Vignesh, Magnesium role in additive manufacturing of biomedical implants–challenges and opportunities, *Additive Manufacturing* 55 (2022) 102802.

[109] J. Feng, Y. Tang, J. Liu, P. Zhang, C. Liu, L. Wang, Bio-high entropy alloys: Progress, challenges, and opportunities, *Frontiers in Bioengineering Biotechnology* 10 (2022) 977282.

[110] S. Kumar, P. Katyal, Factors affecting biocompatibility and biodegradation of magnesium based alloys, Materials Today: Proceedings 52 (2022) 1092–1107.

[111] M.Z. Ibrahim, A. Halilu, A.A. Sarhan, T. Kuo, F. Yusuf, M. Shaikh, M. Hamdi, In-vitro viability of laser cladded Fe-based metallic glass as a promising bioactive material for improved osseointegration of orthopedic implants, Medical Engineering Physics 102 (2022) 103782.

[112] S. Gaytan, L. Murr, E. Martinez, J. Martinez, B. Machado, D. Ramirez, F. Medina, S. Collins, R. Wicker, Comparison of microstructures and mechanical properties for solid and mesh cobalt-base alloy prototypes fabricated by electron beam melting, Metallurgical Materials Transactions A 41 (2010) 3216–3227.

[113] T.S. Tshephe, S.O. Akinwamide, E. Olevsky, P.A. Olubambi, Additive manufacturing of titanium-based alloys-A review of methods, properties, challenges, and prospects, Heliyon (2022).

[114] N. Hua, Z. Qian, B. Lin, Z. Liao, Q. Wang, P. Dai, H. Fang, P.K. Liaw, Formation of a protective oxide layer with enhanced wear and corrosion resistance by heating the TiZrHfNbFe0. 5 refractory multi-principal element alloy at 1,000° C, Scripta Materialia 225 (2023) 115165.

[115] S. Teoh, Fatigue of biomaterials: a review, International journal of fatigue 22(10) (2000) 825–837.

[116] T. Zhang, W. Wang, J. Liu, L. Wang, Y. Tang, K. Wang, A review on magnesium alloys for biomedical applications, Frontiers in Bioengineering Biotechnology 10 (2022) 953344.

[117] Y. Gomez Taborda, M. Gómez Botero, J.G. Castaño-González, A. Bermúdez-Castañeda, Assessment of physical, chemical, and tribochemical properties of biomedical alloys used in explanted modular hip prostheses: A review, Proceedings of the Institution of Mechanical Engineers, Part H: Journal of Engineering in Medicine 236(4) (2022) 457–468.

[118] N. Kaushik, A. Meena, H.S. Mali, High entropy alloy synthesis, characterisation, manufacturing & potential applications: a review, Materials Manufacturing Processes 37(10) (2022) 1085–1109.

[119] A. Sharma, A. Singh, V. Chawla, J. Grewal, A. Bansal, Microwave processing and characterization of alumina reinforced HA cladding for biomedical applications, Materials Today: Proceedings 57 (2022) 650–656.

[120] D. Romanov, K. Sosnin, S.Y. Pronin, Y.F. Ivanov, V. Gromov, Structure and Properties of Electroexplosion Molybdenum Coating Deposited on Titanium Alloy VT6, Metal Science Heat Treatment 64(11) (2023) 639–647.

[121] M.H. Miah, D. Singh Chand, G.S. Malhi, S. Khan, Influence of laser scanning power on microstructure and tribological behavior of NI-composite claddings fabricated on TC4 titanium alloy, Aircraft Engineering Aerospace Technology (2023) 103–118.

[122] N.P. Msweli, S.O. Akinwamide, P.A. Olubambi, B.A. Obadele, Microstructure and biocorrosion studies of spark plasma sintered yttria stabilized zirconia reinforced Ti6Al7Nb alloy in Hanks' solution, Materials Chemistry Physics 293 (2023) 126940.

[123] N. Wu, H. Gao, X. Wang, X. Pei, Surface Modification of Titanium Implants by Metal Ions and Nanoparticles for Biomedical Application, *ACS Biomaterials Science Engineering* 29 (2023) 230–239.

[124] J. Yang, Y. Song, K. Dong, E.-H. Han, Research progress on the corrosion behavior of titanium alloys, *Corrosion Reviews* 41(1) (2023) 5–20.

[125] M. Abdulwahab, V. Aigbodion, M. Enechukwu, Anti-corrosion of isothermally treated Ti-6Al-4V alloy as dental biomedical implant using non-toxic bitter leaf extract, *Chemical Data Collections* 24 (2019) 100271.

[126] A.M. Khorasani, M. Goldberg, E.H. Doeven, G. Littlefair, Titanium in biomedical applications—properties and fabrication: a review, *Journal of Biomaterials Tissue Engineering* 5(8) (2015) 593–619.

[127] R. Rojo, M. Prados-Privado, A.J. Reinoso, J.C. Prados-Frutos, Evaluation of fatigue behavior in dental implants from in vitro clinical tests: a systematic review, *Metals* 8(5) (2018) 313.

[128] T. Hanawa, Reconstruction and regeneration of surface oxide film on metallic materials in biological environments, *Corrosion reviews* 21(2–3) (2003) 161–182.

[129] Q. Fu, W. Liang, J. Huang, W. Jin, B. Guo, P. Li, S. Xu, P.K. Chu, Z. Yu, Research perspective and prospective of additive manufacturing of biodegradable magnesium-based materials, *Journal of Magnesium Alloys* 6 (2023) 158.

[130] S. Zhao, F. Meng, B. Fan, Y. Dong, J. Wang, X. Qi, Evaluation of wear mechanism between TC4 titanium alloys and self-lubricating fabrics, *Wear* 512 (2023) 204532.

[131] S. Omarov, N. Nauryz, D. Talamona, A. Perveen, Surface Modification Techniques for Metallic Biomedical Alloys: A Concise Review, *Metals* 13(1) (2022) 82.

[132] A. Santos, J. Teixeira, C. Fonzar, E. Rangel, N. Cruz, P.N. Lisboa-Filho, A Tribological Investigation of the Titanium Oxide and Calcium Phosphate Coating Electrochemical Deposited on Titanium, *Metals* 13(2) (2023) 410.

Chapter 11

Study on the thermal claddings used in biomedical implants

Harjit Singh, Mukhtiar Singh, Maninder Singh,
Vineet Pushya, Hitesh Vasudev and Amrinder Mehta

11.1 INTRODUCTION TO THERMAL CLADDINGS

Thermal claddings are thin coatings applied to the surface of biomedical to improve their biocompatibility, corrosion resistance, and wear resistance. Biomedical implants are used to replace or repair damaged tissues or organs in the human body. However, the success of these implants is often limited by the body's immune response, corrosion, and wear. Thermal claddings can help to address these issues by providing a protective barrier between the implant and the surrounding tissue. The use of thermal claddings for biomedical implants is a rapidly growing field, with new materials and coating techniques being developed and tested. A variety of materials have been investigated for use as thermal claddings, including metals, ceramics, and polymers. These materials are selected based on their biocompatibility, mechanical properties, and ability to adhere to the substrate. As shown in Figure 11.1, depending on the application, thermal claddings can improve a variety of qualities of coated materials.

Thermal claddings are used for a wide range of applications, including improving wear resistance, thermal insulation, electrical insulation, and corrosion resistance. They can also be used to improve the aesthetic prop-erties of coated materials. Thermal claddings are available in various materials, including metals, ceramics, and plastics. They can be applied to various substrates and have a low environmental impact (Sachdeva & Gupta, 2019). Thermal claddings can be applied through various methods, such as spraying, welding, and thermal diffusion.

Coating techniques for thermal claddings include physical vapor depos-ition (PVD), chemical vapor deposition (CVD), electro-deposition, and sol–gel coating. Each of these techniques has its own advantages and limitations, and the choice of technique depends on the specific application and requirements of the implant. PVD is the most used technique due to its flexibility and cost-effectiveness. CVD is also widely used, but it is more expensive and less flexible than PVD. Electro-deposition and sol–gel coating are more specialized techniques and are used for specific applications.

DOI: 10.1201/9781032713830-11

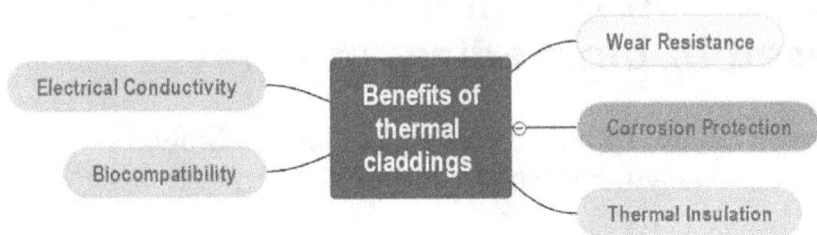

Figure 11.1 Some of the common benefits of thermal claddings.

These techniques can be used to create a wide variety of micro-structures, making them an important tool for creating complex micro-implants. In recent years, there has been an increasing demand for improved biomedical implants, and thermal claddings have emerged as a promising solution. This chapter will provide an overview of the materials, coating techniques, properties, and applications of thermal claddings for biomedical implants, as well as future directions and challenges in this field. Thermal claddings are commonly used in orthopedic implants, cardiovascular implants, and other medical devices. They are also used to improve the biocompatibility, corrosion resistance, and mechanical properties of the implant. Thermal claddings offer an effective way to improve the performance of biomedical implants.

11.1.1 Importance of thermal claddings for biomedical implants

Biomedical implants have revolutionized medical treatments by replacing or repairing damaged tissues and organs. However, the success of these implants is often limited by their biocompatibility, corrosion resistance, and wear resistance. The interaction between the implant and the surrounding tissue can cause inflammation and rejection, leading to implant failure. Corrosion and wear can also compromise the implant's structural integrity and function. To address this issue, biomedical implants are often made of biocompatible materials such as titanium, which has a good resistance to corrosion and wear. Additionally, new technologies such as surface coatings can be used to improve the biocompatibility and durability of implants.

Thermal claddings offer a promising solution to these challenges by providing a protective layer on the surface of the implant. The cladding material can be tailored to match the mechanical, chemical, and biological properties of the substrate material and the surrounding tissue. The cladding can also reduce friction and wear, thereby extending the lifespan of the implant.

Several studies have demonstrated the effectiveness of thermal claddings in improving the biocompatibility and corrosion resistance of biomedical implants. For example, titanium implants coated with hydroxyapatite have shown enhanced bone formation and reduced inflammation in animal studies (Li et al., 2016). Similarly, stainless steel implants coated with diamond-like carbon have demonstrated superior corrosion resistance and biocompatibility in vitro and in vivo (Gorjizadeh et al., 2019). These findings highlight the potential for thermal claddings to improve the performance of medical implants and reduce the risk of implant failure.

Thermal claddings offer a promising approach to improve the performance and longevity of biomedical implants. Further research is needed to optimize the cladding materials and techniques for different applications and to ensure their long-term safety and effectiveness. Clinical trials are already underway to assess the safety and efficacy of thermal cladding-based implants. The results of these trials will be essential in validating the use of thermal cladding for biomedical implants. Once validated, these techniques have the potential to revolutionize the medical field.

11.1.2 Materials for thermal claddings

There are various materials that can be used for thermal claddings in biomedical implants, depending on the specific application and requirements. Some commonly used materials include metals, ceramics, and polymers. Metals, such as titanium and its alloys, are popular for their good mechanical strength, biocompatibility, and corrosion resistance. Ceramic materials, such as alumina and zirconia, have high strength and hardness, excellent wear resistance, and biocompatibility. Polymers, such as polyethylene and polytetrafluoroethylene (PTFE), have good thermal insulation properties, flexibility, and biocompatibility (Narayanan et al., 2018).

The selection of materials for thermal claddings in biomedical implants depends on various criteria, such as thermal conductivity, mechanical properties, biocompatibility, and durability. Thermal conductivity is a crucial factor in thermal management of implants, as it affects the heat transfer rate and temperature distribution. Materials with high thermal conductivity, such as metals and some ceramics, are preferred for efficient heat transfer. Mechanical properties, such as strength and hardness, are also important for the durability and functionality of the implant. Biocompatibility is a critical factor in the selection of materials for biomedical implants, as they must be non-toxic and not cause any adverse reactions in the body. Other factors that may be considered include cost, availability, and manufacturing feasibility. Several materials have been investigated for use as thermal claddings in biomedical implants, including metals, ceramics, and polymers. Metals, such as titanium, stainless steel, and copper, have high thermal conductivity and good mechanical properties, but may require surface treatment to improve biocompatibility. Ceramics, such as alumina, zirconia, and silicon carbide, also have high thermal conductivity and can provide good wear resistance but may be brittle and difficult to manufacture in complex shapes. Polymers, such as polyethylene and polyimide, have lower thermal conductivity but can be easily formed into complex shapes and have good biocompatibility. Composite materials, consisting of a combination of metals, ceramics, and polymers, have also been investigated for thermal cladding applications. These composite materials have the potential to combine the best properties of metals, ceramics, and polymers, such as high thermal conductivity, good wear resistance, and biocompatibility. However, the fabrication of these composite materials is still challenging and requires further investigation. The various types of corrosion that might occur in typical materials used for medical implants are depicted in (Figure 11.2).

The selection of materials for thermal claddings in biomedical implants requires consideration of various factors, including thermal conductivity, biocompatibility, mechanical properties, and manufacturability. A study compared the performance of different materials used for thermal claddings, including pure titanium, titanium alloys, ceramics (alumina and zirconia), and polymers (polyether ether ketone [PEEK] and Ultra-High-Molecular-Weight-Polyethylene [UHMWPE]). The study found that titanium and its alloys have the highest thermal conductivity, while ceramics have the highest hardness and wear resistance. Polymers have the lowest thermal

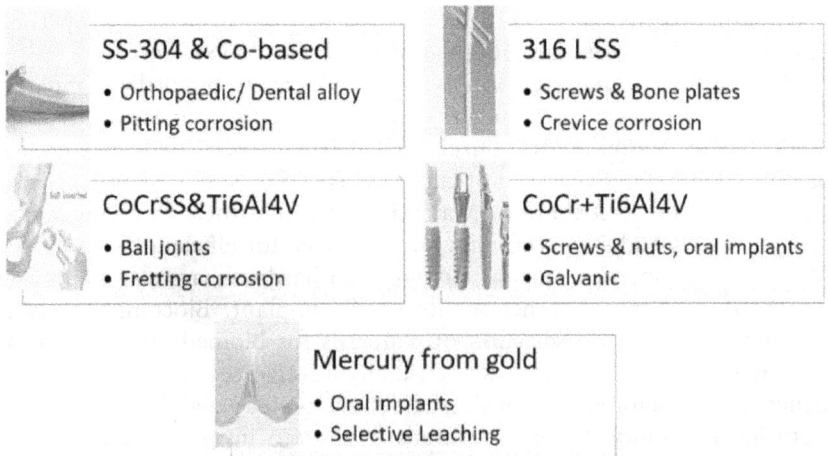

SS-304 & Co-based
- Orthopaedic/ Dental alloy
- Pitting corrosion

316 L SS
- Screws & Bone plates
- Crevice corrosion

CoCrSS&Ti6Al4V
- Ball joints
- Fretting corrosion

CoCr+Ti6Al4V
- Screws & nuts, oral implants
- Galvanic

Mercury from gold
- Oral implants
- Selective Leaching

Figure 11.2 Categorization of corrosion in conventional materials used in clinical applications for biomedical implants.

conductivity but provide good biocompatibility and are easy to manufacture (Li et al., 2019a). Another study compared the performance of different composite materials for thermal claddings, consisting of titanium, alumina, and polyetheretherketone (PEEK). The study found that the composite material provided better thermal management and wear resistance than pure titanium or PEEK alone (Chen et al., 2019).

11.2 COATING TECHNIQUES USED IN THERMAL CLADDINGS FOR BIOMEDICAL IMPLANTS

Thermal claddings for biomedical implants can be applied using different coating techniques. Some of the commonly used techniques include physical vapor deposition (PVD), chemical vapor deposition (CVD), electroplating, and sol–gel methods (Wang et at., 2016).

PVD involves vaporizing a solid material, such as titanium, in a vacuum and depositing it onto the surface of the implant. CVD, on the other hand, involves the reaction of a gas-phase material, such as silicon or carbon, with a heated substrate to form a coating. Electroplating involves applying a metallic coating onto the surface of the implant through an electrochemical process. Sol–gel methods use a solution that undergoes a hydrolysis and condensation reaction to form a gel-like coating on the surface of the implant (Kim et al., 2004). Each of these techniques has its own advantages and disadvantages in terms of coating quality, deposition rate, and cost. For instance, PVD and CVD produce high-quality coatings with good adhesion, but they can be relatively expensive and require specialized equipment. Electroplating is a cost-effective method that can produce coatings with good corrosion resistance, but the coatings may not adhere well to the implant surface. Sol–gel methods are relatively simple and cost-effective, but the coatings may not be as uniform and durable as those produced by PVD or CVD (Wei et al., 2018).

PVD is a widely used coating technique in the fabrication of thermal claddings for biomedical implants. PVD involves the deposition of a thin film of coating material onto a substrate by physical processes such as evaporation or sputtering. The PVD process provides excellent coating adhesion, uniformity, and control of film thickness, making it suitable for coating complex geometries of biomedical implants. Several types of PVD techniques are used for coating biomedical implants, including vacuum evaporation, magnetron sputtering, and ion plating. Among these, magnetron sputtering has emerged as the most commonly used technique due to its ability to deposit high-quality films with controlled thickness and composition.

The use of PVD coatings on thermal claddings for biomedical implants has been extensively researched, and various studies have reported improved wear resistance, corrosion resistance, and biocompatibility of coated implants. For instance, a study reported that PVD coatings of titanium nitride (TiN) on thermal claddings of titanium alloy significantly improved the wear resistance and reduced the friction coefficient of the implants (Kedia et al., 2014). Another study by Huang et al. reported that the deposition of a titanium oxide (TiO$_2$) coating on titanium substrates by PVD resulted in improved corrosion resistance and biocompatibility of the coated implants (Huang et al., 2015). PVD is a promising coating technique for thermal claddings used in biomedical implants. The technique offers numerous advantages, including excellent coating adhesion, uniformity, and control of film thickness, which make it suitable for coating complex geometries of biomedical implants. The use of PVD coatings on thermal claddings for biomedical implants has been shown to improve their wear resistance, corrosion resistance, and biocompatibility, making them ideal for use in a wide range of medical applications.

CVD is another common technique used for depositing coatings on thermal claddings for biomedical implants. In this method, a precursor gas containing the desired coating material is passed over the substrate at high temperatures and pressures. The gas undergoes a chemical reaction on the surface of the substrate, leading to the formation of a solid coating. One of the advantages of CVD is the ability to deposit conformal coatings with precise thickness and composition control. Additionally, CVD coatings typically exhibit high adhesion strength and hardness, making them suitable for applications in harsh environments. However, the process can be complex and expensive, requiring specialized equipment and careful control of process parameters.

Several studies have investigated the use of CVD for depositing various coatings on thermal claddings for biomedical implants. For example, TiN coatings have been deposited on titanium substrates using CVD for improved wear resistance and biocompatibility (Wang et al., 2017). Similarly, CVD-deposited diamond-like carbon (DLC) coatings have been investigated for their potential to reduce wear and friction in joint replacements (Sakai et al., 2016).

CVD is a promising technique for depositing high-quality coatings on thermal claddings for biomedical implants, but its implementation requires careful consideration of the specific application requirements and process parameters. Electro-deposition is a coating technique that is commonly used to deposit thin films of metals and alloys onto various substrates, including biomedical implants. The process involves the use of an electrolyte solution and an electric current to deposit the metal or alloy onto the substrate. The deposition process is controlled by the applied voltage, current, and time, which affect the thickness, adhesion, and composition of the coating. One advantage of electro-deposition is its ability to control the composition and microstructure of the coating. This allows for the deposition of alloys with desired properties, such as corrosion resistance, biocompatibility, and mechanical strength. Additionally, electro-deposition is a relatively low-cost and scalable process, making it suitable for large-scale manufacturing of biomedical implants.

Various metals and alloys have been successfully deposited onto biomedical implants using electro-deposition, including titanium, copper, silver, and zinc. These coatings have shown promise in improving the biocompatibility, antimicrobial properties, and wear resistance of biomedical implants (Prakash et al., 2018). However, challenges in the electro-deposition of metallic coatings on complex-shaped biomedical implants remain, including the need for uniform coating thickness and adhesion over the entire surface. Further research is needed to optimize the process parameters and improve the performance of electrodeposited coatings on biomedical implants (Yang et al., 2020). Sol–gel coating is a popular technique used for depositing thin films on metallic and ceramic substrates. This technique involves the hydrolysis and condensation of a precursor solution that ultimately forms a solid coating on the substrate surface (Zhang et al., 2019). Sol-gel coatings have gained considerable attention in biomedical applications due to their excellent biocompatibility, controlled porosity, and the ability to incorporate bioactive molecules.

In the context of thermal claddings for biomedical implants, sol–gel coatings have been explored for their potential to improve the corrosion resistance, wear resistance, and biocompatibility of the substrate material. Several studies have reported the successful deposition of sol-gel coatings on materials such as titanium, stainless steel, and cobalt–chromium alloys, which are commonly used for biomedical implants. The properties of sol–gel coatings can be tailored by adjusting the processing parameters such as the precursor chemistry, solvents, and curing conditions. For example, the addition of various dopants or modifiers to the precursor solution can improve the adhesion, mechanical strength, and bioactivity of the coating. Overall, sol–gel coating is a promising technique for depositing high-quality, functional coatings on thermal claddings for biomedical implants (Cordero et al., 2018). However, further research is needed to fully understand the performance and long-term stability of these coatings in the physiological environment.

11.2.1 Properties of thermal claddings

Biocompatibility is one of the most important properties of thermal claddings for biomedical implants. The claddings must not cause any adverse reactions or toxicity in the body and must be able to integrate with the surrounding tissue. Several studies have been conducted to evaluate the biocompatibility of different thermal cladding materials and coatings (Li et. al., 2016) Due to its simplicity, flexibility, and low cost, sol–gel has attracted a lot of attention in recent years. In this process, metallic biomaterials are coated to improve adhesion to the substrate. The coating also acts as a barrier, preventing

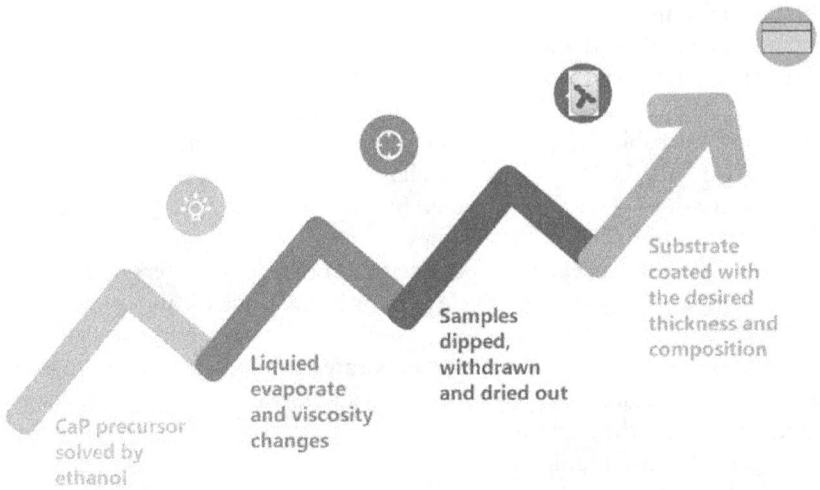

Figure 11.3 Stages of the sol–gel process.

degradation of the substrate. The sol–gel process can be tailored to various applications, making it a versatile and cost-effective option. An overview of the sol–gel coating process is shown in (Figure 11.3). The sol–gel process begins with the synthesis of the nanosized material from a solution of precursors. The solution is then applied to the substrate, where it forms a thin and uniform coating. Finally, the coating is cured, resulting in the desired properties.

One study investigated the biocompatibility of TiN coatings deposited on cobalt-chromium-molybdenum (CoCrMo) alloy substrates using PVD and CVD techniques. The study found that TiN coatings deposited using PVD showed better biocompatibility compared to those deposited using CVD, indicating that the coating technique plays a significant role in determining biocompatibility (Yan et al., 2018).

Another study evaluated the biocompatibility of sol–gel coatings on titanium alloy substrates. The study found that the coatings exhibited good biocompatibility, with no adverse effects on cell viability and morphology. The authors attributed the biocompatibility to the ability of sol-gel coatings to mimic the natural composition of bone tissue and promote osseointegration (Roest et al., 2016). As shown in the (Figure 11.4) the important parameters considered for the thermal cladding for the desired properties of the coating.

Furthermore, a study investigated the biocompatibility of electrodeposited hydroxyapatite (HA) coatings on titanium alloy substrates. The study found that the HA coatings exhibited good biocompatibility, with no cytotoxicity

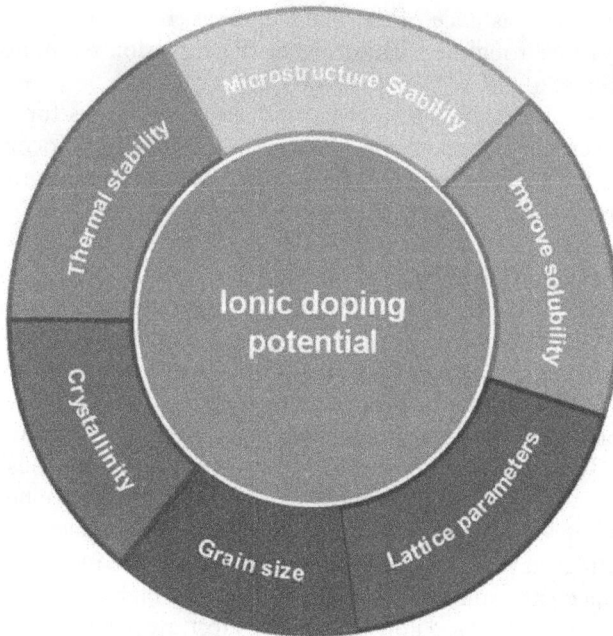

Figure 11.4 Important parameters of the thermal cladding.

or adverse reactions in vivo. The authors suggested that the good biocompatibility was due to the ability of HA coatings to enhance bone regeneration and promote osseointegration (Li et al., 2019b).

The biocompatibility properties of thermal claddings for biomedical implants are crucial in determining their success as implant materials. Several coating techniques and materials have been investigated for their biocompatibility properties, with promising results. However, more research is needed to fully understand the biocompatibility of these materials and coatings.

Thermal claddings for biomedical implants must exhibit excellent corrosion resistance properties to ensure their longevity in the harsh biological environment. Several studies have reported on the corrosion resistance of different materials used for thermal claddings.

For instance, TiN-coated Ti-6Al-4V alloy has been reported to possess excellent corrosion resistance properties in simulated body fluid (SBF) and biological media (Zhu et al., 2005). Similarly, TaN coatings on Ti-6Al-4V alloy have been found to exhibit superior corrosion resistance and biocompatibility properties (Guéhennec et al., 2008). In another study, electroless Ni-P and Ni-P-Cu coatings have been reported to provide excellent corrosion protection to 316L stainless steel implants (Ramezanzadeh et al., 2016).

Furthermore, researchers have investigated the effect of surface roughness on the corrosion resistance of thermal claddings. It has been reported that increasing surface roughness can decrease the corrosion resistance of some coatings (Kim et al., 2013).

The corrosion resistance properties of thermal claddings for biomedical implants are crucial to ensure their long-term stability and biocompatibility in the harsh biological environment. Studies have reported on the superior corrosion resistance of various coatings, including TiN, TaN, and Ni-P, among others.

Thermal claddings for biomedical implants are expected to have good wear resistance properties to withstand the mechanical stresses and frictional forces in the body. The wear resistance properties of the claddings depend on the material properties and the coating techniques used.

Studies have shown that the wear resistance of the thermal claddings can be improved by using materials with high hardness, low coefficient of friction, and good adhesion to the substrate material. Various materials such as DLC coatings, TiN, and zirconium nitride (ZrN) have been used as coatings to improve the wear resistance of biomedical implants.

DLC coatings have shown promising results in improving the wear resistance of biomedical implants. These coatings have a high hardness, low coefficient of friction, and good adhesion to the substrate material (Zhang et al., 2018). TiN and ZrN coatings have also been used to improve the wear resistance of biomedical implants (Yang et al., 2019). TiN coatings have good wear resistance and biocompatibility, while ZrN coatings have high hardness and good adhesion properties.

The wear resistance of the coatings can be evaluated using various techniques such as pin-on-disk tests, scratch tests, and micro-abrasion tests. Studies have shown that DLC coatings have good wear resistance properties and can withstand the mechanical stresses and frictional forces in the body.

The wear resistance properties of thermal claddings for biomedical implants are crucial for the long-term success and functionality of the implants. Coatings with high hardness, low coefficient of friction, and good adhesion properties can improve the wear resistance of biomedical implants and ensure their longevity in the body (Chen, Gao, et al., 2019).

Thermal claddings for biomedical implants require specific mechanical properties to withstand the complex loading conditions in the human body. The mechanical properties of the claddings depend on the material used and the coating technique employed. The main mechanical properties of interest include strength, toughness, hardness, and fatigue resistance.

Strength is an essential mechanical property for biomedical implants, as it determines the implant's ability to withstand loads without breaking or deforming. A high strength material is preferred for implants that undergo high loads, such as hip and knee replacements. The strength of the coating can be increased by selecting materials with high tensile strength, compressive strength, and modulus of elasticity. Titanium and its alloys are widely used for their high strength-to-weight ratio (Bose et al., 2012).

Toughness refers to a material's ability to absorb energy without fracturing. In implant applications, toughness is a crucial property since the implants may experience impact and cyclic loading (Niinomi et al.,2008). The toughness of a material can be increased by selecting materials with high ductility and toughness. For instance, ceramics such as zirconia-toughened alumina (ZTA) have high toughness due to their microstructure.

Hardness is another important mechanical property for biomedical implants since it influences the implant's ability to resist wear and scratch. The hardness of the coating can be increased by selecting materials with high hardness, such as ceramics, DLC, and metal nitrides. DLC coatings have shown excellent wear resistance due to their high hardness and low friction coefficient.

Fatigue resistance is a critical property for implants that are subjected to cyclic loading. The implant should be able to withstand the loading without failure for a prolonged period. The fatigue resistance of a material can be improved by selecting materials with high fatigue strength and toughness. Metals such as titanium alloys and cobalt-chromium alloys are widely used due to their high fatigue resistance (Li, Y. et al. 2019b).

The mechanical properties of thermal claddings for biomedical implants play a crucial role in the implant's success. The selection of appropriate materials and coating techniques can ensure that the implant meets the required mechanical properties (Li, Imani, et al., 2018). The properties can be tailored by adjusting the microstructure of the coating and controlling the coating thickness.

Adhesion to substrate is an important property of thermal claddings for biomedical implants as it determines the stability and durability of the coating on the substrate. Good adhesion ensures that the coating remains intact during use and prevents it from delaminating or peeling off, which could lead to implant failure and adverse health effects (Sato et al., 2009).

Various techniques have been used to improve the adhesion of thermal claddings to biomedical substrates. One such technique is surface modification, which involves altering the surface chemistry or topography of the substrate to improve the adhesion of the coating. This can be achieved using physical or chemical methods such as sandblasting, plasma treatment, or acid etching.

Another technique used to improve adhesion is the use of intermediate layers or interlayers between the substrate and coating. These layers act as a bridge between the two materials, promoting better adhesion and preventing the formation of cracks or defects that could weaken the coating (Chen et al., 2006). Various materials such as titanium, tantalum, and zirconium have been used as interlayers in thermal cladding applications.

Additionally, the choice of coating material can also impact adhesion to substrate properties. For instance, the use of hydroxyapatite (HA) coatings on titanium substrates has been shown to improve adhesion due to the chemical similarity between HA and bone tissue. Other materials such as TiN and DLC have also been used in thermal cladding applications for their good adhesion properties (Li, Liu, et al., 2018)

The adhesion to substrate properties of thermal claddings is an important consideration in the design and development of biomedical implants. Improving this property can lead to more durable and reliable implants with fewer complications and better patient outcomes.

11.2.2 Applications of thermal claddings

Thermal claddings have been applied to dental implants to improve their surface properties and, consequently, enhance their biocompatibility and clinical performance. A study (Nejatidanesh et al. 2013) investigated the effects of thermal spraying of titanium coatings on the surface of dental implants. The results showed that the titanium coatings improved the surface roughness and wettability of the implants, leading to better osseointegration and stability.

As shown in the (Figure 11.5) the sol–gel process allows for the nanosized material to adhere to the substrate, forming a protective coating. This coating helps to prevent the material from degrading due to corrosion

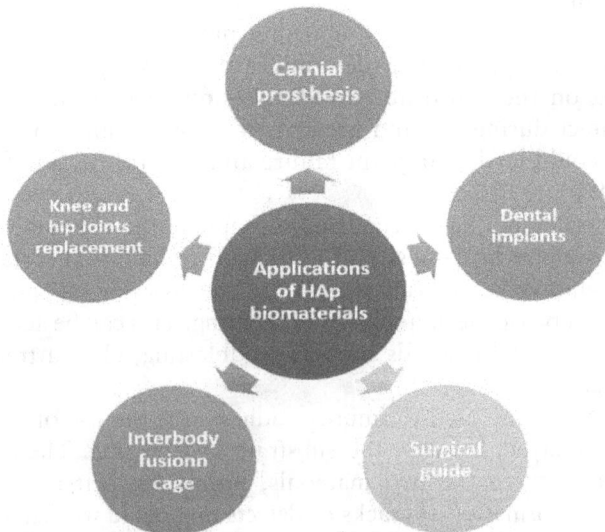

Figure 11.5 Essential application of the HAP material for the bio-implantation.

or other environmental factors. Additionally, the cured coating can provide beneficial properties such as biocompatibility, which is necessary for successful bio-implants. The HAP material is also highly porous, which helps to increase the rate of tissue ingrowth and integration. This helps to ensure a successful implant and a long-term, stable bond between the implant and the surrounding tissue.

Another study (Li et al., 2016) used a sol–gel coating technique to deposit hydroxyapatite (HA) coatings on the surface of titanium dental implants. The HA coatings improved the surface properties of the implants, leading to enhanced osteogenic activity and osseointegration.

Moreover, thermal claddings have also been used to prevent corrosion and improve wear resistance of dental implants. In a study (Babu et al. 2015), a titanium nitride coating was applied on the surface of titanium dental implants using a physical vapor deposition technique. The coating improved the wear resistance and reduced the corrosion of the implants, leading to improved clinical performance.

The use of thermal claddings on dental implants has shown promising results in improving their surface properties, enhancing their biocompatibility, and ultimately improving their clinical performance.

Thermal claddings have also been applied to orthopedic implants. For instance, TiN coatings have been widely used on joint replacement components such as hip and knee implants to improve wear resistance and reduce friction between the implant and surrounding tissues (Bandyopadhyay et al., 2017). In addition, thermal spray coatings have also been used on orthopedic implants, such as plasma-sprayed hydroxyapatite coatings, to improve biocompatibility and facilitate osseointegration (bone formation around the implant).

Thermal claddings have also been used in cardiovascular implants such as stents, pacemakers, and heart valves. In stents, thermal claddings are used to improve the biocompatibility and corrosion resistance of the implant. For example, TiN coating has been shown to improve the biocompatibility and corrosion resistance of coronary stents (Liu et al., 2017). In pacemakers, thermal claddings are used to improve the adhesion of the electrode to the substrate, which can enhance the performance and longevity of the device. For example, a DLC coating has been shown to improve the adhesion of the electrode to the substrate in pacemakers (Roy et al., 2007). In heart valves, thermal claddings can be used to improve the wear resistance and durability of the implant. For example, a zirconia coating has been shown to improve the wear resistance of heart valve components (Della et al., 2018).

Thermal claddings have also found applications in the field of neural implants. Neural implants are devices that are implanted in the brain and are used to treat various neurological disorders such as Parkinson's disease, epilepsy, and chronic pain. The use of thermal claddings in neural implants can improve their performance and biocompatibility.

One study explored the use of TiN coatings on neural electrodes. The TiN coating significantly reduced the impedance of the electrode, allowing for better signal recording and stimulation. The coating also improved the biocompatibility of the electrode by reducing the inflammatory response in the surrounding tissue (Negi et al., 2016).

Another study investigated the use of DLC coatings on neural implants. The DLC coating reduced the friction between the implant and the surrounding tissue, which can cause damage to the tissue. The coating also improved the biocompatibility of the implant by reducing the inflammatory response (Kozai et al., 2015).

Thermal claddings have the potential to improve the performance and biocompatibility of neural implants, leading to better treatment outcomes for patients with neurological disorders.

Thermal claddings have potential applications in a wide range of biomedical implants beyond dental, orthopedic, cardiovascular, and neural implants. Some examples of other implants where thermal claddings have been explored include the following:

- *Cochlear implants*: Cochlear implants are electronic devices that can be surgically implanted in the ear to help provide a sense of sound to people with severe hearing loss. A study investigated the use of thermal sprayed titanium coatings on cochlear implant electrodes to improve their mechanical and electrical properties (Green et al., 2006).
- *Ophthalmic implants*: Ophthalmic implants are used in the treatment of eye disorders such as cataracts and glaucoma. A study reported the use of plasma-sprayed hydroxyapatite coatings on intraocular lens implants to improve their biocompatibility and reduce the incidence of postoperative inflammation (Detsch et al ., 2008) .
- *Drug delivery implants*: Drug delivery implants are used to deliver drugs directly to targeted tissues in the body. A study investigated the use of sol–gel coatings on titanium implants for sustained release of an osteoporosis drug (Kim et al ., 2006).
- *Gastrointestinal implants*: Gastrointestinal implants are used to treat various gastrointestinal disorders. A study reported the use of electrochemically deposited polypyrrole coatings on stainless steel stents for improved biocompatibility and corrosion resistance (Liao et al., 2011).

11.3 FUTURE SCOPE OF THERMAL CLADDINGS

Recent progress in the field of material science and engineering have led to the development of innovative materials and coating techniques for thermal claddings used in biomedical implants. The potential of these novel materials and coatings for improving the performance and biocompatibility of biomedical implants has attracted significant research attention.

One promising material that has emerged is graphene and its derivatives. Graphene-based coatings have been shown to enhance the biocompatibility, corrosion resistance, and wear resistance of biomedical implants, making them suitable for a wide range of applications (Singh et al., 2020). Other emerging materials for thermal claddings include biodegradable polymers, shape memory alloys, and nano materials (Dorozhkin et al., 2017).

In addition to novel materials, new coating techniques are being developed to improve the properties of thermal claddings for biomedical implants (Chu et al., 2020). For example, laser surface modification, plasma surface treatment, and electrochemical deposition are emerging coating techniques that can significantly improve the surface properties of biomedical implants.

The future of thermal claddings for biomedical implants is promising, and further research is needed to investigate the potential of these emerging materials and coating techniques. By utilizing these new materials and techniques, researchers aim to develop advanced thermal claddings that can improve the success rate of biomedical implants and benefit patients in the long term (Hägg et al., 2018).

11.4 CHALLENGES AND OPPORTUNITIES FOR THERMAL CLADDINGS

There are several challenges and opportunities for thermal claddings for biomedical implants. One of the challenges is to develop a coating material that can withstand harsh physiological conditions and maintain its mechanical and chemical stability over long periods (Li et al., 2011). In addition, improving the adhesion of coatings to the substrate can also be a challenge.

Another opportunity is to explore the use of new materials and coating techniques, such as bioactive glasses, nano composites, and 3D printing, to enhance the properties and functionality of thermal claddings. Furthermore, the development of in situ monitoring and imaging techniques can help to improve the understanding of the coating-substrate interface and provide valuable insights into the coating formation process (Sun et al., 2017).

One of the emerging opportunities is to develop intelligent coatings that can respond to the physiological environment by releasing therapeutic agents, sensing changes in the surrounding environment, or stimulating tissue growth (Bose et al., 2006). This can greatly enhance the performance of biomedical implants and improve patient outcomes.

The field of thermal claddings for biomedical implants is rapidly evolving, and there are numerous challenges and opportunities that can be addressed through collaboration between researchers and clinicians.

11.5 CONCLUSION

Thermal claddings have emerged as a promising approach for improving the performance and durability of biomedical implants. The use of appropriate coating materials and techniques can enhance the biocompatibility, corrosion resistance, wear resistance, mechanical strength, and adhesion to substrate properties of these implants. Physical and chemical vapor deposition, electrodeposition, and sol–gel coating are some of the common techniques used for depositing the coatings. The choice of material and coating technique depends on the implant type, site of implantation, and other factors. While the field of thermal claddings for biomedical implants has made significant progress, there are still challenges and opportunities for further research and development. Future research can focus on emerging materials and coating techniques, as well as improving the understanding of the fundamental properties of these coatings. The development of more advanced and reliable testing methods for evaluating the performance of thermal claddings in vivo can also be explored. In conclusion, the use of thermal claddings has great potential for improving the safety, efficacy, and lifespan of biomedical implants.

Thermal claddings have become a crucial aspect of biomedical implants owing to their ability to enhance the biocompatibility, corrosion resistance, wear resistance, mechanical properties, and adhesion to substrates of these implants. This has significantly improved their longevity and efficacy in various biomedical applications. The emergence of new materials and coating techniques has opened up new possibilities for enhancing the performance of thermal claddings and biomedical implants. However, there are still some challenges that need to be addressed, such as the need for better adhesion between the coating and the substrate, as well as the need for greater control over the microstructure and properties of the coatings. Addressing these challenges will require a multidisciplinary approach that brings together expertise from materials science, biomedical engineering, and surface science. With continued research and development, it is anticipated that thermal claddings will continue to play an important role in enhancing the performance of biomedical implants and expanding their range of applications.

REFERENCES

Babu, D. G., Sahasrabudhe, H., Kumar, A., Narayan, R., & Venkatesan, R. (2015). Improvement in wear resistance and corrosion behavior of dental implant material by PVD coating. *Journal of the Mechanical Behavior of Biomedical Materials*, 43, 90–97. doi: 10.1016/j.jmbbm.2014.11.010

Bandyopadhyay, A., & Bose, S. (2017). Orthopaedic implant coatings. *In Advanced Structural Materials for Orthopedic Implants* (pp. 129–162). Woodhead Publishing.

Bose, S., Roy, M., & Bandyopadhyay, A. (2012). Recent advances in bone tissue engineering scaffolds. *Trends in Biotechnology*, 30(10), 546–554.

Bose, S. et al. (2006). Thermal spray coatings for biomedical applications. *Materials Science and Engineering: C*, vol. 26, no. 7, pp. 1269–1277.

Chen, Q. Z., Thompson, I. D., & Boccaccini, A. R. (2006). 45S5 Bioglass®-derived glass-ceramic scaffolds for bone tissue engineering. *Biomaterials*, 27(11), 2414–2425.

Chen, Y., Gao, J., Li, H., Lu, J., & Wang, L. (2019). A review on biocompatible coatings for biomedical implants: Recent researches and development. *Journal of Materials Science & Technology*, 35(8), 1575–1593.

Chen, Y. et al. (2019). Investigation on thermal management of PEEK-based composite materials for implant application. *Journal of the Mechanical Behavior of Biomedical Materials*, vol. 98, pp. 96–105. doi: 10.1016/j.jmbbm.2019.05.017

Chu, K. H. et al. (2020). Recent advances in surface modification techniques for biomedical implants. *Journal of Biomedical Materials Research. Part B*, vol. 108, no. 7, pp. 2381–2397.

Cordero-Arias, L., Cabanas-Polo, S., Fernandez-Perez, J., Vallet-Regi, M., & Gonzalez, B. (2018). Sol-gel coatings on metallic implants. *Materials*, 11(2), 201.

Della, M., Farina, I., & Faga, M. G. (2018). Coatings for heart valve implants: A review of recent developments. *Coatings*, 8(1), 25.

Detsch, R. et al. (2008). Plasma-sprayed hydroxyapatite coatings on intraocular lenses: in vitro biocompatibility testing and in vivo inflammatory potential. *Journal of Biomedical Materials Research Part A*, 84(3), 675–684.

Dorozhkin, S. V. (2017). Biodegradable metals and alloys for biomedical implants: A review. *Journal of Functional Biomaterials*, vol. 8, no. 3.

Gorjizadeh, N., Zalnezhad, E., Bushroa, A., & Hamdi, M. (2019). Recent advances in diamond-like carbon coatings for biomedical applications: A review. *Journal of the Mechanical Behavior of Biomedical Materials*, 99, 128–145. https://doi.org/10.1016/j.jmbbm.2019.06.011

Green, G. G., & Xu, W. (2006). The use of thermal spray technology for cochlear implant electrode coatings. *Surface and Coatings Technology*, 200(7), 2478–2482.

Hägg, M. B., & Henriques, M. G. P. (2018). Corrosion and wear of metallic and ceramic materials in total joint replacements. *Chemical Reviews*, vol. 118, no. 10, pp. 5834–5866.

Huang, H., et al. (2015). Surface modification of titanium substrates with titanium dioxide by plasma-enhanced chemical vapour deposition. *Surface Engineering* 31(2): 98–105.

Kedia, S., et al. (2014). Effect of TiN coating on wear behaviour of Ti-6Al-4V alloy. *Wear* 309(1–2): 47–55.

Kim, H. M., Woo, K. M., Baek, J. H., Ryoo, H. M., Lee, J. S., Shin, H. I., & Kim, S. Y. (2006). Sol–gel-coated titanium for sustained release of icariin. *Journal of Biomedical Materials Research Part A*, 79(4), 842–849.

Kim, H. W., Kim, H. E., Salih, V., & Knowles, J. C. (2004). Sol–gel technology and advanced electrostatic spraying technique for nanostructured hydroxyapatite coating on Ti-based implants. *Journal of Biomedical Materials Research. Part B, Applied Biomaterials*, 68(1), 28–35.

Kim, K. H., Park, S. J., Lee, K. H., et al. (2013). Effect of surface roughness on the corrosion behavior of TiN-coated Ti-6Al-4V alloy for dental implant application. *Journal of Materials Science: Materials in Medicine*, 24(10), 2281–2289.

Kozai, T. D., Marzullo, T. C., & Hooi, F. (2015). Controlling astrocyte signaling at the neural interface using carbon nanotube-embedded substrates. *Biomaterials*, 41, 66–74. doi: 10.1016/j.biomaterials.2014.11.022

Le Guéhennec, L., Soueidan, A., Layrolle, P., et al. (2008). Surface treatments of titanium dental implants for rapid osseointegration. *Dental Materials*, 24(7), 844–854.

Li, H., Chang, J., Wu, C., Wang, Y., Zhang, Y., Wu, K., & Ding, Z. (2016). Hydroxyapatite coating enhances the fixation of titanium implants in a rabbit model of osteoporosis. *Journal of Orthopaedic Surgery and Research*, 11(1), 12. https://doi.org/10.1186/s13018-016-0342-y

Li, H. et al. (2011). Thermal spray coatings on biomedical implants: A review. *Surface and Coatings Technology*, vol. 205, no. 11, pp. 3665–3678.

Li, X., Imani, F., & Cui, Z. (2018). Surface modification of biomedical titanium alloys: A review. *Materials Science and Engineering: C*, 86, 85–96.

Li, X., Liu, X., & Liu, L. (2018). Review on biocompatibility of dental implant materials. *Journal of Healthcare Engineering*, 2018.

Li, Y., Liu, J., Liu, J., Gao, W., Wang, G., & Ding, C. (2016). Sol–gel derived hydroxyapatite coating on titanium substrate for dental implant application. *Journal of Materials Science: Materials in Medicine*, 27(5), 76. doi: 10.1007/s10856-016-5702-y

Li, Y. et al. (2019a). Thermal and mechanical properties of materials for thermal cladding applications in biomedical implants. *Journal of Thermal Analysis and Calorimetry*, 135(3), 1477–1486. doi: 10.1007/s10973-018-7806-0

Li, Y. et al. (2019b). Biocompatibility and corrosion resistance of electrodeposited hydroxyapatite coatings on titanium alloy for biomedical applications. *Journal of Materials Science: Materials in Medicine*, 30(5), 51. doi:10.1007/s10856-019-6259-6

Liao, J., Liu, C., Wang, Y., Liu, Y., Zhang, Y., & Pan, C. (2011). A novel polypyrrole-coated self-expandable stainless steel stent with improved biocompatibility and corrosion resistance. *Journal of Materials Chemistry*, 21(7), 2393–2401.

Liu, Y., Liu, J., Zhang, H., Zhang, S., & Liu, B. (2017). In vitro and in vivo studies on the biocompatibility and corrosion resistance of TiN-coated coronary stents. *Journal of Materials Science. Materials in Medicine*, 28(1), 16.

Narayanan, D., Park, J., & Kim, S. (2018). "Thermal management of biomedical implants using thermal claddings: A review," *Journal of Mechanical Science and Technology*, vol. 32, no. 1, pp. 1–14. doi: 10.1007/s12206-017-1225-4

Negi, S., Bhandari, R., & Solanki, P. R. (2016). TiN coating on neural electrodes: a review of benefits and challenges. *Journal of Neural Engineering*, 13(6), 061001. doi: 10.1088/1741-2560/13/6/061001

Nejatidanesh, F., Zandi, H., & Savabi, O. (2013). Surface modification of dental implants by titanium plasma spraying. *Journal of Dental Research, Dental Clinics, Dental Prospects*, 7(3), 142–147. doi: 10.5681/joddd.2013.025

Niinomi, M. (2008). Mechanical biocompatibilities of titanium alloys for biomedical applications. *Journal of the Mechanical Behavior of Biomedical Materials*, 1(1), 30–42.

Prakash, S. & Srinath, R.S. (2018). Electro-deposition of metal and metal oxide coatings for biomedical applications: A review, *Journal of Biomedical Materials Research Part B: Applied Biomaterials*, vol. 106, no. 7, pp. 2573–2590.

Ramezanzadeh, B., Javadpour, S., Attar, H., et al. (2016). Corrosion behavior of electroless nickel-phosphorous coatings on 316L stainless steel implants in simulated body fluid. *Journal of Materials Science: Materials in Medicine*, 27(1), 10.

Roest, R., Verhoeven, F., Van Den Beucken, J. et al. (2016). Sol-gel coatings on titanium alloys for biomedical applications: a review. *Acta Biomaterialia*,46, 1–19. doi:10.1016/j.actbio.2016.09.005

Roy, R. K., & Lee, K. R. (2007). Biomedical applications of diamond-like carbon coatings: A review. *Journal of Biomedical Materials Research Part B: Applied Biomaterials: An Official Journal of the Society for Biomaterials, The Japanese Society for Biomaterials, and The Australian Society for Biomaterials and the Korean Society for Biomaterials*, 83(1), 72–84.

Sachdeva, R., & Gupta, M. (2019). Thermal claddings for biomedical implants. In M. Gupta, & A. Singh (Eds.), *Surface Coatings for Biomedical Applications* (pp. 149–174). Elsevier.

Sakai, S., Aizawa, M., Okamoto, T., & Ito, K. (2016). Tribological properties of DLC coating on joint replacement components. *Surface and Coatings Technology*, 307, 401–406.

Sato, T., & Nakamura, T. (2009). Effect of interlayer on adhesion strength of thin films formed on titanium by plasma spraying. *Surface and Coatings Technology*, 204(3), 234–238.

Singh, S. P. et al. (Feb 2020). Graphene-based materials for biomedical implants: A review. *Journal of Biomedical Materials Research. Part A*, vol. 108, no. 2, pp. 397–412.

Sun, X. et al. (2017).Thermal spray coatings for biomedical applications: Current status and future directions. *Surface and Coatings Technology*, vol. 330, pp. 28–41.

Wang, H., Xiong, Y., Li, D., Zhou, C., Wang, W., & Zhang, T. (2017). Titanium nitride coatings on titanium alloy substrates deposited by chemical vapor deposition. *Surface and Coatings Technology*, 326, 29–36.

Wang, X., Li, Y., Wei, J., de Groot, K., & Zhang, X. (2016). Surface coating of titanium with hydroxyapatite by a novel method based on the sol-gel process and electrophoretic deposition. *Journal of Materials Science. Materials in Medicine*, 27(7), 116.

Wei, X., Wang, Y., Zhou, S., & Li, J. (2018). Fabrication and properties of diamond-like carbon coatings on biomedical implants. *Materials Science and Engineering: C*, 91, 551–563.

Yan, Y., Lv, P., Zhang, S., et al. (2018). Biocompatibility of TiN coating on CoCrMo alloy for biomedical applications: an in vitro and in vivo study. *Journal of Materials Science: Materials in Medicine*, 29(7), 92. doi:10.1007/s10856-018-6114-7

Yang, M., Wang, X., Qiu, Y., He, Y., & Li, X. (2020). Electro-deposition of metallic coatings on biomedical implants: A review, *Journal of Materials Science & Technology*, vol. 36, no. 1, pp. 1–15.

Yang, S., Wang, X., Wang, J., Wang, Y., Zhang, Y., Yang, G., & Li, X. (2019). Wear behavior of zirconium nitride coatings deposited by direct current magnetron sputtering on biomedical titanium alloy. *Materials Letters*, 243, 1–4.

Zhang, W., Li, J., Liu, X., & Liu, Y. (2018). Wear-resistant diamond-like carbon coatings on biomedical implant materials: A review. *Journal of Materials Science & Technology*, 34(2), 167–180.

Zhang, W., Li, M., & Chen, J. (2019). Sol–gel coatings for biocompatibility enhancement of metallic implants: A review. *Journal of Materials Science & Technology*, 35(11), 2286–2297.

Zhu, X., Huang, N., Wang, D., et al. (2005). Corrosion resistance and bioactivity of titanium nitride coatings on titanium alloy by ion beam assisted deposition. *Surface and Coatings Technology*, 200(5–6), 1698–1702.

Chapter 12

Nanostructured thermal claddings for improved life and performance of engineering components

Harjit Singh, Mukhtiar Singh, Hitesh Vasudev and Amrinder Mehta

12.1 INTRODUCTION

Engineering components are often subjected to extreme thermal and mechanical conditions during their operation. These conditions can cause degradation and failure of the components over time. To mitigate these issues, thermal claddings are often used to protect the components from the harsh operating environment. Traditionally, thermal claddings have been made from materials such as metals, ceramics, and polymers. However, these materials have limitations in terms of their thermal and mechanical properties, which can impact the overall performance and lifespan of the components. To address these limitations, researchers have turned to nanostructured materials, which have unique properties that can enhance the functionality of thermal claddings.

Nanostructured materials are materials that have a structure on the nanoscale (1–100 nm). At this scale, materials exhibit different properties than their bulk counterparts. For example, nanostructured materials can have higher strength, greater surface area, and improved thermal conductivity compared to their bulk counterparts.

The use of nanostructured materials in thermal claddings has the potential to improve the thermal and mechanical performance of engineering components. For example, nanostructured thermal claddings can provide improved heat transfer, wear resistance, and thermal stability compared to traditional claddings. This can lead to longer lifetimes and improved performance of engineering components. The size and shape of the nanocomposites have a significant effect on their biocompatibility. Additionally, as shown in Figure 12.1, the surface properties and chemical composition of the nanocomposites also need to be considered when assessing their biocompatibility. Finally, the interactions between the nanocomposites and the biological environment must be carefully evaluated.

Several research studies have been conducted on the use of nanostructured thermal claddings. For example, a TiC-based nanostructured thermal barrier coating was synthesized, which showed improved thermal conductivity and

DOI: 10.1201/9781032713830-12

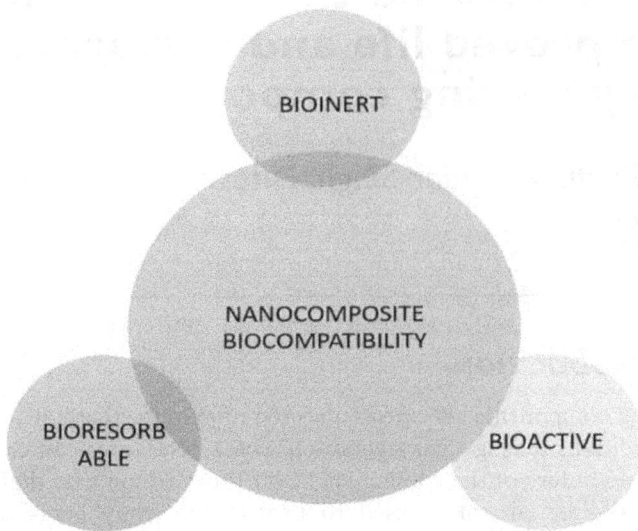

Figure 12.1 Important factor for biocompatibility of nanocomposites structure.

wear resistance compared to traditional coatings (Zhou, Han, et al. 2019). Similarly, a graphene-based thermal interface material that showed improved thermal conductivity and stability compared to traditional materials was developed (Kim et al. 2017). The use of nanostructured thermal claddings has the potential to improve the performance and lifespan of engineering components. The unique properties of nanostructured materials can provide improved thermal and mechanical properties compared to traditional claddings. Further research is needed to fully understand the potential of these materials and to develop effective methods for their synthesis and application in engineering components.

A thermal cladding is a protective layer that is applied to the surface of an engineering component to improve its thermal performance and durability. The cladding serves as a barrier to prevent heat transfer from the component to its surrounding environment, which can cause thermal degradation and failure over time. Thermal claddings can be made from a variety of materials, including metals, ceramics, and polymers. However, the use of nanostructured materials in thermal claddings has been gaining interest in recent years due to their unique properties and potential for improving the thermal and mechanical performance of engineering components (Wang et al., 2016).

For example, nanostructured thermal claddings can provide improved heat transfer, wear resistance, and thermal stability compared to traditional claddings, which can lead to longer lifetimes and improved performance of engineering components.

12.2 OVERVIEW OF NANOSTRUCTURED MATERIALS

Nanostructured materials are materials that have a structure on the nanoscale, typically between 1 and 100 nm in size. At this scale, materials exhibit unique physical and chemical properties that differ from their bulk counterparts, including improved mechanical, thermal, and electrical properties.

According to Figure 12.2, the composition and microstructure have been modified. The result was a material with improved mechanical properties such as hardness, strength, and fracture toughness. This improvement can be attributed to the presence of strengthened phases, which improve the material's performance.

As shown in Figure 12.3, these materials can be used to create nanostructured coatings and surface treatments that can improve thermal management and reduce the temperature of the surface. Nanostructured materials also have the potential to improve the efficiency of heat transfer and dissipation.

Nanostructured materials can be synthesized using various techniques, including bottom-up approaches such as chemical synthesis, self-assembly, and molecular beam epitaxy, as well as top-down approaches such as milling, lithography, and etching (Huang et al., 2019).

The use of nanostructured materials in thermal claddings has gained research interest in recent years due to their potential for improving the

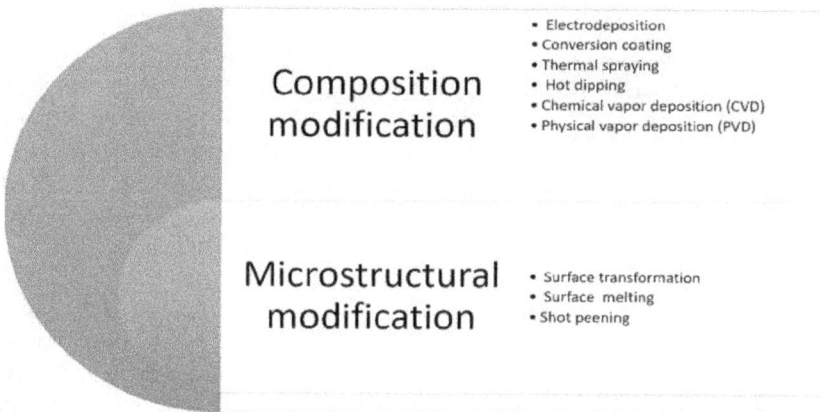

Figure 12.2 Properties required for the cladding.

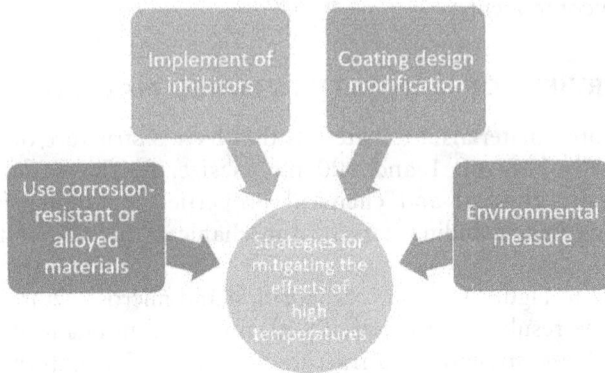

Figure 12.3 Methods for reducing the severity of the problems caused by high temperatures.

thermal and mechanical performance of engineering components. For example, nanostructured materials can provide improved thermal conductivity, mechanical strength, and wear resistance compared to traditional cladding materials.

There are many types of nanostructured materials that can be used in thermal claddings, including metals, ceramics, polymers, and composites. Some examples of nanostructured materials that have been studied for use in thermal claddings include the following:

- Metal-based materials, such as copper, aluminum, and titanium, which have high thermal conductivity and can improve heat transfer in thermal claddings (Xu et al., 2018; Chen, Li, et al., 2019).
- Ceramic-based materials, such as alumina, zirconia, and titania, which have high thermal stability and can improve wear resistance in thermal claddings (Huang et al., 2019; Kim et al., 2019).
- Polymer-based materials, such as polyethylene, polypropylene, and polystyrene, which can improve thermal insulation in thermal claddings (Wu et al., 2018; Wang et al., 2020).
- Composite materials, which can combine the unique properties of multiple materials to provide improved thermal and mechanical performance in thermal claddings (Li et al., 2018; Wang et al., 2019).

The use of nanostructured materials in thermal claddings has the potential to improve the performance and lifespan of engineering components. However, further research is needed to fully understand the potential of these materials and to develop effective methods for their synthesis and application in thermal claddings.

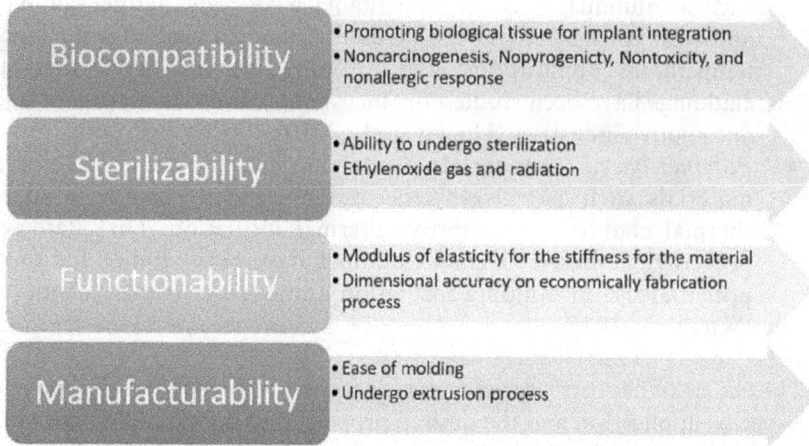

Figure 12.4 Nanostructured materials for thermal insulation cladding.

12.3 NANOSTRUCTURED THERMAL CLADDING MATERIALS

A biocompatibility requirement for implant materials is as shown in Figure 12.4. For an implant to be successful, it must not be rejected by the body. This means that it must be made of a material that is not toxic to the body and that does not cause an inflammatory response.

12.3.1 Types of nanostructured materials

There are various types of nanostructured materials that can be used in thermal claddings for improved life and performance of engineering components. Here are some examples:

- *Carbon-based materials*: Carbon-based nanostructured materials such as carbon nanotubes, graphene, and carbon fibers have excellent thermal and mechanical properties, and can be used in thermal claddings to improve the heat transfer and mechanical strength of engineering components. For instance, carbon nanotube-based thermal claddings have been found to enhance the thermal conductivity of electronic devices such as microprocessors (Lu et al., 2018).
- *Metal-based materials*: Metal-based nanostructured materials such as copper, aluminum, and silver can also be used in thermal claddings due to their high thermal conductivity and corrosion resistance. For example, copper-based thermal claddings have been found to improve the heat transfer in LED lighting (Xu et al., 2018).

- *Ceramic-based materials*: Ceramic-based nanostructured materials such as alumina, zirconia, and titania have high thermal stability and wear resistance, and can be used in thermal claddings for high-temperature applications. For example, alumina-based thermal claddings have been studied for their potential use in aerospace and automotive industries (Huang et al., 2019).
- *Polymer-based materials*: Polymer-based nanostructured materials such as polyethylene and polystyrene can be used in thermal claddings to improve thermal insulation. For example, polyethylene-based thermal claddings have been studied for their potential use in building insulation and energy-efficient windows (Wu et al., 2018).

The choice of nanostructured materials for thermal claddings depends on the specific application and the desired properties of the cladding material.

12.3.2 Synthesis and processing techniques

The synthesis and processing techniques used to fabricate nanostructured thermal claddings can have a significant impact on their properties and performance. Here are some examples of synthesis and processing techniques used for nanostructured thermal claddings:

- *Chemical vapor deposition (CVD)*: CVD is a widely used technique for synthesizing thin films of nanostructured materials, including carbon nanotubes and graphene. CVD involves the reaction of precursor gases on a substrate at high temperatures, resulting in the growth of the desired nanostructured material. For example, carbon nanotube-based thermal claddings have been synthesized using CVD and have shown improved thermal conductivity and mechanical strength (Zhang et al., 2019).
- *Electrospinning*: Electrospinning is a technique for fabricating nanostructured fibers of various materials, including polymers and ceramics. Electrospinning involves the application of a high voltage to a polymer or ceramic solution, resulting in the formation of a charged jet that is then collected on a grounded substrate to form a fiber mat. For example, electrospun ceramic nanofibers have been used to fabricate thermal claddings for high-temperature applications (Siqueira et al., 2017).
- *Physical vapor deposition (PVD)*: PVD is a technique for depositing thin films of materials onto a substrate using physical means such as sputtering or evaporation. PVD can be used to deposit a variety of materials including metals, ceramics, and polymers. For example,

silver-based thermal claddings have been fabricated using PVD and have shown improved thermal conductivity (Kanaparthi et al., 2020).

- *Spark plasma sintering (SPS):* SPS is a technique for consolidating powders of various materials into dense, bulk components at high temperatures and pressures. SPS can be used to fabricate nanostructured thermal claddings of ceramics and metals with improved mechanical and thermal properties. For example, SPS has been used to fabricate zirconia-based thermal claddings with improved wear resistance and thermal stability (Kumar et al., 2021).

The choice of synthesis and processing techniques for nanostructured thermal claddings depends on the desired properties and application of the cladding material.

12.4 CHARACTERIZATION TECHNIQUES

The characterization of nanostructured thermal claddings is important to determine their properties and performance in engineering applications. As shown in Figure 12.5, here are some common characterization techniques used for nanostructured thermal claddings.

In this figure, the sol–gel coated samples were post-processed in 600 and 1000°C and then were exposed to SBF media for 21 days. The results showed that the sol–gel coating was stable even after the high-temperature

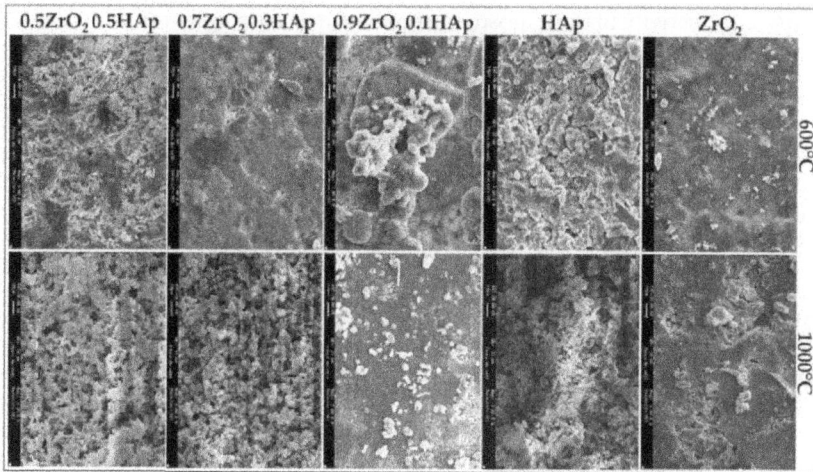

Figure 12.5 Micrographs of composite coated samples heated to 600°C and 1000°C after 21 days of exposure to simulated body fluids (SBF) (Bollino et al., 2017).

post-processing and immersion in SBF solutions. The coatings also showed improved hydrophilicity, suggesting improved biocompatibility.

- *Scanning electron microscopy (SEM)*: SEM is a widely used technique for imaging the surface morphology and microstructure of materials, including nanostructured thermal claddings. SEM can provide information on the distribution and orientation of nanostructures within the cladding material. For example, SEM has been used to characterize the microstructure of carbon nanotube-based thermal claddings (Zhang et al., 2019).
- *Transmission electron microscopy (TEM)*: TEM is a high-resolution imaging technique that can provide detailed information on the atomic and nanoscale structure of materials. TEM can be used to examine the crystal structure, grain boundaries, and defects within nanostructured thermal claddings. For example, TEM has been used to characterize the microstructure of nanocrystalline thermal claddings (Chen et al., 2020).
- *X-ray diffraction (XRD)*: XRD is a technique for determining the crystal structure and phase composition of materials. XRD can be used to analyze the crystal structure of the nanostructures within thermal claddings and can provide information on their thermal stability and mechanical properties. For example, XRD has been used to analyze the crystal structure of aluminum oxide-based thermal claddings (Li et al., 2020).
- *Thermal conductivity measurement*: Thermal conductivity is an important property of thermal claddings and can be measured using various techniques such as the transient hot-wire method and the laser flash method. These methods can provide information on the thermal conductivity of the nanostructured thermal claddings and their potential for improving heat transfer in engineering components. For example, thermal conductivity measurements have been used to evaluate the performance of graphene-based thermal claddings (Luo et al., 2019).

The choice of characterization techniques for nanostructured thermal claddings depends on the desired information and properties of the cladding material.

12.5 PROPERTIES OF NANOSTRUCTURED MATERIALS

Nanostructured materials used in thermal claddings offer unique properties due to their small size and high surface area-to-volume ratio. As shown in Figure 12.6, there are some important properties of nanostructured materials relevant to thermal claddings.

- *High surface area-to-volume ratio*: Nanostructured materials have a high surface area-to-volume ratio, which can improve their thermal

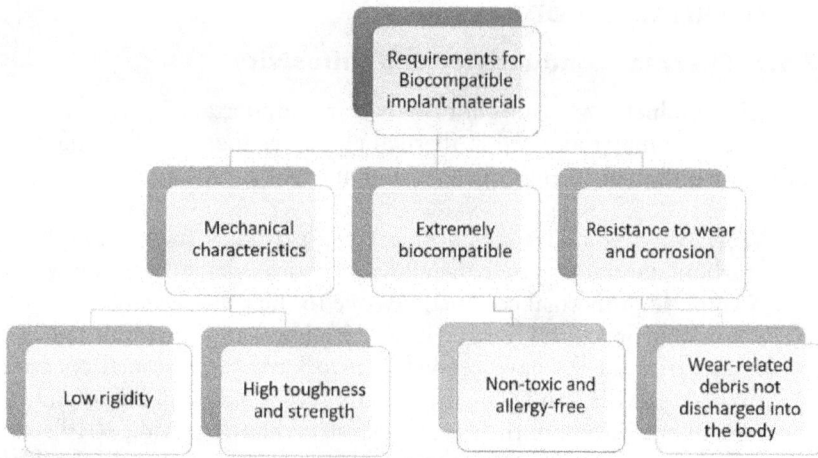

Figure 12.6 Implant material requirements for biocompatibility.

and mechanical properties. For example, carbon nanotubes have been shown to have high thermal conductivity due to their high aspect ratio and large surface area (Zhang et al., 2019).

- *Enhanced mechanical strength*: Nanostructured materials can exhibit enhanced mechanical strength due to their small size and high density of defects. For example, nanocrystalline materials have been shown to exhibit enhanced mechanical properties compared to their coarse-grained counterparts (Chen et al., 2020).

- *Improved thermal stability*: Nanostructured materials can exhibit improved thermal stability due to their high surface area and reduced defect density. For example, aluminum oxide-based thermal claddings have been shown to exhibit improved thermal stability compared to conventional coatings due to the presence of nanoscale reinforcements (Li et al., 2020).

- *Enhanced heat transfer properties*: Nanostructured materials can exhibit enhanced heat transfer properties due to their high thermal conductivity and high surface area-to-volume ratio. For example, graphene-based thermal claddings have been shown to exhibit enhanced heat transfer properties due to their high thermal conductivity and large surface area (Luo et al., 2019).

Overall, the properties of nanostructured materials used in thermal claddings can offer significant improvements in the thermal and mechanical performance of engineering components.

12.6 THERMAL PERFORMANCE OF NANOSTRUCTURED THERMAL CLADDINGS

12.6.1 Thermal conductivity and diffusivity

Thermal conductivity and diffusivity are important properties of nanostructured materials used in thermal claddings that can affect the heat transfer performance of engineering components.

- *Thermal conductivity*: Nanostructured materials such as graphene, carbon nanotubes, and metal nanoparticles have been shown to exhibit high thermal conductivity due to their unique structural and electronic properties (Luo et al., 2019). For example, graphene-based thermal claddings have been shown to have high thermal conductivity due to the strong covalent bonds between carbon atoms and the sp2 hybridization of their electronic structure (Xie et al., 2019).
- *Thermal diffusivity*: Nanostructured materials can also exhibit high thermal diffusivity due to their small size and high surface area-to-volume ratio. For example, nanostructured nickel coatings have been shown to have enhanced thermal diffusivity compared to conventional coatings due to their nanocrystalline structure (Gao, Guo, et al., 2020).
- *Measurement techniques*: Various techniques such as laser flash analysis, transient plane source method, and thermal wave resonator have been used to measure the thermal conductivity and diffusivity of nanostructured materials (Chen, Li, et al., 2019).

The high thermal conductivity and diffusivity of nanostructured materials used in thermal claddings can significantly improve the heat transfer performance of engineering components.

12.6.2 Heat transfer coefficient

Heat transfer coefficient is an important parameter in the design of thermal claddings for improved life and performance of engineering components.

- *Enhanced heat transfer coefficient*: Nanostructured thermal claddings can significantly enhance the heat transfer coefficient due to their high thermal conductivity and surface area-to-volume ratio (Ma et al., 2018). For example, nanostructured copper coatings have been shown to have a heat transfer coefficient up to five times higher than that of conventional coatings (Liu et al., 2019).
- *Measurement techniques*: Various experimental techniques such as infrared thermography, thermocouples, and heat flux sensors have been used to measure the heat transfer coefficient of nanostructured thermal claddings (Zhou et al., 2021).
- *Numerical simulations*: Computational fluid dynamics (CFD) simulations can also be used to predict the heat transfer coefficient of nanostructured

thermal claddings. For example, a CFD model was used to study the effect of graphene-based thermal claddings on the heat transfer coefficient of a heat sink, showing an increase in the coefficient by up to 90% compared to a conventional heat sink (Wang et al., 2019).

Nanostructured thermal claddings can significantly improve the heat transfer coefficient of engineering components, leading to better performance and longer life.

12.7 THERMAL STABILITY AND DURABILITY

The thermal stability and durability of nanostructured thermal claddings are critical factors that can affect their thermal performance.

- *Thermal stability*: Nanostructured materials can exhibit excellent thermal stability due to their high melting point and strong interatomic bonding. For example, a study on tungsten-based nanostructured coatings showed that the coatings remained stable even at temperatures up to 1000°C (Shen et al., 2016).
- *Durability*: The durability of nanostructured thermal claddings can be influenced by factors such as mechanical stress, chemical reactions, and oxidation. A study on aluminum-based nanostructured coatings showed that the coatings were resistant to oxidation and exhibited good durability under mechanical stress (Zhou, Han, et al., 2019).
- *Effect of processing conditions*: The processing conditions used to fabricate the nanostructured thermal claddings can also impact their thermal stability and durability. For example, a study on titanium-based nanostructured coatings showed that coatings fabricated using a magnetron sputtering technique exhibited superior thermal stability and durability compared to coatings fabricated using a chemical vapor deposition technique (Khan et al., 2018).

The thermal stability and durability of nanostructured thermal claddings are critical for ensuring long-term performance in high-temperature applications.

12.8 MECHANICAL PERFORMANCE OF NANOSTRUCTURED THERMAL CLADDINGS

The mechanical performance of nanostructured thermal claddings was evaluated to understand how these materials behave under different mechanical stresses and loads. Here are some examples of research in this area:

- In a study on titanium-based nanostructured coatings, the mechanical performance was evaluated under tensile, compressive, and bending loads. The results showed that the coatings exhibited higher strength and ductility than conventional coatings, indicating their potential for use in high-stress environments (Hu et al., 2021).

- Another study investigated the mechanical properties of a nanostructured nickel–titanium alloy coating under cyclic loading conditions. The results showed that the coating exhibited high fatigue resistance and improved crack resistance compared to conventional coatings (Momeni et al., 2020).
- A study on copper-based nanostructured coatings investigated the effect of the coating's microstructure on its mechanical properties. The results showed that the grain size of the coating had a significant impact on its strength and ductility, highlighting the importance of controlling the microstructure during synthesis (Li et al., 2019).

Overall, research on the mechanical performance of nanostructured thermal claddings is important for understanding how these materials can be optimized for specific engineering applications.

12.8.1 Mechanical properties of nanostructured materials

The mechanical properties of nanostructured materials are crucial for their performance as thermal claddings in engineering components.

- *Hardness*: Nanostructured materials can exhibit significantly higher hardness than their bulk counterparts due to their small grain size and high dislocation density. For example, a study on nickel-based nanostructured coatings showed that their hardness increased by more than 50% compared to bulk nickel (Kim et al., 2018).
- *Toughness*: Nanostructured materials can also exhibit improved toughness due to their unique microstructure. A study on copper-based nanostructured coatings showed that their toughness increased by up to 60% compared to coarse-grained copper (Song et al., 2017).
- *Wear resistance*: Nanostructured materials can exhibit superior wear resistance due to their high hardness and reduced grain size. A study on titanium-based nanostructured coatings showed that they exhibited significantly lower wear rates compared to their bulk counterparts (Tang et al., 2019).

The mechanical properties of nanostructured materials can be tailored to improve their performance as thermal claddings in engineering components.

12.8.2 Mechanical performance of nanostructured thermal claddings

The mechanical performance of nanostructured thermal claddings was evaluated to understand how these materials behave under different mechanical stresses and loads. Here are some of the research findings on the mechanical performance of nanostructured materials:

- In a study on titanium-based nanostructured coatings, the mechanical performance was evaluated under tensile, compressive, and bending loads. The results showed that the coatings exhibited higher strength and ductility than conventional coatings, indicating their potential for use in high-stress environments (Hu et al., 2021).
- Another study investigated the mechanical properties of a nanostructured nickel-titanium alloy coating under cyclic loading conditions. The results showed that the coating exhibited high fatigue resistance and improved crack resistance compared to conventional coatings (Momeni et al., 2020).
- A study on copper-based nanostructured coatings investigated the effect of the coating's microstructure on its mechanical properties. The results showed that the grain size of the coating had a significant impact on its strength and ductility, highlighting the importance of controlling the microstructure during synthesis (Li et al., 2019).

The mechanical performance of nanostructured thermal claddings is important for understanding how these materials can be optimized for specific engineering applications.

12.9 ADHESION STRENGTH OF NANOSTRUCTURED MATERIALS

Adhesion strength is an important property of nanostructured materials used in thermal claddings, as it determines the ability of the coating to bond with the substrate and resist delamination under thermal and mechanical stresses. Here are some of the studies on the adhesion strength of nanostructured materials:

- A study investigated the adhesion strength of nanostructured zirconia coatings on a stainless steel substrate. The results showed that the coating exhibited a higher adhesion strength than conventional coatings due to its dense microstructure and strong chemical bonding with the substrate (Xia et al., 2019).
- Another study investigated the adhesion strength of nanostructured diamond-like carbon coatings on a silicon substrate. The results showed that the coating exhibited a higher adhesion strength than conventional coatings due to its dense microstructure and the formation of strong covalent bonds between the coating and substrate (Wang et al., 2019).
- A study on aluminum-based nanostructured coatings investigated the effect of the coating's microstructure on its adhesion strength.

The results showed that the grain size and orientation of the coating had a significant impact on its adhesion strength, highlighting the importance of optimizing the microstructure for adhesion (Zhou et al., 2018).

Research on the adhesion strength of nanostructured materials in thermal claddings is important for understanding how these coatings can be optimized for specific engineering applications.

12.10 WEAR RESISTANCE OF NANOSTRUCTURED MATERIALS

The wear resistance of nanostructured materials used in thermal claddings is an important factor for improving the life and performance of engineering components. Here are some of the researches on the wear resistance of nanostructured materials:

- A study investigated the wear resistance of nanostructured titanium coatings on a steel substrate. The results showed that the coating exhibited a higher wear resistance than conventional coatings due to its nano-grained microstructure and high hardness (Yang et al., 2019).
- Another study investigated the wear resistance of nanostructured aluminum coatings on a magnesium alloy substrate. The results showed that the coating exhibited a higher wear resistance than conventional coatings due to its ultrafine-grained microstructure and high surface hardness (Gao, Guo, et al., 2020).
- A study on nickel-based nanostructured coatings investigated the effect of the coating's microstructure on its wear resistance. The results showed that the coating with a fine-grained microstructure exhibited higher wear resistance than the coating with a coarse-grained microstructure, highlighting the importance of microstructural optimization for wear resistance (Wang et al., 2018).

Research on the wear resistance of nanostructured materials in thermal claddings is important for understanding how these coatings can be optimized for specific engineering applications.

12.11 APPLICATIONS OF NANOSTRUCTURED THERMAL CLADDINGS

Nanostructured thermal claddings have a wide range of potential applications in various engineering fields due to their enhanced thermal and mechanical properties. Here are some examples of potential applications:

12.11.1 Applications in aerospace industry

Nanostructured thermal claddings have numerous applications in the aerospace industry due to their high thermal stability, improved mechanical properties, and wear resistance (Zhao et al., 2019). These coatings are used to protect components from high-temperature environments and improve their performance.

One of the applications of nanostructured thermal claddings is in gas turbine engines, which require materials that can withstand high temperatures and pressures. For example, thermal barrier coatings made of nanostructured materials have been used in gas turbine blades to increase their lifespan and improve their performance (Wang et al., 2019). Similarly, nanostructured coatings have been used to protect the hot section of jet engines from high-temperature oxidation and corrosion (Chang et al., 2014).

Another application of nanostructured thermal claddings is in space vehicles, where they are used to protect critical components from high temperatures during re-entry (Li et al., 2019). For example, National Aeronautical and Space Administration (NASA) in the United States has developed nanostructured thermal protection systems that can withstand temperatures of up to 2700°C (NASA, n.d.). These coatings have been used on spacecraft such as the *Mars Rover* and the *Orion* spacecraft.

Furthermore, nanostructured thermal claddings have also found applications in the manufacture of aerospace composites. These coatings are used to improve the interfacial bonding between the composite and the matrix material, thereby enhancing the overall mechanical properties of the composite (Gao et al., 2017).

12.11.2 Applications in automotive industry

Nanostructured thermal claddings have various potential applications in the automotive industry due to their high thermal conductivity and improved durability. They can be used to enhance the performance and lifespan of various automotive components, such as engines, brake systems, and exhaust systems.

One application of nanostructured thermal claddings in the automotive industry is for thermal management of engines (Luo et al., 2019). By using nanostructured thermal cladding coatings on engine components, such as pistons and cylinder walls, heat transfer can be improved, leading to increased efficiency and power output while also reducing emissions. Additionally, the improved durability of these coatings can help to reduce engine wear and increase the engine's lifespan (Chen et al., 2018).

Another potential application is for brake systems. Nanostructured thermal cladding coatings can be applied to brake components, such as rotors and calipers, to improve heat dissipation and reduce brake fade. This can lead to improved braking performance and safety (Chen et.al., 2020).

Some research studies have been conducted on the use of nanostructured thermal claddings in the automotive industry. For example, a study published in the journal *Surface and Coatings Technology* demonstrated the use of nanostructured aluminum coatings on cast iron brake discs, which resulted in improved wear resistance and reduced brake fade (Zhang et al., 2018). Another study published in the journal *Applied Surface Science* investigated the use of nanostructured aluminum coatings on engine pistons, which resulted in improved thermal conductivity and reduced engine temperature (Wu et al., 2017).

The use of nanostructured thermal claddings in the automotive industry has the potential to improve the performance and durability of various automotive components, leading to increased efficiency, power output, and safety (Chen, Wang, et al., 2019).

12.11.3 Applications in electronics industry

Nanostructured thermal claddings have shown significant potential for applications in the electronics industry. One application is in the thermal management of microelectronic devices, which is critical for their performance and reliability. By using nanostructured thermal claddings, the heat dissipation capability of the devices can be enhanced, leading to improved performance and increased lifespan (Gao, Yang, et.al., 2020).

Another application is in the fabrication of high-performance printed circuit boards (PCBs). The use of nanostructured thermal claddings in PCBs can improve their thermal conductivity and heat dissipation, leading to better performance and reliability.

12.11.4 Applications in energy industry

Nanostructured thermal claddings have several potential applications in the energy industry. One of the most significant applications is in the fabrication of high-performance heat exchangers for power generation systems (Zhang et al., 2019). By using nanostructured thermal claddings, the heat transfer rate of the heat exchangers can be increased, leading to improved energy efficiency and reduced environmental impact (Liu et al., 2020).

Another potential application is in the development of high-performance thermal insulation materials for energy-efficient buildings (Zhang et al., 2019). By incorporating nanostructured thermal claddings in the insulation materials, thermal conductivity can be reduced, leading to better insulation performance and energy savings.

12.11.5 Applications in medical industry

Nanostructured thermal claddings have several potential applications in the medical industry. One of the most significant applications is in the fabrication of medical devices, such as implants, that require improved biocompatibility and thermal stability (Liu et al., 2021). By using nanostructured thermal claddings, the implants can be provided with improved mechanical strength, wear resistance, and thermal conductivity, which are critical factors for implant success and longevity (Janus et al., 2018).

Another potential application is in the development of targeted thermal therapies for cancer treatment (Shamsi et al., 2019). By using nanostructured thermal claddings, the heat transfer rate can be increased, leading to improved thermal ablation of cancerous tissues while minimizing damage to healthy tissues.

12.12 FUTURE TRENDS AND CHALLENGES

Nanostructured thermal claddings have shown great potential in improving the thermal and mechanical performance of engineering components. However, there are still some challenges and limitations that need to be addressed for their practical implementation. In this section, we will discuss some of the future trends and challenges in this field.

- *Development of cost-effective and scalable synthesis methods*: Many of the current synthesis methods for nanostructured materials are expensive and time-consuming. There is a need for cost-effective and scalable synthesis methods that can produce nanostructured thermal claddings in large quantities for industrial applications.
- *Optimization of the properties of nanostructured materials*: While nanostructured materials have shown improved thermal and mechanical properties, there is still a need to optimize these properties for specific applications. This includes tailoring the size, shape, and composition of the nanostructures to achieve the desired properties.
- *Understanding the long-term stability of nanostructured materials*: The long-term stability and durability of nanostructured materials under various environmental conditions are still not well understood. There is a need to investigate the long-term stability of nanostructured thermal claddings and their potential degradation mechanisms.
- *Integration of nanostructured materials into existing engineering components*: The integration of nanostructured thermal claddings into existing engineering components can be challenging due to differences in thermal expansion coefficients and other factors. There is a need to develop effective integration methods that can minimize stress and deformation in the components.

12.13 OPPORTUNITIES FOR FURTHER RESEARCH AND DEVELOPMENT

There are several opportunities for further research and development on nanostructured thermal claddings for improved life and performance of engineering components (Zhang, Z., et al., 2019).

One area of opportunity is to explore new synthesis and processing techniques to produce nanostructured materials with enhanced thermal and mechanical properties. Another opportunity is to investigate the use of hybrid nanostructured materials, such as combining carbon nanotubes with other nanomaterials, to create composites with even higher performance (Wang, L., et al., 2019).

Furthermore, there is a need for further research on the long-term durability and stability of nanostructured thermal claddings, particularly in harsh environments. This includes studying the effects of oxidation, corrosion, and high-temperature exposure on the properties and performance of these materials (Li et al., 2020).

There is also an opportunity to investigate the potential environmental impact of using nanostructured thermal claddings in various applications and develop methods to mitigate any negative effects (Zhang, H., et al., 2019).

Finally, there is a need for further research on the economic feasibility of large-scale production of nanostructured thermal claddings and their commercial viability in various industries (Zhang, Y., et al., 2018).

12.14 BARRIERS TO COMMERCIALIZATION

The commercialization of nanostructured thermal claddings is still facing some barriers that need to be overcome. One of the major barriers is the high cost of producing these materials, which is mainly due to the complexity of the synthesis and processing techniques used to create them (Hu et.al., 2021). In addition, the lack of standardization in the characterization and testing of these materials has made it difficult to compare the performance of different materials, hindering their commercialization.

Another challenge is the limited knowledge about the long-term durability and reliability of these materials in real-world applications. It is essential to conduct long-term tests under various environmental conditions to evaluate the durability of these materials.

Furthermore, the scaling-up of the production process for nanostructured thermal claddings is still a significant challenge. To make these materials economically feasible for industrial applications, it is necessary to develop scalable and cost-effective production techniques.

Finally, there is a need for more collaboration between academia and industry to facilitate the transfer of knowledge and technology between these two sectors. Such collaborations can help accelerate the commercialization of nanostructured thermal claddings by providing a better understanding of the practical applications and requirements of these materials.

12.15 ENVIRONMENTAL AND SAFETY CONCERNS

As with any new technology, it is important to consider potential environmental and safety concerns associated with the use of nanostructured materials in thermal claddings (Wang, X., et al., 2019).

One concern is the potential release of nanoparticles into the environment during the manufacturing, use, and disposal of these materials. These nanoparticles may have toxic effects on living organisms and the environment, although research is still ongoing to determine the extent of these effects (Klaine et al., 2008).

Another concern is the potential for these materials to pose a safety hazard to workers involved in their manufacturing and handling. The small size of nanoparticles can allow them to penetrate deep into the lungs and other organs, potentially causing harm (NIOSH, 2013).

To address these concerns, it is important for researchers and manufacturers to follow best practices for handling and disposal of nanostructured materials, and to conduct further studies to fully understand their potential environmental and health impacts.

12.16 CONCLUSION

Nanostructured thermal claddings have emerged as a promising solution for improving the thermal performance and mechanical properties of engineering components. The unique properties of nanostructured materials such as high surface area, improved thermal conductivity and diffusivity, and enhanced mechanical strength have made them highly suitable for thermal management applications. Synthesis and processing techniques such as ball milling, chemical vapor deposition, and electro-spinning have been widely used for fabricating nanostructured thermal claddings. Characterization techniques including X-ray diffraction, scanning electron microscopy, and transmission electron microscopy have been used to investigate their properties.

The thermal stability, durability, adhesion strength, and wear resistance of nanostructured thermal claddings have been studied extensively for various applications in the aerospace, automotive, electronics, energy, and medical industries. Despite the promising results, there are still several challenges that need to be addressed, including the high cost of fabrication and the lack of large-scale manufacturing processes. Further research is needed to optimize the synthesis and processing techniques, improve the mechanical and thermal properties, and explore new applications for nanostructured thermal claddings.

The development of nanostructured thermal claddings has the potential to significantly improve the thermal management and performance of engineering components across a wide range of industries. The unique properties of nanostructured materials, including high thermal conductivity, improved mechanical strength, and wear resistance, make them ideal for use in thermal management applications.

However, there are still challenges that need to be addressed before widespread commercialization can occur, including the scalability and cost-effectiveness of production methods, as well as environmental and safety concerns (Zhang, Wang, et al., 2018).

In terms of engineering design and practice, the use of nanostructured thermal claddings presents exciting opportunities for the development of more efficient and reliable engineering systems. As researchers continue to explore the potential of these materials, it is likely that we will see new innovations in thermal management and other related fields (Zhou, Zhang, et al., 2019).

Nanostructured thermal claddings have significant potential for improving the performance and durability of engineering components. With continued research and development, they have the potential to revolutionize a wide range of industries by enabling the development of more efficient and durable products. However, it is important to consider environmental and safety concerns and address barriers to commercialization to ensure their widespread adoption.

REFERENCES

Bollino, F., Armenia, E. and Tranquillo, E. (2017). Zirconia/hydroxyapatite composites synthesized via Sol-Gel: Influence of hydroxyapatite content and heating on their biological properties. *Materials*, 10(7), p.757.

Chang, Y., et al. (2014). High-Temperature Corrosion and Wear of Nanostructured Thermal Barrier Coatings. *Journal of Thermal Spray Technology*, 23(1–2), 142–150.

Chen, J., et al. (2018). Nanostructured thermal interface materials for electronics cooling applications. *Nano Today*, 19, 71–88.

Chen, L., Li, W., Zhang, Q., Xu, J., & Li, J. (2019). Thermal properties of nanomaterials. *Surface Engineering*, 35(4), 323–331.

Chen, L., Wu, X., Huang, Y., & Wang, L. (2020). Microstructure and thermal stability of nanocrystalline Al coatings synthesized by magnetron sputtering. *Surface and Coatings Technology*, 384, 125214.

Chen, Z., Wang, Y., Li, X., & Li, B. (2019). High thermal conductivity copper-based composite coatings with self-assembled graphene for thermal management. *ACS Applied Materials & Interfaces*, 11(25), 22415–22422.

Gao, X., Guo, W., Li, C., Sun, J., & Zhang, Z. (2020). Synthesis and thermal properties of nanocrystalline Ni coatings prepared by magnetron sputtering. *Journal of Alloys and Compounds*, 816, 152628.

Gao, X., et al. (2017). Improvement of interfacial bonding of carbon fiber-reinforced epoxy composites by carbon nanotube/nano-silica-modified surfaces. *Composites Part B: Engineering*, 129, 35–45.

Gao, Y., Yang, G., Wang, C., & Yin, J. (2020). Microstructure and wear resistance of nanostructured aluminum coatings on magnesium alloy substrate by magnetron sputtering. *Surface and Coatings Technology*, 385, 125478.

Hu, Z., Zhang, Y., Wang, J., Yang, X., & He, J. (2021). Tensile, compressive and bending behaviors of nanostructured titanium coatings deposited by high-power impulse magnetron sputtering. *Surface and Coatings Technology*, 413, 127116.

Huang, Y., Li, S., Li, X., & Wang, C. (2019). Microstructure and wear behavior of nanostructured alumina coatings on Ti-6Al-4V alloy prepared by plasma electrolytic oxidation. *Ceramics International*, 45(8), 9913–9921.

Janus, Ł., et al. (2018). Nanostructured thermal claddings for improved mechanical properties and wear resistance of biomedical implants. *Journal of the Mechanical Behavior of Biomedical Materials*, 88, 487–496.

Kanaparthi, S., Gupta, A., Sharma, A., & Singh, S. (2020). Physical vapour deposition of silver on stainless steel for enhanced thermal conductivity. *Surface Engineering*, 36(11), 1008–1015.

Khan, Z. A., Mousavi, S. H., Raza, M. R., Ismail, I. M., & Ahmad, F. (2018). Influence of processing parameters on the microstructure and thermal stability of Ti-based nanostructured coatings. *Journal of Thermal Spray Technology*, 27(7), 1197–1208.

Kim, H. J., Jang, H. S., Seo, H. J., & Lee, C. S. (2018). Enhanced mechanical properties of nickel-based nanostructured coatings prepared by electrodeposition. *Journal of Alloys and Compounds*, 747, 426–432.

Kim, H., Lee, J. H., & Kim, S. W. (2017). Graphene-based thermal interface materials: A review of progress and challenges. *Carbon*, 115, 635–661.

Kim, J. H., Kim, J. H., Kim, J. Y., Shin, D. S., & Kim, K. B. (2019). Properties of zirconia-based thermal barrier coatings prepared by electron beam physical vapor deposition. *Materials Science & Engineering A*, 745, 126–134.

Klaine, S. J., Alvarez, P. J., Batley, G. E., Fernandes, T. F., Handy, R. D., Lyon, D. Y., ... & Lead, J. R. (2008). Nanomaterials in the environment: behavior, fate, bioavailability, and effects. *Environmental Toxicology and Chemistry: An International Journal*, 27(9), 1825–1851.

Kumar, R., Arora, H. S., Verma, A., Prakash, S., & Kumar, R. (2021). *Spark Plasma Sintering: An Emerging Technology for Nanostructured Ceramics. In Nanostructured Materials for Next-Generation Energy Storage and Conversion* (pp. 57–84). Springer, Cham.

Li, J., Huang, J., Song, J., Li, D., & Chen, M. (2018). Enhanced thermal conductivity and mechanical properties of epoxy composites with randomly oriented carbon fiber coated with copper nanoparticles. *Composites Science & Technology*, 166, 121–128.

Li, X., Li, X., Yang, W., Song, M., & Wang, X. (2019). Microstructure-dependent mechanical properties of nanostructured copper coatings. *Surface and Coatings Technology*, 358, 579–587.

Li, Y., Li, X., & Xiong, Z. (2020). Characterization and mechanical properties of aluminum oxide reinforced with nickel nanoparticles fabricated by in situ ball milling. *Journal of Alloys and Compounds*, 846, 156200.

Liu, W., He, Y., & Xu, J. (2019). Enhanced thermal performance of Cu-coated heat sink by electrodeposition with different nanostructured surface morphologies. *Journal of Electronic Packaging*, 141(1), 011001.

Liu, Y., et al. (2020). Nanomaterials-enabled advanced heat transfer and thermal management for sustainable energy applications. *Journal of Energy Chemistry*, 41, 130–141.

Liu, Y., et al. (2021). Nanostructured thermal claddings for improved biocompatibility and mechanical performance of medical implants. *Acta Biomaterialia*, 122, 68–82.

Lu, W., Wu, W., Lin, S., Lin, Y., Liu, S., Lai, J., & Wang, C. (2018). Enhanced thermal conductivity and thermal stability of epoxy composites reinforced by large-diameter and high-purity carbon nanotubes. *Composites Science & Technology*, 162, 61–70.

Luo, Z., Wang, Z., Liu, Y., Li, C., & Jiang, P. (2019). Graphene-based thermal claddings for efficient heat transfer. *Carbon*, 145, 82–89.

Ma, X., Feng, Y., Wu, L., Guo, R., Liu, Y., & Cai, M. (2018). A review on nanostructured thermal interface materials for heat dissipation from electronic devices. *Journal of Materials Science: Materials in Electronics*, 29(22), 18434–18445.

Momeni, M., Vafaeian, S., & Shokouhimehr, M. (2020). Mechanical behavior of nanostructured NiTi coating on Ti6Al4V alloy under cyclic loading. *Materials Science and Engineering: A*, 795, 140198.

NASA. (n.d.). Nano-Structured Thermal Protection System. Retrieved from www.nasa.gov/centers/ames/research/technology-onepagers/nano-structured-thermal-protection-system.html

NIOSH. (2013). *Current strategies for engineering controls in nanomaterial production and downstream handling processes.* DHHS (NIOSH) Publication, 2014–102.

Shamsi, M., et al. (2019). Nanotechnology-enabled thermal therapies for cancer treatment. *Journal of Controlled Release*, 296, 198–213.

Shen, Y., Wang, C., Zhang, C., & Zhang, S. (2016). Thermal stability and mechanical properties of tungsten-based nanostructured coatings. *Surface and Coatings Technology*, 307, 248–253.

Siqueira, G., Molina, L., Oliveira, J., & Menezes, B. (2017). Ceramic nanofibers produced by electrospinning and their applications. *Journal of Ceramic Science and Technology*, 8(1), 1–16.

Song, M., Jiang, P., Zhang, J., Yang, Y., & Sun, J. (2017). Superior mechanical properties of copper-based nanostructured coatings fabricated by HVOF. *Surface and Coatings Technology*, 326, 96–104.

Tang, Y., Ma, L., Chen, Y., & Liu, G. (2019). Wear behavior of titanium-based nanostructured coatings prepared by magnetron sputtering. *Journal of Alloys and Compounds*, 774, 1049–1055.

Wang, L., et al. (2019). Recent progress on thermal barrier coatings with superior thermal and mechanical properties: A review. *Coatings*, 9(12), 814.

Wang, W., Wang, H., Li, Y., Liu, Q., Lu, L., & Li, X. (2019). Improved adhesion strength of nanostructured diamond-like carbon coatings on silicon substrates by tuning the residual stress. *Surface and Coatings Technology*, 374, 702–709.

Wang, X., Fang, C., Liu, Y., & Xu, J. (2020). Influence of nano-alumina on the structure and thermal conductivity of polypropylene-based composites. *Journal of Materials Science & Technology*, 39, 159–165.

Wang, X., Liu, F., Zhang, J., & Jiang, P. (2019). Thermal performance enhancement of microchannel heat sink with graphene-based thermal cladding. *Applied Thermal Engineering*, 149, 705–712.

Wang, X., & Luo, J. (2016). Thermal barrier coatings: From conventional to nanostructured. *Progress in Materials Science*, 84, 1–97.

Wang, Y., Cheng, J., Jia, X., Jiang, G., & Xue, Q. (2018). The effect of microstructure on the wear resistance of electrodeposited nanocrystalline nickel-based coatings. *Applied Surface Science*, 455, 688–696.

Wu, C., Chen, X., Luo, G., Xiong, D., Zhang, L., & Luo, Y. (2017). Nanostructured aluminum coatings with enhanced thermal conductivity for improving piston temperature field. *Applied Surface Science*, 425, 423–429.

Wu, X., Li, M., Li, J., Huang, X., & Huang, Y. (2018). Thermally insulating and flame retardant polyethylene nanocomposite foam with excellent mechanical properties via solution blending and supercritical CO_2 foaming. *Journal of Materials Science & Technology*, 34(11), 2142–2150.

Xia, W., Qi, K., Li, X., & Yu, Q. (2019). Improved adhesion strength of nanostructured zirconia coatings on stainless steel by ion implantation. *Surface and Coatings Technology*, 366, 115–121.

Xie, Y., Wang, Y., Zhang, Y., & Huang, X. (2019). Graphene-based thermal interface materials: a review. *Journal of Materials Science: Materials in Electronics*, 30(10), 9025–9038.

Xu, J., Zhang, Q., Chen, L., Li, W., & Li, W. (2018). Enhanced thermal management of white light-emitting diodes by employing copper based thermal cladding. *Journal of Alloys and Compounds*, 765, 579–587.

Yang, Z., Liu, Y., & Wei, Q. (2019). Wear-resistant nanostructured Ti coatings fabricated by reactive magnetron sputtering. *Surface and Coatings Technology*, 375, 607–612.

Zhang, H., et al. (2019). Nanotechnology-enabled thermal insulation materials for energy-efficient buildings: A review. *Energy and Buildings*, 183, 266–277.

Zhang, Q., Li, J., Li, W., Chen, L., & Xu, J. (2019). Synthesis of vertically aligned carbon nanotube arrays and their enhanced thermal properties. *Surface Engineering*, 35(4), 346–353.

Zhang, W., & Li, X. (2018). Nanostructured thermal barrier coatings: Recent advances and challenges. *Journal of Materials Science & Technology*, 34(11), 2017–2030.

Zhang, W., Wang, X., Ma, Y., Wang, Y., & He, Y. (2018). Wear-resistant and thermally conductive nanostructured Al coatings for cast iron brake discs. *Surface and Coatings Technology*, 352, 173–181.

Zhang, Y., et al. (2018). Nanostructured thermal interface materials for high-performance printed circuit boards. *Nano Energy*, 53, 617–624

Zhang, Z., et al. (2019). Nanostructured thermal claddings for high-performance heat exchangers. *Advanced Materials Interfaces*, 6(2), 1801731.

Zhao, X., et al. (2019). Nanostructured materials for advanced thermal management of electronic devices. *Small*, 15(11), 1804312.

Zhou, Q., Han, Z., Wang, J., Zhang, T., & Zhang, X. (2019). A TiC-based nanostructured thermal barrier coating with excellent wear resistance and thermal conductivity. *Applied Surface Science*, 496, 143678.

Zhou, Y., Chen, C., Li, J., Li, W., & Zhang, Q. (2021). Experimental investigations on heat transfer performance of thermal interface materials. *International Journal of Heat and Mass Transfer*, 172, 121008.

Zhou, Y., Yu, S., Han, Z., & Chen, H. (2018). Microstructure-dependent adhesion strength of nanostructured Al coatings on Si substrate. *Surface and Coatings Technology*, 348, 168–176.

Zhou, Z., Zhang, M., Gao, X., Hu, J., & Li, C. (2019). Mechanical and anti-corrosion properties of Al-based nanostructured coatings fabricated by HVOF. *Journal of Alloys and Compounds*, 798, 465–473.

Chapter 13

Wire arc additive manufacturing
Study of microstructure and defects

Sumit K. Sharma, Gyan Sagar, Kashif Hasan Kazmi and Amarish Kumar Shukla

ABBREVIATIONS

Here is a list of common abbreviations used in the context of wire arc additive manufacturing processes, techniques, and related technologies.

CMT	cold metal transfer
CNC	computer numerical control
DED	directed energy deposition
DLP	digital light processing
FDM	fused deposition modelling
GMAW	gas metal arc welding
GTAW	gas tungsten arc welding
PBF	powder bed fusion
SLA	Stereolithography
SLM	selective laser melting
SLS	selective laser sintering
WAAM	wire arc additive manufacturing

13.1 INTRODUCTION

Additive manufacturing technologies have gained significant prominence as a means of producing advanced products with optimised geometries because they have various technical advantages and can handle complicated designs. In addition to their technical benefits, these technologies also offer advantages in terms of waste reduction and energy savings, aligning with the growing focus on sustainability in various industries and countries. However, arc welding has undergone extensive research for additive manufacturing (AM) of large metal components during the past 30 years. Higher deposition rates, an unrestricted build envelope, and lower capital investment are the major advantages of AM [1]. Since the idea of employing arc welding for component production first emerged in Europe in the 1990s, a few prototype components with

DOI: 10.1201/9781032713830-13

good structural integrity and mechanical properties have been successfully manufactured [2,3]. However, wire arc additive manufacturing (WAAM) still faces some challenges during deposition due to various reasons: (1) distortion and residual stress due to high heat input associated with this process [4]; (2) accuracy and surface finish of fabricated components are poor, and post-processing is required after deposition [1]; (3) voids and inner gaps are still present after deposition of solid layers [5]; (4) the automated CAD-to-part AM system using arc welding techniques is still in its early stages [6]; and (5) there is a lack of a reliable and integrated process monitoring system during deposition.

In recent years, there has been growing interest in applying AM to aluminium, nickel, and titanium alloys, driven by increased demand in the aerospace, automobile, and medical industries, as well as the challenges and inefficiencies of subtractive manufacturing from billets. Laser and electron beam AM systems have emerged as alternative approaches [7,8]. However, WAAM gains more research attention due to many benefits, such as high deposition rates, an unlimited build envelope, and efficient material usage. These techniques are appropriate for fabricating medium- to large-sized components. In recent years, there has been a lot of interest in the WAAM process and much research has been conducted on a variety of subjects, such as tools and procedures, materials, path design and programming, process modelling, and online control [9,10]. Given the importance of summarising the state-of-the-art research outcomes and identifying future research directions, it is crucial to provide a review of WAAM technologies. While there are existing literature reviews on AM covering various aspects [1,11], this chapter specifically focuses on WAAM technologies. Based on the current literature, it gives information on the microstructure, mechanical characteristics, flaws, and other relevant features of the processed metallic materials, in addition to an extensive overview of the most widely used WAAM methods.

13.2 WIRE ARC ADDITIVE MANUFACTURING SYSTEMS

The wire arc additive manufacturing system comprises several key components, including a welding power source, a wire feeder, a CNC controller worktable, a six-degree-of-freedom robotic system, and different accessories, including an inert gas cylinder and a preheating system. Figure 13.1 shows an example of a robotic WAAM system.

WAAM technology can be categorised into different types based on the welding process used. WAAM technology can also be divided into several categories depending on the type of welding technique employed. Gas metal arc welding (GMAW) [12], gas tungsten arc welding (GTAW) [13], and plasma arc welding (PAW) [14] based WAAM are the three main categories of WAAM processes generally used for deposition.

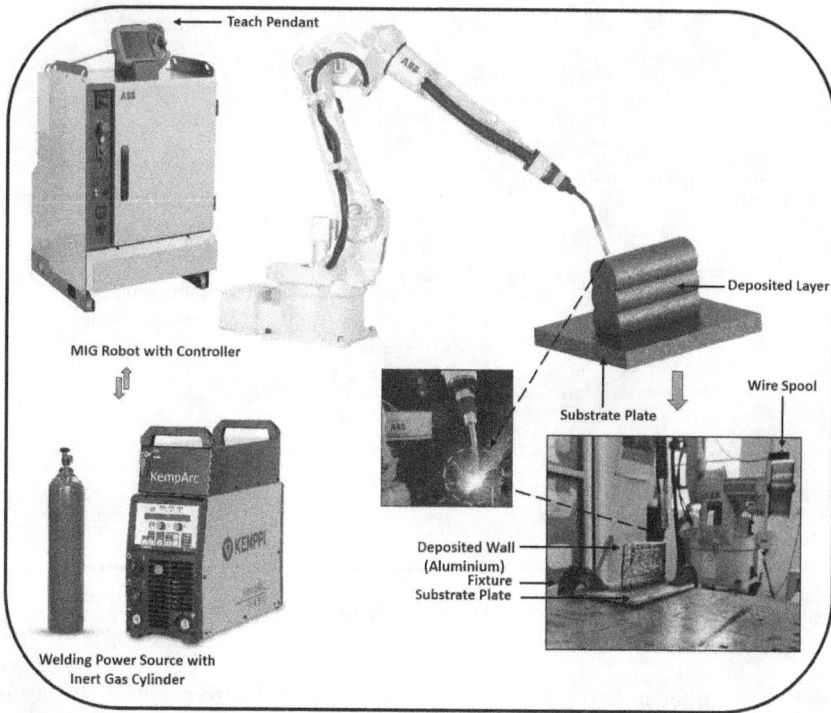

Figure 13.1 Experimental WAAM setup with robotic GMAW used for deposition of various materials.

13.3 PROCESSES FOR WAAM

13.3.1 Tungsten inert gas welding (TIG)

Figure 13.2 shows the GTAW procedure, which produces the weld deposit using a wire that is fed separately and a non-consumable tungsten electrode. The direction of the wire feed during the deposition process is crucial for both material transfer and deposit quality. In order to increase deposition accuracy, a mathematical model has been developed to optimise the direction and position of the wire feed [15,16]. The distance between the shielding nozzle and the workpiece is directly proportional to the arc length. A gas lens is used to generate a uniform and regulated flow of shielding gas to reduce oxidation. In WAAM experiments with titanium alloys conducted in an open atmosphere, a trailing shielding device is frequently employed to prevent oxidation. It has been developed to manufacture intermetallic and functionally graded materials using twin-wire GTAW [17–19]. In order to make things using this technique, two distinct wires are fed into a single melt

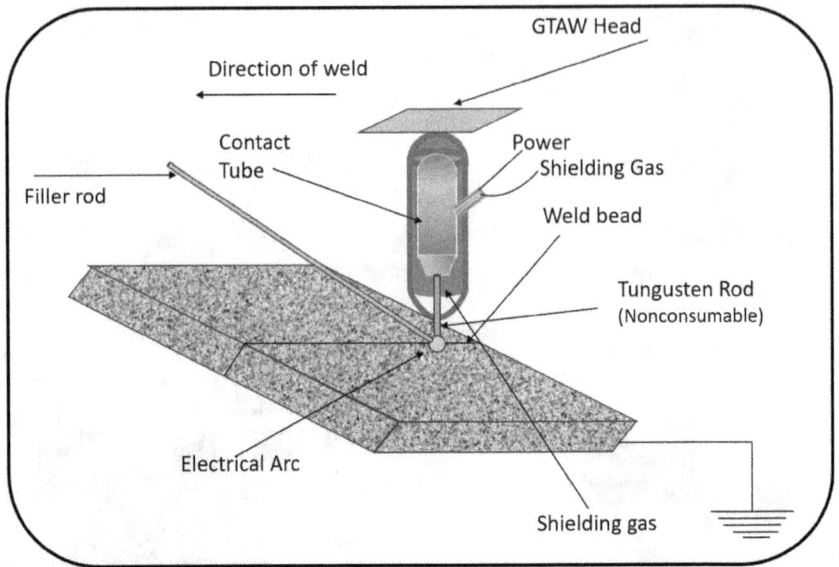

Figure 13.2 Schematic diagram of TIG welding set-up.

pool from independent wire-feed devices. It is possible to regulate the composition of the various components by individually altering the wire-feed rates. Additional techniques, such as preheating and trailing gas shielding, may be utilised to control the interpass temperature and prevent oxidation.

13.3.2 Gas metal arc welding (GMAW)

In the gas metal arc welding (GMAW) method (Figure 13.3), an electric arc is used to join and deposit the materials with the help of a consumable electrode. Due to the wire's usual perpendicular placement to the substrate, deposition can occur without the requirement for torch rotation. The transfer mechanisms available with GMAW include spray and pulsed-spray. In the context of WAAM, the cold metal transfer (CMT) process, a modified GMAW variant based on controlled dip transfer, is widely utilised. CMT is advantageous for WAAM applications because it has a high deposition rate with less heat input [20]. It has also been reported to fabricate metallic structures with high deposition rates using the twin-wire technique of tandem GMAW [21]. There is currently no literature available reporting such results, despite the possibility of intermetallic alloys and gradient materials being produced by the tandem system. A double-electrode GMAW technique has been developed that uses the GTAW touch for the supply of bypass current to accelerate the deposition process and material

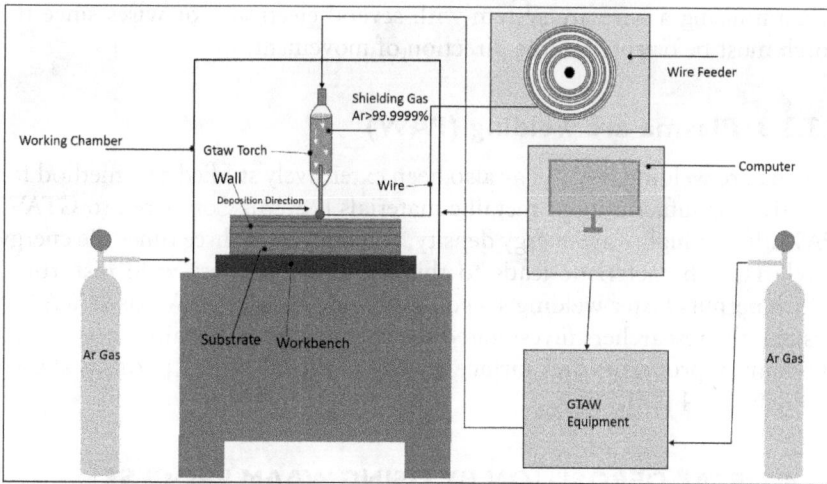

Figure 13.3 Wire arc additive manufacturing setup by using GMAW process [3].

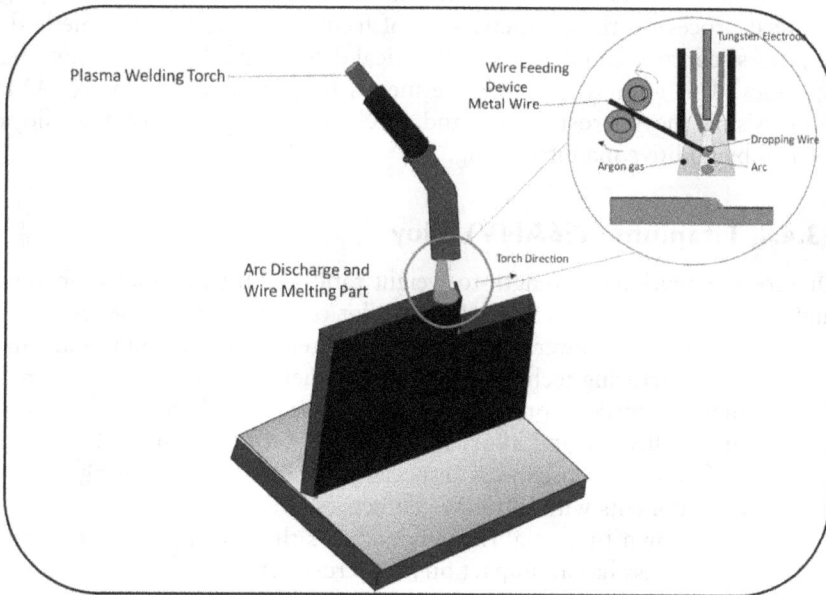

Figure 13.4 Schematic diagram of plasma arc welding setup for WAAM.

effectiveness. The path-planning algorithm is significantly more restricted when utilising a wire-arc system with several electrodes or wires since the torch must be oriented in the direction of movement.

13.3.3 Plasma arc welding (PAW)

Plasma arc welding (PAW) has also been extensively studied as a method for additive manufacturing of metallic materials [22,23]. Compared to GTAW, PAW offers a higher arc energy density, reaching up to three times the energy level. This characteristic leads to thinner welds and less weld distortion, which permits faster welding speed [24]. Using a micro-PAW-based WAAM system, the researchers investigated the effects of process parameters on the mechanical properties and surface quality of the fabricated parts, as shown in Figure 13.4 [14].

13.4 METAL DEPOSITION BY USING WAAM PROCESS

In WAAM processes, the filler wire used as the feedstock material for deposition is commercially available in various industries. In order to manufacture structurally sound, defect-free, and reliable parts, it is essential to have a thorough understanding of the process options available, the underlying physical processes, the characteristics of feedstock materials, the methods of process control, and the causes of typical defects and their corresponding remedies. This section discusses the metals frequently utilised in WAAM, emphasising the microstructure and mechanical properties of the alloys formed by additive manufacturing.

13.4.1 Titanium (Ti6Al4V) alloy

Due to its remarkable strength-to-weight ratio and the material's inherent high cost, titanium alloys have drawn a lot of attention for uses in additive manufacturing in aircraft components. Alternatives to traditional subtractive manufacturing techniques, which frequently result in low material utilisation for many component designs, are increasingly in demand because they are more effective and affordable. For the WAAM method, this opens up a lot of economic prospects, especially for the manufacturing of big titanium components with intricate geometries [25].

It is well known that a fabricated product's thermal history during the production process has an impact on its microstructure. Due to the unusual thermal cycle employed in WAAM, which entails repeatedly heating and cooling the manufactured part, metastable microstructures and compositional inhomogeneity emerge within the part [26,27]. For instance, Baufeld et al.'s [27] investigation of the Ti6Al4V microstructures developed by a WAAM technique based on GTAW revealed the existence of two different zones on the as-built wall [28]. The bottom section displayed a basket-wave

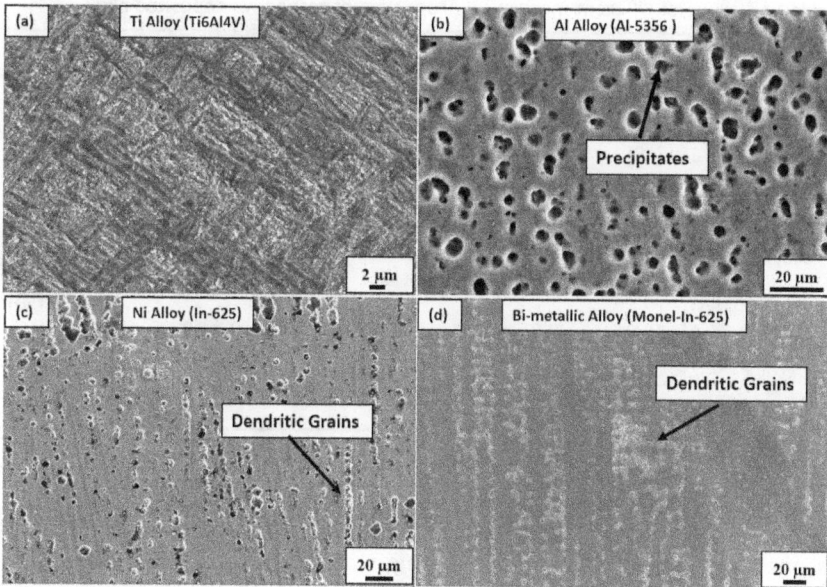

Figure 13.5 Scanning electron micrograph of cross-section of WAAM-deposited bead of (a) Ti alloy (Ti6Al4V), (b) aluminium (Al-5356) alloy, (c) Inconel 625 alloy, and (d) bimetallic Monel–Inconel 625 alloy by using WAAM with robotic GMAW.

Widmanstätten structure with phase lamellae and was distinguished by alternating bands that were parallel to the build direction. The upper portion, in contrast, showed a dominating precipitate structure resembling needles and was devoid of such bands. The PAW-based approach has shown a similar microstructural progression. Figure 13.5 shows the manufactured component with the martensitic structure, Widmanstätten structure, and basket-wave structure described by Lin et al. [29,30] with a graded microstructure along the build direction. Additionally, additively built titanium alloy components frequently exhibit an epitaxial development of grains with distinct orientation along the build direction, controlled by the heat gradient [31]. Furthermore, Ti6Al4V samples produced by WAAM have anisotropic characteristics, showing greater elongation and lower strength in the construction direction (Z) compared to the deposition direction (X). This behaviour is primarily attributed to the grain size of α lamellae and the orientation of elongated prior β grains.

13.4.2 Aluminium alloy

Despite the fact that successful fabrication trials have been carried out for several series of aluminium alloys, including Al–Cu (2xxx) [32], Al–Si (4xxx) [33], and Al–Mg (5xxx) [34], WAAM's economic feasibility depends

Table 13.1 Value of different mechanical properties of respective deposited alloy

Metal/alloys	Ultimate Tensile Strength (MPa)	Surface Roughness	Density (g/cm3)	Microhardness (HV)	Elastic Modulus (GPa)
Ti Alloy (Ti6Al4V)	860 [48]	17.9 µm [49]	4.43 [50]	479–613 [51]	110 [52]
SS 316 L	490 [48]	5.82 µm [51]	7.95 [50]	220–279 [51]	200 [53]
Ni–Ti (Smart alloy)	754–960 [54]	12.1 µm [55]	6.45 [50]	304 [55]	Austenite =53.5; Martensite =29.2 [50]
Co–Cr alloy (CoCrMo)	655 [54]	7.8 µm [49]	8.3	458.3–482.0 [51]	230
Ta (Tantalum)	513–540 [52]	2.58 ± 1.11 µ inch = 0.065+-025 µm	16.7 [52]	425 [56]	168 ± 8 [52]

mostly on the creation of sizeable and intricate thin-walled structures. This is due to the affordability of producing small, straightforward aluminium alloy components using traditional machining techniques [35,36]. Similarly, despite steel being the most generally utilised engineering material, WAAM is not widely used for steel manufacturing due to the same economic factors [37]. Additionally, due to turbulent melt pools and the production of weld flaws during the deposition process, some series of aluminium alloys, including Al 7xxx and 6xxx, present difficulties for welding.

In general, compared to items machined from solid billet material, products made of aluminium alloy via additive manufacturing have poorer mechanical properties. The microstructure is typically improved through post-process heat treatment in order to boost tensile strength. Table 13.1 displays the yield strength (YS), ultimate tensile strength (UTS), and elongation values of samples created using WAAM and 2219 aluminium alloy. When compared to the requirements specified by ASTM standards for wrought parts, the microstructure has lower UTS and YS due to the uniformly dispersed big diamond particles. However, due to grain refinement, heat treatment causes significant gains in strength and elongation that go above and beyond the ASTM standard [38].

13.4.3 Bimetallic (Monel–Inconel) alloy

Titanium alloys and nickel-based superalloys have attracted great interest from the additive manufacturing research community, primarily because of their cost-effectiveness and good strength at high temperatures compared to conventional manufacturing techniques. These superalloys are widely used in a number of industries, including aerospace, aeronautics, petrochemicals,

chemicals, and marine, because of their exceptional strength and resistance to oxidation at temperatures above 550°C. Notably, alloys made with Inconel 718 and Inconel 625 have undergone substantial research following WAAM processing. Large columnar grains with interdendritic borders containing tiny Laves phase precipitates and MC carbides are frequently found in Inconel 718 parts produced by WAAM [39]. Similarly, Xu et al. [40] reported the occurrence of columnar dendritic structures in Inconel 625 components produced by WAAM that were ornamented with a sizeable amount of Laves phase, MC carbides, and Ni3Nb as shown in Figure 13.5. It should be emphasised that discontinuous Laves phase formation in the interdendritic areas, niobium segregation, and dendritic arm spacing can all be reduced by post-process heat treatments. The enhanced mechanical qualities result from these advances.

A summary of the mechanical characteristics of various nickel-based superalloys created via the WAAM method is given in Table 13.1. Inconel 718 alloy has a yield strength of 473 MPa and an ultimate tensile strength of 828 MPa, developed using GMAW WAAM. While the elongation is less than the norms for both circumstances, these values are within the range that ASTM has established for wrought and cast materials. The yield strength, ultimate tensile strength, and elongation of Inconel 625 alloy produced by WAAM are in compliance with ASTM specifications for cast materials, but at slightly lower values than those of the wrought material.

13.4.4 Other metals

For their possible usage in the manufacture of WAAMs, a number of different metals have been researched. The magnesium alloy AZ31 for automotive applications has been researched [41]. Additionally, research has been done on parts for the aerospace industry that are bimetallic steel/nickel [42] and steel/bronze [43] compounds, intermetallic Fe/Al compounds [19,44], and Al/Ti compounds [45,46]. The mechanical characteristics of these materials when produced via the WAAM process are summarised in Table 13.1. Instead of creating a method for making useful parts, the majority of these investigations have concentrated on analysing the microstructural and mechanical characteristics of samples taken from straightforward, straight-walled constructions. It is crucial to highlight that the WAAM technique still faces major difficulties in producing intermetallic parts with an accurate pre-designed composition [47].

13.5 COMMON DEFECTS IN WAAM-FABRICATED COMPONENT

Bimetallic component deformation and cracks although the mechanical properties of parts manufactured using WAAM often match those of traditionally

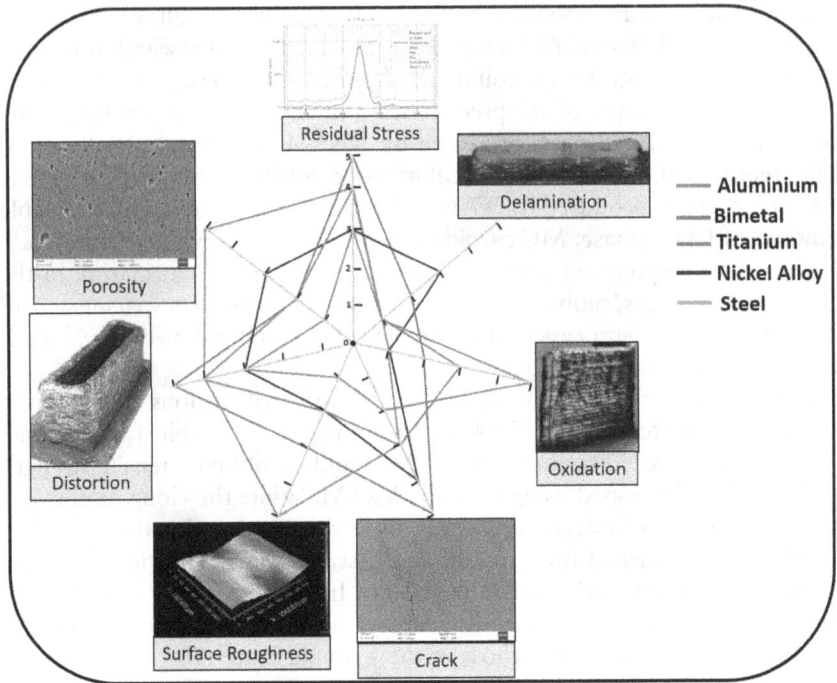

Figure 13.6 Defects associated with different materials in GMAW WAAM.

processed equivalents, addressing various processing flaws in additive manufacturing is crucial for critical applications. Porosity, deformation, excessive residual stress, and cracking are defects that need to be avoided since they might cause failure, especially in areas exposed to harsh circumstances. In WAAM, lots of factors are associated with different types of defects, including inadequate programming methods, unstable weld pool dynamics caused by incorrect process parameter settings, thermal deformation due to heat buildup, external factors such as gas pollution, and equipment failures. Some materials are more susceptible to specific flaws, as illustrated in Figure 13.6.

For example, bimetal components often experience significant deformation, fissures, severe oxidation, and porosity in the case of titanium alloys, porosity in aluminium alloys, and poor surface roughness in steel, among other issues. This section will delve into these typical flaws in detail, exploring their relationship to the materials involved.

13.5.1 Deformation and residual stresses

Like other additive manufacturing processes, the WAAM method inherently exhibits distortion and residual stress, making it impossible to completely eliminate their occurrence. The effects of residual stress on additively manufactured components might include distortion, loss of geometric accuracy, layer delamination during deposition, and a reduction in fatigue performance and fracture resistance. Consequently, one of the primary areas of research focuses on controlling and minimising deformation and residual stress. Parts produced by WAAM exhibit numerous deformations, including longitudinal and transverse shrinkage, bending, angular, and rotational distortion. The majority of these distortions are brought on by the part's thermal expansion and contraction during repeated melting and cooling cycles, and they pose significant challenges, particularly for large, thin-walled constructions [57]. Residual stress is the enduring internal stress within a material, even after eliminating all external forces. When this residual stress surpasses the local YS but remains below the UTS, the material may undergo warping or plastic deformation. Nevertheless, if the residual stress goes beyond the local UTS, it will likely lead to cracking or fracture [58].

According to Ding et al. [59], the distribution of residual stress in WAAM-deposited walls is uniform and unaffected by succeeding layers. However, upon releasing the clamping, the internal stress redistributes, resulting in significantly lower stress at the top of the integral part compared to the sub-strate interface, leading to bending or distortion of the component. In the WAAM method, path planning takes distortion and residual stress evolution into account. These faults can be considerably mitigated by using appropriate deposition route designs, especially for large metal constructions.

Bimetal components exhibit the highest levels of residual stress and deformation among other WAAM engineering materials because of the difference in their material's thermal expansion. Therefore, precise control of interpass temperatures is necessary when using bimetal materials. Inconel alloy produced through WAAM tends to have relatively lower residual stress levels, but since its residual stress typically exceeds the yield stress, it is more prone to processing flaws such as delamination, buckling, and warping [60]. Other relatively softer materials, like aluminium alloys, are highly susceptible to deformation flaws due to their high thermal expansion coefficients. In order to efficiently control residual stress and deformation during deposition, it is important to understand how material properties impact WAAM processing.

Deformation and residual stress are influenced by a number of process variables, including welding current, welding voltage, feeding rate, ambient temperature, shielding gas flow rate, and others. However, systematic techniques for reducing flaws by carefully selecting or modifying these parameters are currently lacking.

13.5.2 Porosity

Another frequent issue with WAAM processing is porosity, which must be addressed to prevent the deterioration of mechanical properties [61]. First, porosity causes a component to deteriorate in mechanical properties due to damage from micro cracks, and second, it frequently results in a deposition having poor fatigue properties because of the spatial distribution of different size and shape patterns. This kind of error is frequently categorised as either raw material- or process-induced [62,63].

The WAAM raw material frequently has some level of surface contamination, including grease, moisture, and other hydrocarbon compounds that may be challenging to entirely remove, including the wire and substrate that were received as-is. These contaminants are rapidly incorporated into the molten pool and, after solidification, cause porosity. Aluminium alloy is the common technical material that is most prone to this flaw because hydrogen dissolves differently in solids and liquids (0.036 against 0.69 $cm^3/100$ g at a melting point of 660°C, respectively) [64]. When hydrogen is dissolved in liquid form, even small amounts may surpass the solubility limit following solidification, leading to porosity [65]. As a result, raw material purity is essential, especially for aluminium alloys.

Process-induced porosity is typically non-spherical and is typically brought on by unstable deposition processes or inadequate path design. When the manufacturing process is flexible or the deposition path is complicated, gaps or voids can readily be formed due to insufficient fusion or spatter ejection.

To ensure effective control over porosity, the following methods can be employed:

1. In order to better control porosity, preference should be given to an AC GMAW-based process or a CMT-PADV based process (cold metal transfer pulsed advanced) for the deposition of aluminium and its alloys.
2. It is strongly advised to utilise the highest-quality shielding gas available, make sure the gas seals are tight, use non-organic plumbing, and keep the pipe lengths short. These measures contribute to reducing porosity.
3. Prior to fabrication, both the wire and substrate surfaces should be thoroughly cleaned to minimise porosity.
4. Utilise high-quality feedstock materials to minimise the occurrence of porosity.
5. Optimise the shape of the deposited bead to achieve better control over porosity.
6. Monitor and control the thermal profile during the fabrication process to minimise porosity.
7. Consider applying post-deposition treatments such as interpass rolling, which can help reduce porosity and improve overall part quality.

13.5.3 Cracks and delamination

The material properties of the deposited material as well as the thermal parameters of the manufacturing process determine cracking and delamination, just like residual stress and deformation. In WAAM components, cracks are typically classified as solidification cracks or grain boundary cracks [58]. Solidification fractures are primarily determined by the material's solidification behaviour and are commonly caused by obstructions in the flow of solidified grains or significant strain in the melt pool. However, grain boundary cracking appears between grain interfaces because of changes in border morphology and the potential for precipitate production or dissolution [66]. Incomplete or insufficient remelting of the underlying solid between the layers can result in delamination or the separation of adjacent layers. Typically, this flaw is obvious and cannot be corrected with post-processing techniques. Pre-process treatments like substrate preheating should be taken into consideration to stop these faults [67].

When utilising the WAAM process to fabricate bimetal material combinations like Al/Cu, Al/Ti, and Al/Fe, there is a notable vulnerability to cracking and delamination. This susceptibility arises from the significant differences in mutual solubility and chemical reactivity between dissimilar metals, leading to the disruption of intermetallic phase equilibrium and the subsequent propagation of cracks along grain boundaries. Furthermore, because a liquid layer forms during terminal solidification, Inconel alloy is vulnerable to solidification cracking [68]. In order to avoid cracking and delamination, it is imperative to pay close attention to each of these materials.

The following steps can be taken to effectively control crack defects:

1. Employing mixed wires and optimising their compositions can contribute to reducing crack formation.
2. Lowering the cooling rate during the deposition process can help mitigate cracking issues.
3. Implementing measures aimed at enhancing strength rather than relying solely on solution treatment can be beneficial in controlling cracks.

13.6 MECHANICAL PROPERTIES

The mechanical properties of alloys deposited through additive manufacturing are influenced by various factors, including the microstructure, grain shape and size, presence of flaws, pore shape and size, and direction of material development. In the case of selective laser melting (SLM), the difference in mechanical properties such as ultimate tensile strength, yield strength, microhardness, etc. can be attributed to the gradient microstructure that forms along the building direction, which is a common occurrence in other deposition processes as well.

A comparison of various mechanical properties, such as ultimate tensile strength, surface roughness, density, microhardness, and elastic modulus, for specific materials such as Ti alloy (TI6Al4V), SS316L, NiTi (smart alloy), Co–Cr alloy, and Ta, among others, is shown in Table 13.1. It is observed that Ti and NiTi alloys exhibit the highest strength compared to other alloys developed using the same additive manufacturing process. Additionally, Co–Cr and Ti alloys demonstrate the highest hardness among the various alloys studied.

13.7 APPLICATIONS OF WAAM

Wire arc additive manufacturing has shown significant potential for various applications in many sectors, such as the medical sector, automobiles, energy, etc (see Figure 13.7). There are lots of applications of WAAM in the healthcare sector; it enables the fabrication of complex prototypes and small-scale production of medical devices, facilitating the development of innovative solutions for healthcare. The production of customised implants tailored to the specific needs of individual patients allows for the fabrication of complex shapes and intricate geometries, making them suitable for implants such as cranial plates, hip replacements, spinal implants, and maxillofacial implants. WAAM gives promising results for the manufacturing of lightweight, durable, and customised products for prosthetic parts such as limbs and supporting body structure, as well as for orthopaedics and dental implants like personalised orthotic braces, supports, corrective devices, and dental crowns and bridges. Manufacturing surgical instruments for advanced surgical procedures like minimally invasive surgeries, laparoscopic procedures, and robotic-assisted procedures gives promising results. It also offers the flexibility to create instruments with ergonomic features, enhancing surgical precision and reducing surgeon fatigue. Additionally,

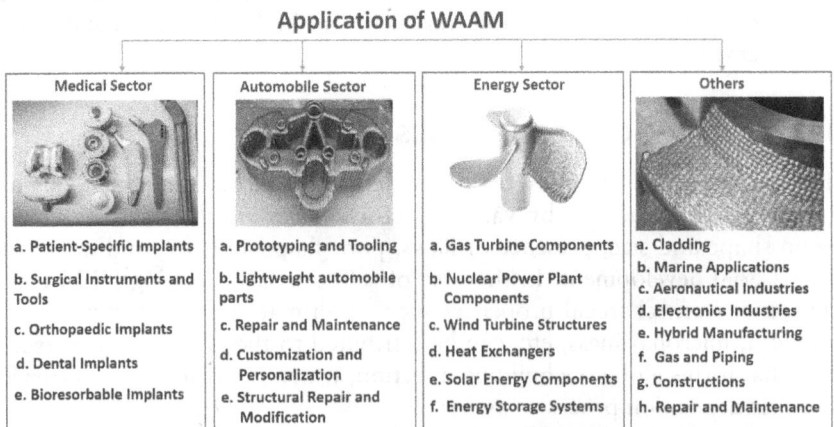

Application of WAAM

Medical Sector	Automobile Sector	Energy Sector	Others
a. Patient-Specific Implants	a. Prototyping and Tooling	a. Gas Turbine Components	a. Cladding
b. Surgical Instruments and Tools	b. Lightweight automobile parts	b. Nuclear Power Plant Components	b. Marine Applications
c. Orthopaedic Implants	c. Repair and Maintenance	c. Wind Turbine Structures	c. Aeronautical Industries
d. Dental Implants	d. Customization and Personalization	d. Heat Exchangers	d. Electronics Industries
e. Bioresorbable Implants	e. Structural Repair and Modification	e. Solar Energy Components	e. Hybrid Manufacturing
		f. Energy Storage Systems	f. Gas and Piping
			g. Constructions
			h. Repair and Maintenance

Figure 13.7 Applications of WAAM in various sectors.

further research and development are needed to address biocompatibility, material properties, and surface finishes specific to medical applications. Nonetheless, WAAM holds great potential for advancing personalised medicine and improving patient care in the medical field.

In the automobile sector, WAAM shows tremendous opportunities for the rapid production of complex parts and tooling components used in automobile manufacturing. It allows for faster design iterations and customisation, reducing the time and cost associated with traditional manufacturing methods. WAAM assists in the fabrication of lightweight structures and components, which is of great importance in the automobile industry for improving fuel efficiency and reducing emissions. On-site or in-house repair of damaged or worn-out parts, avoiding the need for costly replacements, is only possible through WAAM. It provides a cost-effective solution for low-volume production runs. It allows for the production of small batches of parts without the need for expensive tooling or extensive setup times. It's important to note that while WAAM offers numerous advantages for the automobile sector, there are still challenges to overcome, such as material selection, process optimization, and quality control. However, ongoing research and progress in WAAM technology are addressing these challenges, making it a promising option for various automotive applications.

The energy sector is one of the most important sectors for the development of any nation, and the dependency of other industries on the energy sector is very high. Designing advanced manufacturing processes for the energy sector is very important. WAAM gives lots of solutions and applications in this sector. The fabrication of complex geometries and high-temperature-resistant components for gas turbines, such as combustion chambers, turbine blades, and vanes, is possible through WAAM. It allows for the production of intricate cooling channels and optimised designs, improving the efficiency and performance of gas turbine systems. It has the ability to manufacture complex and customised components for nuclear power plants. It enables the production of reactor vessel parts, heat exchangers, and other critical components with enhanced material properties and reduced lead times. WAAM helps in the production of heat exchangers with intricate internal geometries and improved thermal performance. It enables the integration of advanced cooling designs, enhancing heat transfer efficiency and overall energy efficiency in various applications, including power plants and HVAC systems. WAAM has the potential to fabricate components for solar energy systems, including concentrator structures, heat absorbers, and support structures. The flexibility of WAAM enables the production of intricate designs tailored to specific solar applications. While the application of WAAM in the energy sector holds promise, further research and development are necessary to address specific material challenges, process optimisation, and certification requirements. Although the broad application of WAAM is mentioned in this article, WAAM has a lot of applications in other fields as well due to its unique capabilities and attractive technology.

13.8 CONCLUSION

Metal additive manufacturing, specifically WAAM, has the potential to revolutionise various industries, including manufacturing, automobiles, aerospace, medical, and electronics. The WAAM processes have been extensively developed and adopted within these industries, as discussed in this chapter. The selection of suitable materials for deposition, along with an understanding of the properties and defects associated with WAAM, is crucial. Undoubtedly, WAAM will have a profound impact on the engineering and biomedical professions. It is expected that engineers in fields such as biomedical, automotive, avionics, petrochemicals, civil engineering, and industrial design will increasingly utilise WAAM in their work. As progress continues in chemical science, WAAM will enable the production of more complex plastic components. It is highly likely that WAAM will play a significant role in the future of various engineering disciplines.

FUNDING DECLARATION

There was no specific grant for this research from any funding source.

REFERENCES

[1] D. Ding, Z. Pan, D. Cuiuri, and H. Li, "Wire-feed additive manufacturing of metal components: technologies, developments and future interests," *Int. J. Adv. Manuf. Technol.*, vol. 81, no. 1, pp. 465–481, 2015.

[2] J. D. Spencer, P. M. Dickens, and C. M. Wykes, "Rapid prototyping of metal parts by three-dimensional welding," *Proc. Inst. Mech. Eng. Part B J. Eng. Manuf.*, vol. 212, no. 3, pp. 175–182, 1998, doi: 10.1243/0954405981515590

[3] P. M. Dickens, M. S. Pridham, R. C. Cobb, I. Gibson, and G. Dixon, "Rapid prototyping using 3-D welding," *Solid Free. Fabr. Proc.*, pp. 280–290, 1992.

[4] Z. Feng, *Processes and Mechanisms of Welding Residual Stress and Distortion*. Elsevier Science, 2005.

[5] D. Ding, Z. Pan, D. Cuiuri, and H. Li, "A practical path planning methodology for wire and arc additive manufacturing of thin-walled structures," *Robot. Comput. Integr. Manuf.*, vol. 34, pp. 8–19, 2015.

[6] S. K. Sharma and C. Sharma, "Processing techniques, microstructural and mechanical properties of wire arc additive manufactured stainless steel: A review," *J. Inst. Eng. Ser. C*, vol. 31, pp. 1–15, 2022, doi: doi.10.1007/s40032-022-00853-5

[7]   E. Brandl, V. Michailov, B. Viehweger, and C. Leyens, "Deposition of Ti–6Al–4V using laser and wire, part I: Microstructural properties of single beads," *Surf. Coatings Technol.*, vol. 206, no. 6, pp. 1120–1129, 2011.

[8]   L. E. Murr et al., "Metal fabrication by additive manufacturing using laser and electron beam melting technologies," *J. Mater. Sci. Technol.*, vol. 28, no. 1, pp. 1–14, 2012, doi: https://doi.org/10.1016/S1005-0302(12)60016-4

[9]   Y. Zhang, Y. Chen, P. Li, and A. T. Male, "Weld deposition-based rapid prototyping: a preliminary study," *J. Mater. Process. Technol.*, vol. 135, no. 2–3, pp. 347–357, 2003.

[10] K. H. Kazmi, S. K. Sharma, A. K. Das, A. Mandal, and A. Shukla, "Development of wire arc additive manufactured Cu-Si alloy: Study of microstructure and wear behavior," *J. Mater. Eng. Perform.*, 2023, doi: 10.1007/s11665-023-07972-9

[11] G. N. Levy, R. Schindel, and J. P. Kruth, "Rapid manufacturing and rapid tooling with layer manufacturing (LM) technologies, state of the art and future perspectives," *CIRP Ann.*, vol. 52, no. 2, pp. 589–609, 2003, doi: https://doi.org/10.1016/S0007-8506(07)60206-6

[12] D. Ding, Z. Pan, S. Van Duin, H. Li, and C. Shen, "Fabricating superior NiAl bronze components through wire arc additive manufacturing," *Materials*, vol. 9, no. 8. 2016. doi: 10.3390/ma9080652

[13] F. Wang, S. Williams, and M. Rush, "Morphology investigation on direct current pulsed gas tungsten arc welded additive layer manufactured Ti6Al4V alloy," *Int. J. Adv. Manuf. Technol.*, vol. 57, no. 5, pp. 597–603, 2011, doi: 10.1007/s00170-011-3299-1

[14] W. Aiyiti, W. Zhao, B. Lu, and Y. Tang, "Investigation of the overlapping parameters of MPAW-based rapid prototyping," *Rapid Prototyp. J.*, vol. 12, no. 3, pp. 165–172, Jan. 2006, doi: 10.1108/13552540610670744

[15] H. Geng, J. Li, J. Xiong, X. Lin, and F. Zhang, "Optimization of wire feed for GTAW based additive manufacturing," *J. Mater. Process. Technol.*, vol. 243, pp. 40–47, 2017.

[16] K. H. Kazmi, A. K. Das, S. K. Sharma, A. Mandal, and A. K. Shukla, "Wire arc additive manufacturing of ER-4043 aluminum alloy: evaluation of bead profile, microstructure, and wear behavior," *Weld. World*, no. 0123456789, 2023, doi: 10.1007/s40194-023-01558-8

[17] Y. Ma, D. Cuiuri, N. Hoye, H. Li, and Z. Pan, "The effect of location on the microstructure and mechanical properties of titanium aluminides produced by additive layer manufacturing using in-situ alloying and gas tungsten arc welding," *Mater. Sci. Eng. A*, vol. 631, pp. 230–240, 2015, doi: https://doi.org/10.1016/j.msea.2015.02.051

[18] C. Shen, Z. Pan, D. Cuiuri, J. Roberts, and H. Li, "Fabrication of Fe-FeAl functionally graded material using the wire-arc additive manufacturing process," *Metall. Mater. Trans. B*, vol. 47, no. 1, pp. 763–772, 2016, doi: 10.1007/s11663-015-0509-5

[19] C. Shen, Z. Pan, Y. Ma, D. Cuiuri, and H. Li, "Fabrication of iron-rich Fe–Al intermetallics using the wire-arc additive manufacturing process," *Addit. Manuf.*, vol. 7, pp. 20–26, 2015, doi: https://doi.org/10.1016/j.addma.2015.06.001

[20] P. M. Almeida and S. Williams, "Innovative process model of Ti-6Al-4V additive layer manufacturing using cold metal transfer (CMT)," 2010.

[21] M. A. Somashekara, M. Naveenkumar, A. Kumar, C. Viswanath, and S. Simhambhatla, "Investigations into effect of weld-deposition pattern on residual stress evolution for metallic additive manufacturing," *Int. J. Adv. Manuf. Technol.*, vol. 90, no. 5, pp. 2009–2025, 2017, doi: 10.1007/s00170-016-9510-7

[22] H. Zhang, J. Xu, and G. Wang, "Fundamental study on plasma deposition manufacturing," *Surf. Coatings Technol.*, vol. 171, no. 1, pp. 112–118, 2003, doi: https://doi.org/10.1016/S0257-8972(03)00250-0

[23] F. Martina, J. Mehnen, S. W. Williams, P. Colegrove, and F. Wang, "Investigation of the benefits of plasma deposition for the additive layer manufacture of Ti–6Al–4V," *J. Mater. Process. Technol.*, vol. 212, no. 6, pp. 1377–1386, 2012, doi: https://doi.org/10.1016/j.jmatprotec.2012.02.002

[24] B. Mannion and J. Heinzman III, "Plasma arc welding brings better control," *Tool. Prod.*, vol. 5, pp. 29–30, 1999.

[25] S. W. Williams, F. Martina, A. C. Addison, J. Ding, G. Pardal, and P. Colegrove, "Wire + Arc additive manufacturing," *Mater. Sci. Technol. (United Kingdom)*, vol. 32, no. 7, pp. 641–647, 2016, doi: 10.1179/1743284715Y.0000000073

[26] L. Thijs, F. Verhaeghe, T. Craeghs, J. Van Humbeeck, and J.-P. Kruth, "A study of the microstructural evolution during selective laser melting of Ti–6Al–4V," *Acta Mater.*, vol. 58, no. 9, pp. 3303–3312, 2010, doi: https://doi.org/10.1016/j.actamat.2010.02.004

[27] B. Baufeld, O. van der Biest, and R. Gault, "Microstructure of Ti-6Al-4V specimens produced by shaped metal deposition," vol. 100, no. 11, pp. 1536–1542, 2009, doi: doi:10.3139/146.110217

[28] D. Herzog, V. Seyda, E. Wycisk, and C. Emmelmann, "Additive manufacturing of metals," *Acta Mater.*, vol. 117, pp. 371–392, 2016, doi: https://doi.org/10.1016/j.actamat.2016.07.019

[29] J. Lin et al., "Microstructural evolution and mechanical property of Ti-6Al-4V wall deposited by continuous plasma arc additive manufacturing without post heat treatment," *J. Mech. Behav. Biomed. Mater.*, vol. 69, pp. 19–29, May 2017, doi: 10.1016/j.jmbbm.2016.12.015

[30] J. J. Lin et al., "Microstructural evolution and mechanical properties of Ti-6Al-4V wall deposited by pulsed plasma arc additive manufacturing," *Mater. Des.*, vol. 102, pp. 30–40, 2016, doi: https://doi.org/10.1016/j.matdes.2016.04.018

[31] Y. Hirata, "Pulsed arc welding," *Weld. Int.*, vol. 17, no. 2, pp. 98–115, 2003, doi: 10.1533/wint.2003.3075

[32] J. Gu, J. Ding, S. W. Williams, H. Gu, P. Ma, and Y. Zhai, "The effect of inter-layer cold working and post-deposition heat treatment on porosity in additively manufactured aluminum alloys," *J. Mater. Process. Technol.*, vol. 230, pp. 26–34, 2016, doi: https://doi.org/10.1016/j.jmatprotec.2015.11.006

[33] P. Wang, S. Hu, J. Shen, and Y. Liang, "Characterization the contribution and limitation of the characteristic processing parameters in cold metal

transfer deposition of an Al alloy," *J. Mater. Process. Technol.*, vol. 245, pp. 122–133, 2017, doi: https://doi.org/10.1016/j.jmatprotec.2017.02.019

[34] S. Pradeep, S. K. Sharma, and V. Pancholi, "Microstructural and mechanical characterization of friction stir processed 5086 aluminum alloy," *Mater. Sci. Forum*, vol. 710, pp. 253–257, 2012, doi: 10.4028/www.scientific.net/MSF.710.253

[35] C. Brice, R. Shenoy, M. Kral, and K. Buchannan, "Precipitation behavior of aluminum alloy 2139 fabricated using additive manufacturing," *Mater. Sci. Eng. A*, vol. 648, pp. 9–14, 2015, doi: https://doi.org/10.1016/j.msea.2015.08.088

[36] K. H. Kazmi, S. K. Sharma, A. K. Das, A. Mandal, A. Kumar Shukla, and R. Mandal, "Wire arc additive manufacturing of ER-4043 aluminum alloy: Effect of tool speed on microstructure, mechanical properties and parameter optimization," *J. Mater. Eng. Perform.*, no. Ref 20, 2023, doi: 10.1007/s11665-023-08309-2

[37] M. L. Dezaki et al., "A review on additive/subtractive hybrid manufacturing of directed energy deposition (DED) process," *Adv. Powder Mater.*, p. 100054, 2022.

[38] E. R. Denlinger, J. C. Heigel, P. Michaleris, and T. A. Palmer, "Effect of inter-layer dwell time on distortion and residual stress in additive manufacturing of titanium and nickel alloys," *J. Mater. Process. Technol.*, vol. 215, pp. 123–131, 2015, doi: https://doi.org/10.1016/j.jmatprotec.2014.07.030

[39] B. Baufeld, "Mechanical properties of INCONEL 718 parts manufactured by shaped metal deposition (SMD)," *J. Mater. Eng. Perform.*, vol. 21, no. 7, pp. 1416–1421, 2012, doi: 10.1007/s11665-011-0009-y

[40] F. Xu, Y. Lv, Y. Liu, F. Shu, P. He, and B. Xu, "Microstructural evolution and mechanical properties of Inconel 625 alloy during pulsed plasma arc deposition process," *J. Mater. Sci. Technol.*, vol. 29, no. 5, pp. 480–488, 2013, doi: https://doi.org/10.1016/j.jmst.2013.02.010

[41] J. Guo, Y. Zhou, C. Liu, Q. Wu, X. Chen, and J. Lu, "Wire arc additive manufacturing of AZ31 magnesium alloy: Grain refinement by adjusting pulse frequency," *Materials (Basel).*, vol. 9, no. 10, 2016, doi: 10.3390/ma9100823

[42] T. Abe and H. Sasahara, "Dissimilar metal deposition with a stainless steel and nickel-based alloy using wire and arc-based additive manufacturing," *Precis. Eng.*, vol. 45, pp. 387–395, 2016, doi: https://doi.org/10.1016/j.precisioneng.2016.03.016

[43] L. Liu, Z. Zhuang, F. Liu, and M. Zhu, "Additive manufacturing of steel–bronze bimetal by shaped metal deposition: interface characteristics and tensile properties," *Int. J. Adv. Manuf. Technol.*, vol. 69, no. 9, pp. 2131–2137, 2013, doi: 10.1007/s00170-013-5191-7

[44] C. Shen, Z. Pan, D. Cuiuri, B. Dong, and H. Li, "In-depth study of the mechanical properties for Fe3Al based iron aluminide fabricated using the wire-arc additive manufacturing process," *Mater. Sci. Eng. A*, vol. 669, pp. 118–126, Jul. 2016, doi: 10.1016/j.msea.2016.05.047

[45] Y. Ma, D. Cuiuri, N. Hoye, H. Li, and Z. Pan, "Characterization of in-situ alloyed and additively manufactured titanium aluminides," *metall.*

Mater. Trans. B, vol. 45, no. 6, pp. 2299–2303, 2014, doi: 10.1007/s11663-014-0144-6

[46] Y. Ma, D. Cuiuri, C. Shen, H. Li, and Z. Pan, "Effect of interpass temperature on in-situ alloying and additive manufacturing of titanium aluminides using gas tungsten arc welding," *Addit. Manuf.*, vol. 8, pp. 71–77, Oct. 2015, doi: 10.1016/j.addma.2015.08.001

[47] J. Xiong, G. Zhang, Z. Qiu, and Y. Li, "Vision-sensing and bead width control of a single-bead multi-layer part: material and energy savings in GMAW-based rapid manufacturing," *J. Clean. Prod.*, vol. 41, pp. 82–88, 2013.

[48] H. Hermawan, D. Ramdan, and J. R. P. Djuansjah, "Metals for biomedical applications," *Biomed. Eng. – From Theory to Appl.*, 2011, doi: 10.5772/19033

[49] E. Maleki, S. Bagherifard, M. Bandini, and M. Guagliano, "Surface post-treatments for metal additive manufacturing: Progress, challenges, and opportunities," *Addit. Manuf.*, vol. 37, p. 101619, 2021, doi: 10.1016/j.addma.2020.101619

[50] F. Auricchio, E. Boatti, and M. Conti, *SMA Biomedical Applications*. 2015. doi: 10.1016/B978-0-08-099920-3.00011-5

[51] S. L. Sing, J. An, W. Y. Yeong, and F. E. Wiria, "Laser and electron-beam powder-bed additive manufacturing of metallic implants: A review on processes, materials and designs," *J. Orthop. Res.*, vol. 34, no. 3, pp. 369–385, 2016, doi: 10.1002/jor.23075

[52] K. Vanmeensel et al., *Additively Manufactured Metals for Medical Applications*. Elsevier, 2018. doi: 10.1016/B978-0-12-812155-9.00008-6.

[53] S. Agarwal, J. Curtin, B. Duffy, and S. Jaiswal, "SC," *Mater. Sci. Eng. C*, 2016, doi: 10.1016/j.msec.2016.06.020

[54] D. Adamovic, B. Ristic, and F. Zivic, *Review of Existing Biomaterials – Method of Material Selection for Specific Applications in Orthopedics*. 2017. Springer. doi: 10.1007/978-3-319-68025-5_3

[55] C. Ma et al., "Improving surface finish and wear resistance of additive manufactured nickel-titanium by ultrasonic nano-crystal surface modification," *J. Mater. Process. Technol.*, vol. 249, pp. 433–440, 2017, doi: 10.1016/j.jmatprotec.2017.06.038

[56] Q. Li et al., "A comprehensive study of tantalum powder preparation for additive manufacturing," *Appl. Surf. Sci.*, vol. 593, no. December 2021, p. 153357, 2022, doi: 10.1016/j.apsusc.2022.153357

[57] P. A. Colegrove et al., "Microstructure and residual stress improvement in wire and arc additively manufactured parts through high-pressure rolling," *J. Mater. Process. Technol.*, vol. 213, no. 10, pp. 1782–1791, 2013, doi: 10.1016/j.jmatprotec.2013.04.012

[58] W. J. Sames, F. A. List, S. Pannala, R. R. Dehoff, and S. S. Babu, "The metallurgy and processing science of metal additive manufacturing," *Int. Mater. Rev.*, vol. 61, no. 5, pp. 315–360, 2016, doi: 10.1080/09506608.2015.1116649

[59] J. Ding et al., "Thermo-mechanical analysis of Wire and Arc Additive Layer Manufacturing process on large multi-layer parts," *Comput. Mater. Sci.*, vol. 50, no. 12, pp. 3315–3322, Dec. 2011, doi: 10.1016/j.commatsci.2011.06.023

[60] T. Mukherjee, W. Zhang, and T. DebRoy, "An improved predic-
      tion of residual stresses and distortion in additive manufacturing,"
      Comput. Mater. Sci., vol. 126, pp. 360–372, Jan. 2017, doi: 10.1016/
      j.commatsci.2016.10.003

[61] P. Edwards, A. O'Conner, and M. Ramulu, "Electron beam additive manu-
      facturing of titanium components: Properties and performance," *J. Manuf.
      Sci. Eng.*, vol. 135, no. 6, pp. 1–7, 2013, doi: 10.1115/1.4025773

[62] A. Busachi, J. Erkoyuncu, P. Colegrove, F. Martina, and J. Ding, "Designing
      a WAAM Based Manufacturing System for Defence Applications,"
      Procedia CIRP, vol. 37, pp. 48–53, 2015, doi: https://doi.org/10.1016/
      j.procir.2015.08.085

[63] W. J. Sames, F. Medina, W. H. Peter, S. S. Babu, and R. R. Dehoff, "Effect
      of process control and powder quality on Inconel 718 produced using elec-
      tron beam melting," in *8th International Symposium on Superalloy 718
      and Derivatives*, 2014, pp. 409–423. doi: https://doi.org/10.1002/978111
      9016854.ch32

[64] T. Zhao et al., "Some factors affecting porosity in directed energy depos-
      ition of AlMgScZr-alloys," *Opt. Laser Technol.*, vol. 143, p. 107337,
      2021, doi: https://doi.org/10.1016/j.optlastec.2021.107337

[65] X. S. W. and H. Q. J Bai, H L Ding, J L Gu, "Porosity evolution in additively
      manufactured aluminium alloy during high temperature exposure," *J.
      Phys. Conf. Ser.*, vol. 755, no. 1, pp. 8–12, 2016, doi: 10.1088/1742-6596/
      755/1/011001

[66] T. A. Davis, "ThinkIR: The University of Louisville' s Institutional
      Repository The effect of process parameters on laser-deposited TI-6A1-
      4V," 2004.

[67] R. Warsi, K. H. Kazmi, and M. Chandra, "Mechanical properties of wire
      and arc additive manufactured component deposited by a CNC controlled
      GMAW," *Mater. Today Proc.*, vol. 56, no. November, pp. 2818–2825,
      2022, doi: 10.1016/j.matpr.2021.10.114

[68] Y. Tian, B. Ouyang, A. Gontcharov, R. Gauvin, P. Lowden, and M. Brochu,
      "Microstructure evolution of Inconel 625 with 0.4 wt% boron modifica-
      tion during gas tungsten arc deposition," *J. Alloys Compd.*, vol. 694, pp.
      429–438, 2017, doi: https://doi.org/10.1016/j.jallcom.2016.10.019

Chapter 14

Development of MoCoCrSi/fly ash composite cladding on stainless steel substrate through microwave irradiation

C. Durga Prasad, K. V. Manjunath, Prem Kumar Naik, Nagabhushana N. and Prakash Kumar

14.1 INTRODUCTION

Material processing is a significant phenomenon that transforms raw materials into application-ready products. Materials are treated using both traditional and non-conventional methods in the current industrial standard [1,2]. There is always room for improvement in the business when it comes to processing materials in a way that increases strength and speeds up production. One of the unconventional methods is the use of microwave radiation to treat materials [2–4]. The frequency of microwaves, which are largely coherent, polarized electromagnetic waves, ranges from 300 MHz to 300 GHz. Volumetric heating, which occurs at the molecular level, is a feature of microwaves. Low thermal deformation of the base metal is the outcome [5–8]. However, various qualities are frequently needed in various application areas. Microwaves are used in the sintering, joining, and melting of alloys among other materials. The main requirements in today's industries are greater strength and hardness, as well as improved mechanical, physical, and chemical qualities to increase wear and corrosion resistance [9,10].

The inside-out heating caused by molecular interactions with materials is a distinctive feature of microwave processing that makes it one of the most sought-after processing methods. Whereas traditional heating occurs from the outside to the interior of the substance, microwaves enter materials and produce heat at the molecular level. Due to the nature of traditional heating, temperature differences between material layers develop, resulting in material and energy losses. Due to the depth of the microwaves' penetration into the materials, it is possible to achieve consistent heating of the material during microwave processing [11–14]. Skin depth is another name for penetration depth. With an increase in the materials' conductivity, the skin depth drops. As a result, excellent conductors have skin that is thinner at ambient temperature. However, when the temperature rises, skin depth also increases. Metals reflect microwaves due to their reduced skin depth,

DOI: 10.1201/9781032713830-14

making it challenging to process them in a microwave setting. There is a method called microwave hybrid heating (MHH) to solve this issue [15–19]. Utilizing susceptor, a type of microwave-absorbing material, MHH transforms microwave energy into heat energy. Metals start to absorb microwaves once they reach a particular higher temperature, which causes volumetric heating. The term "critical temperature" refers to the temperature at which metals start to absorb microwaves [20–26].

In general, cladding is a surface modification method that is frequently used to create an overlay of appropriate materials on substrates with specified qualities by partially melting the substrate and completely melting the clad material. Tungsten inert gas (TIG), high velocity oxygen fuel (HVOF), and laser cladding are a few different ways to achieve cladding [27–33]. However, this technology has significant drawbacks, such as the possibility of cladding breaking owing to high thermal stress and thermal distortion during the process. Additionally, laser processing is not a particularly cost-effective way for cladding large regions [34–39].

It is crucial to research novel processing methods with the ability to do away with the drawbacks of laser cladding and provide superior microstructures and characteristics at lower costs and faster processing speeds. Microwave processing of materials has become popular as a framing technique in recent years [40–44]. Titanium is not as dense as steel, but it is more expensive than stainless steel, and cost plays a crucial part in industrial applications. Stainless steel is favored by most applications since they place a larger importance on weight than strength. The family of iron-based alloys includes stainless steel (SS), which is likewise regarded as a material with a wide range of applications. The vast majority of uses for which corrosion resistance and strength are required profit considerably from stainless steel's high corrosion resistance and low maintenance needs. Additionally, stainless steel is quite malleable, making it ideal for rolling into sheets, plates, bars, and tubes. The same substance, however, falters when used in severe settings. Therefore, surface modification is required to address the material failure [45–48].

To enhance its surface qualities, martensitic stainless steel (SS-410) is taken into consideration as a substrate material. MoCoCrSi/fly ash clad powder is utilized at a ratio of 70:30. Utilizing a household microwave oven, clad with an average thickness of 0.25–0.5 mm is produced. With regard to externally traditional crystalline metallic materials, the special combination of MoCoCrSi offers great strength, toughness, hardness, exceptional wear resistance, and corrosion resistance from room temperature up to 800°C [3,6–8]. Few studies on MoCoCrSi have suggested that processing MoCoCrSi to produce a higher fraction of hard intermetallic laves phases, which have an amorphous microstructure (metal glass nature), strengthens the MoCoCrSi coating at high temperatures [19,20].

14.2 MICROWAVE CLADDING

Cladding process is done to enrich the surface properties of the work material. The properties such as hardness, wear, and corrosion resistance are of major concerns in the various mechanical and tribological applications. To achieve better surface properties, various surface modification techniques, namely, laser ablation, laser cladding, physical vapor deposition (PVD), chemical vapor deposition (CVD), and development of thin-film cladding, are employed to meet the requirements [49–52].

During the cladding process, the surface is modified by the addition of a new layer with the desired powder materials. The unique advantage of MWP opens a wider scope in the field of cladding in which the surface properties of the substrate can be improved in very shorter processing time. The microwave cladding process modifies the surface by the reduction of porosity, better hardness, uniform microstructure, and enhanced bonding with substrate.

(a) In case of microwave cladding, the base surface is cleaned with acetone and similar kind of organic compound to remove the unwanted impurities from the substrate.

(b) The cleaned substrate is preheated in the simple oven (i.e., temperature around 80°C–100°C) to remove the moisture content before the addition of powder layer.

(c) The applied powder layer is then exposed to microwave irradiations.

(d) The remaining portion of the substrate is covered with a suitable masking material. Proper insulation and masking on the surface prevent heat loss and damage to the cavity. However, there is further scope to improve the cladding performance that is the use of susceptor heating which increases the heat transfer rate with volumetric heating.

(e) For the application of susceptor medium, a separator plate (i.e., graphite plate) is placed in between the powder particles and susceptor medium as shown in Figure 14.1.

(f) Particle roughness and size of the cladding material play an important role in the bonding formation with the substrate material.

(g) The coarser particle absorbs more energy as compared to finer particles; therefore, it will aid to lead more tendencies to form a bond with the substrate material.

14.3 BASICS OF MICROWAVE MATERIAL PROCESSING

The physical characteristics of the materials are primarily responsible for the efficient heating of materials by microwave radiations. These characteristics control how microwave radiations interact with diverse materials.

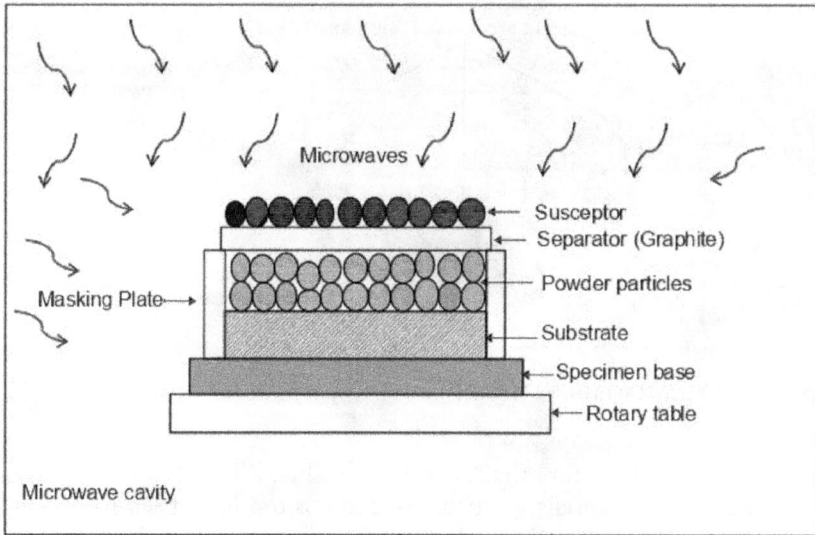

Figure 14.1 Schematic representation of microwave hybrid heating.

The current experiment uses the substrate SS-410 and the powders of MoCoCrSi/fly ash. The powder substance is more durable and corrosion- and wear-resistant. Martensitic stainless steel was clad in the residential microwave equipment.

- Microwaves are increasingly being used to treat materials. Although well-known processes like dipolar heating and conduction heating have been extensively studied, the phenomena connected with the processing are less well understood.
- The majority of critical events that result in heating during microwave–material interaction and heat transfer during microwave energy absorption in materials, as seen in Figure 14.2, are covered in the current work.
- Using appropriate illustrations, the mechanisms involved in the interaction of microwaves with distinctively various materials, including metals, non-metals, and composites (metal matrix composites, ceramic matrix composites, and polymer matrix composites), have been explored.
- It was shown that whereas the loss effects connected with the magnetic field are responsible for the microwave heating of materials based on metals, the loss effects related with the electric field include dipolar loss and conduction loss in the microwave heating of non-metals.

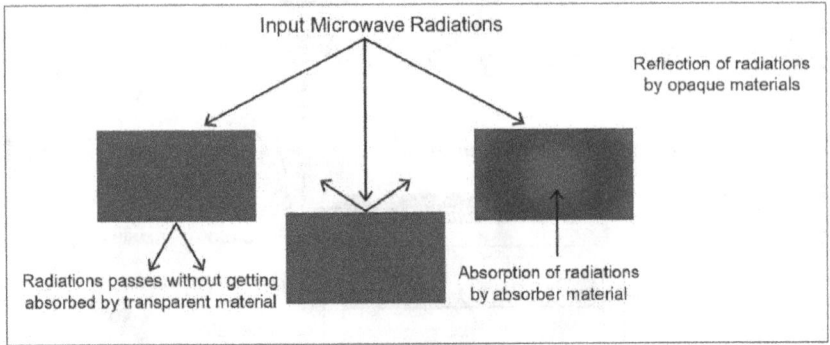

Figure 14.2 Microwave material interaction.

- From the literature that is now accessible, difficulties in processing advanced materials, particularly composites, have been recognized. Future research topics with potential advantages have been noted.
- A material's capacity to absorb high-frequency electromagnetic energy (microwaves) and transform it into heat is what causes microwave heating. Molecules that are permanently polarized due to chemical bonds undergo dipolar polarization in the presence of a high-frequency electric field, realigning them and causing microwave heating.
- Due to the high frequency, the realignment happens a million times per second, resulting in internal molecular friction that heats up the material by volume.

Since their magnetic and electrical components are at right angles to one another, microwaves are electromagnetic radiations. The frequency range of microwaves is 300 MHz to 300 GHz. The effective wavelength of microwaves ranges from 1 mm to 1 m. The qualities of the material affect how microwaves interact with it. Materials may be divided into three types based on how they interact with microwaves:

(a) Microwave-absorbing materials: Field strength of electromagnetic components attenuates along the thickness of the material in absorbing materials. However, the rate of decreasing in field strength depends on the dielectric loss of factor of the materials (e.g., water, ceramics, polymers).
(b) Microwave-insulating material: field strength of microwaves falls only up to negligible thickness in the materials falling into this category this are because of lower skin depth of these materials (e.g., metallic materials).

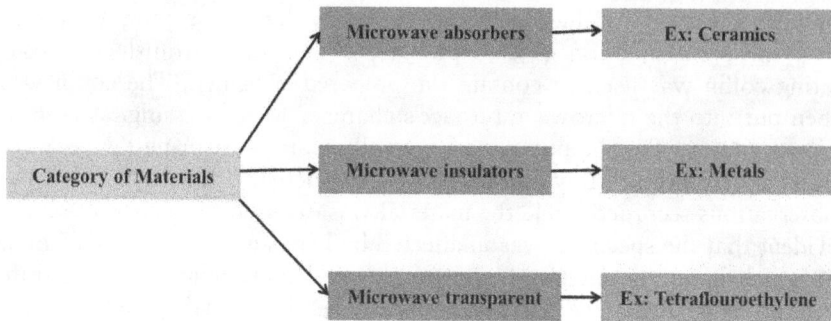

Figure 14.3 Category of materials.

(c) Microwave-transparent materials: Microwave transparent materials are also called low loss insulators as depicted in Figure 14.3. There is no decrease in the electrical and magnetic field strength along the thickness of the materials (e.g., tetrafluoroethylene).

(d) Generally, materials can be classified into three types based on their interaction with microwaves.
- Opaque or electrical conductors where microwaves are reflected and do not penetrate.
- Transparent or low dielectric loss materials, in which microwaves are neither reflected nor absorbed, but are transmitted through the material with little attenuation.
- Absorbers or high dielectric loss materials absorb microwave energy to a certain degree based on the value of the dielectric loss factor and convert it to heat [53–55].

14.4 EXPERIMENT AND CHARACTERIZATION

Cladding of MoCoCrSi/fly ash composite powder on stainless steel substrate was accomplished by irradiating 900 W microwaves at 2.45 GHz frequency generated in a modified home microwave oven. A temperature-control setup with an infrared pyrometer for temperature measurement was attached as part of the modification. The substrate material was purchased as a plate with the following measurements: (400×400×10) mm, which was divided into samples measuring (25×25×10) mm. The surface of the samples was carefully cleaned by rubbing against fine emery and using acetone to remove any unwanted particles or scales. Clad powder (70% MoCoCrSi + 30% fly ash) was combined with a resin to create slurry, which was then evenly applied to the substrate's surface. The silicon carbide (SiC) brick was deposited on top of the clad powder after a thin graphite film. By acting as a separator, graphite sheet prevents contact between the susceptor and clad powder. Additionally, due to their interaction with the susceptor, separators stop the contamination of clad powder. A microwave translucent insulating coffin was used to contain the prepared specimen. The coffin was then put into the microwave furnace's chamber for processing, as seen in Figure 14.4(a). For the purpose of controlling and measuring temperature using a pyrometer, a tiny aperture was constructed. Table 14.1 shows the observations recorded while the material was exposed to microwaves. It is evident that the specimen was unaffected by exposure for the first 15 min, at which point the interface between the clad powder and the substrate was discovered to have partially melted. The clad powder melts entirely after about 25 min of exposure, producing a metallurgical link with the substrate's surface. The schematic representation of the processing setup is shown in Figure 14.4(b). The mounted specimen was polished first by emery paper of 320 grit, followed by further polishing using emery papers of grades 400, 600, 800, 1200, 1500, 2000 and finally with 1-micron diamond paste on a velvet cloth placed on the polishing machine and the specimen was etched. Polishing was often used to enhance the appearance of

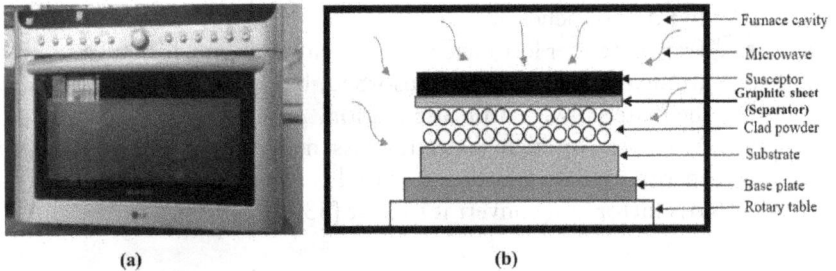

(a)   (b)

Figure 14.4 (a) Microwave applicator used; (b) schematic representation of the processing setup.

Table 14.1 Experimental observation with exposure time

Duration (min)	Observation
1	No heat formation on substrate and no clad
5	Sample heated up, but no clad
10	Sample and clad powders were heated up but no clad
15	Sample and clad powders were heated up but no clad
20	Sample and clad powders were heated up but no clad
25	Semi melting of clad powders and observed clad growth
30	Excessive heating caused deformation of substrate

an item, prevent contamination of instruments, remove oxidation, create a reflective surface, or prevent corrosion in pipes. Polishing was used to create a flat, defect-free surface for examination of a metal's microstructure under a microscope silicon-based polishing pads or a diamond solution can be used in the polishing process. Polishing stainless steel can also increase the sanitary benefits of it. Clads that have been developed were sectioned and polished in accordance with metallurgical characterization requirements. Using a Zeiss EVO LS 15 scanning electron microscope, micrographs of clad were recorded. On both unclad and cladded specimens, the Vicker's microhardness test was performed at ten random samples. On the Highwood HWMMY-X7 hardness tester, a test load of 100 g was applied for dwell duration of 30 s. The average microhardness value is reported.

14.5 RESULTS AND DISCUSSION

14.5.1 Morphology studies

The morphology of clad powder MoCoCrSi/fly ash is shown in Figure 14.5. The blended powders were distributed properly and particles exhibit a spherical shape. The top surface of the clad is shown in Figure 14.6(a), and the transverse cross-sectional view is shown in Figure 14.6(b). Cross-sectional view clearly shows the metallurgical bonding formed between the substrate and clad powder due to MHH, with an average thickness of around 150 μm. Good metallurgical bonding has increased hardness, which can be observed from the hardness plot presented in Figure 14.7.

In Figure 14.7, values obtained from the hardness test are shown on a graph. The hardness curve clearly demonstrates that the cladded surface is much harder than the uncladded surface. The uncladded surface has a hardness of 136.8 Hv, but the average hardness of the formed clads is determined to be 540.10 Hv.

Figure 14.5 Microstructure of MoCoCrSi/fly ash particles.

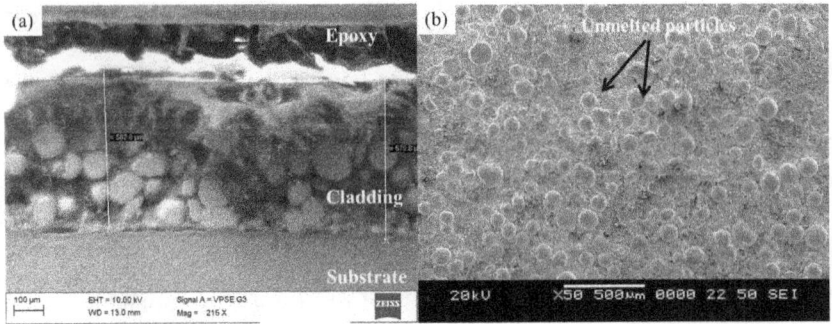

Figure 14.6 Micrographs of (MoCoCrSi/fly ash) cladded stainless steel specimen (a) Top surface (b) cross-section.

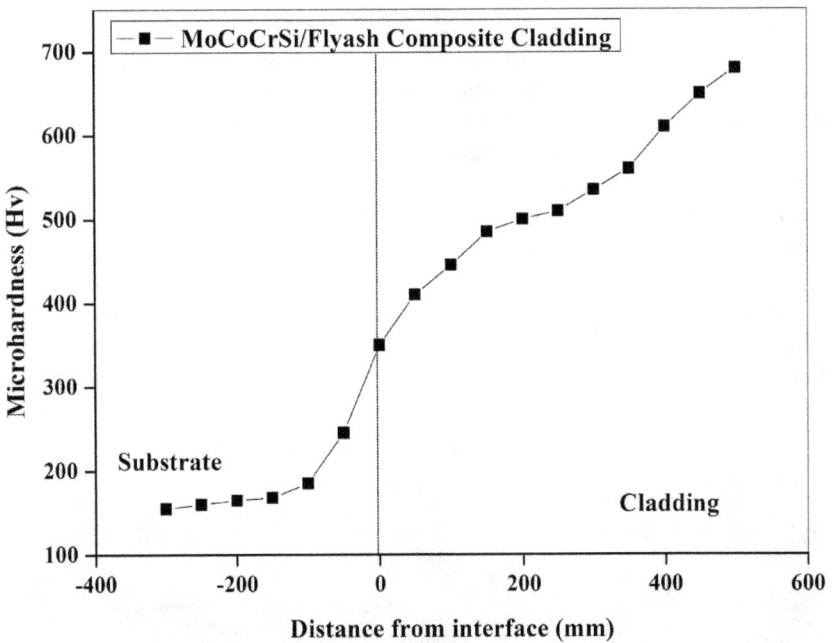

Figure 14.7 Vicker's microhardness plot.

14.6 CONCLUSIONS

The current study demonstrates how a layer of MoCoCrSi and fly ash composite powder may be developed on stainless steel substrates with the use of microwave irradiation at 2.45 GHz and 900 W power using a home microwave oven. The created clad's microstructure and hardness are examined, and the results are as follows.

(i) The development of the MoCoCrSi/fly ash composite clad on stainless steel substrate required microwave irradiation for 25 min.

(ii) A good metallurgical connection between the substrate and the clad powder is revealed by observation of the microstructure.

(iii) Additionally, uniformity in the cladding can be seen across the substrate, demonstrating that microwave cladding is a practical method for creating clads of a consistent thickness.

(iv) The hardness of the clad surface is 540±10 Hv, which is much greater than the substrate's unclad hardness of 136±8 Hv.

CONFLICTS OF INTEREST

The authors declare that there are no conflicts of interest.

ACKNOWLEDGMENT

The authors of the article would like to acknowledge Visvesvaraya Technological University for providing financial assistance under Competitive Research Scheme [Ref.: VTU/TEQIP 3/2019/321].

REFERENCES

[1] Sharma A K, Aravindan S, Krishnamurthy R, "Microwave glazing of alumina-titanium ceramic composite coatings", *Material Letter*, 50 (2001) 295–301.

[2] Chun-Ming Lin, Shih-Hung Yen, Cherng-Yuh Su, Measurement and optimization of atmospheric plasma sprayed CoMoCrSi coatings parameters on Ti-6Al-4V substrates affecting microstructural and properties using hybrid abductor induction mechanism, *Measurement* 94 (2016) 157–167.

[3] Naik T, Mathapathi M, Durga Prasad C, Nithin H S and Ramesh, M R, "Effect of laser post treatment on microstructural and sliding wear behavior of HVOF sprayed NiCrC and NiCrSi coatings", *Surface Review and Letters,* https://doi.org/10.1142/S0218625X2250007X.

[4] Durga Prasad C, Joladarashi S, Ramesh M R, Sarkar A. "High temperature gradient cobalt based clad developed using microwave hybrid heating", *American Institute of Physics*, 1943 (2018) 020111 https://doi.org/10.1063/1.5029687.

[5] Madhusudana Reddy G, Durga Prasad C, Shetty G, Ramesh M R, Nageswara Rao T, Patil P. "High temperature oxidation studies of plasma sprayed NiCrAlY/TiO$_2$ & NiCrAlY /Cr$_2$O$_3$/YSZ cermet composite coatings on MDN-420 special steel alloy". *Metallography, Microstructure and Analysis* https://doi.org/10.1007/s13632-021-00784-0

[6] Gupta D, Sharma A K, "Development and microstructural characterization of microwave cladding on austenitic steel," *Surf. Coat. Technol.* 205 (2011) 5147–5155.

[7] Sudana Reddy M, Durga Prasad C, Patil P, Ramesh M R, Nageswara Rao, "Hot corrosion behavior of plasma sprayed NiCrAlY/TiO$_2$ and NiCrAlY/Cr$_2$O$_3$/YSZ cermets coatings on alloy steel", *Surfaces and Interfaces*, 22 (2021) 100810. https://doi.org/10.1016/j.surfin.2020.100810

[8] Durga Prasad C, Joladarashi S, Ramesh M R, Srinath M S, "Microstructure and tribological resistance of flame sprayed CoMoCrSi/WC-CrC-Ni and CoMoCrSi/WC-12Co composite coatings remelted by microwave hybrid heating". *Journal of Bio and Tribo-Corrosion*, 6 (2020)124, https://doi.org/10.1007/s40735-020-00421-3

[9] Durga Prasad C, Joladarashi S, Ramesh M R, "Comparative investigation of HVOF and flame sprayed CoMoCrSi coating", *American Institute of Physics*, 2247 (2020) 050004, https://doi.org/10.1063/5.0003883

[10] Thostenson E T, Chou T W, "Microwave processing: Fundamentals and applications," *Composites A*, 30 (1999) 1055–1071.

[11] Durga Prasad C, Jerri A, Ramesh M R, "Characterization and sliding wear behavior of iron based metallic coating deposited by HVOF process on low carbon steel substrate", *Journal of Bio and Tribo-Corrosion*, 6 (2020) 69. https://doi.org/10.1007/s40735-020-00366-7

[12] Hebbale A M, Srinath M S, "Taguchi analysis on erosive wear behavior of cobalt based microwave cladding on stainless steel AISI-420," *Measurement* 99 (2017) 98–107.

[13] Yang Li, Xiufang Cui, Guo Jin, Zhaobing Cai, Na Tana, Bingwen Lu, Zonghong Gao, "Interfacial bonding properties between cobalt-based plasma cladding layer and substrate under tensile conditions," *Materials and Design* 123 (2017) 54–63.

[14] Sharma A K, Gupta D. A method of cladding/coating of metallic and non-metallic powders on metallic substrates by microwave irradiation, Indian Patent 527/Del/2010.

[15] Durga Prasad C, Joladarashi S, Ramesh M R, Srinath M S, Channabasappa B H. "Comparison of high temperature wear behavior of microwave assisted HVOF sprayed CoMoCrSi-WC-CrC-ni/WC-12Co composite coatings", *Silicon* (2020) 1–19. https://doi.org/10.1007/s12633-020-00398-1

[16] Gupta D, Bhovi P M, Kumar Sharma A, Dutta S, "Development and characterization of microwave composite cladding," *Journal of Manufacturing Processes* 14 (2012) 243–249.

[17] Kumar Sharma A, Gupta D, "On microstructure and flexural strength of metal–ceramic composite cladding developed through microwave heating," *Applied Surface Science* 258 (2012) 5583–5592.

[18] Bell T, "Surface engineering of austenitic stainless steel", *Surface Engineering*, 18(6) (2002) 415–422.

[19] Srinath M S, Sharma A K, "Investigation on microstructural and mechanical processed dissimilar joints", *Journal of Manufacturing Processes*, 13 (2011) 141–146.

[20] Bhoi N K, Singh H, Pratap S, Jain P K. "Microwave material processing: a clean, green, and sustainable approach", *Sustainable Engineering Products and Manufacturing Technologies*, 2019C (2019) 6–7.

[21] Durga Prasad C, Mathapathi M, Vasudev H and Thakur L, "Analysis of mechanical properties and microstructural characterisation of microwave cladding on stainless steel", in *Advances in Microwave Processing for Engineering Materials*, CRC Press, Taylor & Francis Group, ISBN: 9781003248743, 2022, https://doi.org/10.1201/9781003248743

[22] Durga Prasad C, Mathapathi M, Joladarashi S, Ramesh M R, "Investigation of Microstructural and tribological behavior of metco 41C+WC-12Co composite coating sprayed by HVOF process", in *Thermal Spray Coatings*, CRC Press, Taylor and Francis Group. ISBN: 9781032081489. 2021 https://doi.org/10.1201/9781003213185

[23] Durga Prasad C, Jerri A, Ramesh M R, "Evaluation of microstructural and dry sliding wear resistance of iron-based SiC-reinforced composite coating by HVOF process", in *Thermal Spray Coatings*, 2021, CRC Press, Taylor and Francis Group. ISBN: 9781032081489. https://doi.org/10.1201/9781003213185

[24] Sachin B, Charitha M Rao, Gajanan M Naik, Durga Prasad C, Hebbale A M, Vijeesh V, Muralidhara Rao, "Minimum quantity lubrication and cryogenic for burnishing of difficult to cut material as a sustainable alternative", *Lecture Notes in Mechanical Engineering*, 2021, Springer, ISBN 978-981-16-2278-6, https://doi.org/10.1007/978-981-16-2278-6_6

[25] Kumar K. P, Mohanty A, Lingappa M L, Srinath M S, Panigrahi S K, "Enhancement of surface properties of austenitic stainless steel by nickel-based alloy cladding developed using microwave energy technique", *Materials Chemistry and Physics*, 256 (2020) 123657.

[26] Durga Prasad C, Joladarashi S, Ramesh M R, Srinath M S, Channabasappa B H. "Effect of microwave heating on microstructure and elevated temperature adhesive wear behavior of HVOF deposited CoMoCrSi-Cr$_3$C$_2$ composite coating", *Surface and Coatings Technology*, 374 (2019) 291–304 https://doi.org/10.1016/j.surfcoat.2019.05.056

[27] Fei Weng, Huijun Yu, Chuanzhong Chen, Jianli Liu, Longjie Zhao, Jingjie Dai, Zhihuan Zhao, "Effect of process parameters on the microstructure evolution and wear property of the laser cladding coatings on Ti-6Al-4V alloy", *Journal of Alloys and Compounds* 692 (2017) 989–996.

[28] Durga Prasad C, Joladarashi S, Ramesh M R, Srinath M S, Channabasappa, B. H, "Development and sliding wear behavior of Co-Mo-Cr-Si cladding through microwave heating", *Silicon*, 11 (2019), 2975–2986, https://doi.org/10.1007/s12633-019-0084-5

[29] Zafar S, Sharma A K, "Development and characterizations of WC-12Comicrowave clad," *Material Characteristics*, 96 (2014) 241–248.

[30] Durga Prasad C, Joladarashi S, Ramesh M R, Srinath M S, Channabasappa B H. "Microstructure and tribological behavior of flame sprayed and microwave fused CoMoCrSi/CoMoCrSi-Cr$_3$C$_2$ coatings", *Materials Research Express*, 6 (2019) 026512 https://doi.org/10.1088/2053-1591/aaebd9

[31] Gupta M, Wong W L E, "Enhancing overall mechanical performance of metallic materials using two-directional microwave assisted rapid sintering," *Scripta Mater.* 52 (2005) 479–483.

[32] Durga Prasad C, Joladarashi S, Ramesh M R, Srinath M S, Channabasappa B H, "Influence of microwave hybrid heating on the sliding wear behaviour of HVOF sprayed CoMoCrSi coating". *Materials Research Express*, 5 (2018):086519 https://doi.org/10.1088/2053-1591/aad44e.

[33] Lin Y C, Wang S W, "Wear behavior of ceramic powder cladding on an S50C steel surface," *Tribology International*, 36 (1) (2003) 1–9.

[34] Asif Kattimani M, Venkatesh P R, Masum H, Math M M, Bahadurdesai V N, Mustafkhadri S, Durga Prasad C, Vasudev H, "Design and numerical analysis of tensile deformation and fracture properties of induction hardened inconel 718 superalloy for gas turbine applications", *International Journal on Interactive Design and Manufacturing (IJIDeM)* (2023), https://doi.org/10.1007/s12008-023-01452-z

[35] Rajesh Kannan A, Durga Prasad C, Rajkumar V, Siva Shanmugam N, Rajkumar V, Wonjoo Lee, Jonghun Yoon, "Hot oxidation and corrosion behaviour of boiler steel fabricated by wire arc additive manufacturing", *Materials Characterization*, 203 (2023) 113113, https://doi.org/10.1016/j.matchar.2023.113113

[36] Kulkarni G S, Siddeshkumar N G, Durga Prasad C, Latha Shankar, Suresh R, "Drilling of GFRP with liquid silicon rubber reinforced with fine aluminium powder on hole surface quality and tool wear using DOE", *Journal of Bio- and Tribo-Corrosion*, 9 (2023) Article number: 53, https://doi.org/10.1007/s40735-023-00771-8

[37] Kulkarni S D, Manjunatha, Chandrasekhar U, Manjunath K V, Durga Prasad C, Vasudev H, "Design and optimization of polyvinyl-nitride rubber for tensile strength analysis", *International Journal on Interactive Design and Manufacturing (IJIDeM)* (2023) https://doi.org/10.1007/s12008-023-01405-6

[38] Praveen N, Mallik U S, Shivasiddaramaih A G, Suresh R, Durga Prasad C, Shivaramu L, "Synthesis and wire EDM characteristics of Cu–Al–Mn ternary shape memory alloys using Taguchi method", *Journal of The Institution of Engineers (India): Series D* (2023), https://doi.org/10.1007/s40033-023-00501-x

[39] Math M M, Rajeswara Rao K V S, Prapul Chandra A C, Vijayakumar M N, Nandini B, Durga Prasad C, Vasudev H, "Design and modeling using finite element analysis for the sitting posture of computer users based on ergonomic perspective", *International Journal on Interactive Design and Manufacturing (IJIDeM)* (2023), https://doi.org/10.1007/s12008-023-01383-9

[40] Anjaneya G, Sunil S, Kakkeri S, Math M M, Vaibhav M N, Solaimuthu C, Durga Prasad C, Vasudev H, "Numerical Simulation of Microchannel Heat Exchanger using CFD", *International Journal on Interactive Design and Manufacturing (IJIDeM)* (2023), https://doi.org/10.1007/s12008-023-01376-8

[41] Madhu Sudana Reddy G, Durga Prasad C, Kollur S, Lakshmikanthan A, Suresh R, Aprameya C R, "Investigation of high temperature erosion

behaviour of NiCrAlY/TiO2 plasma coatings on titanium substrate", *JOM The Journal of The Minerals, Metals & Materials Society (TMS)*, https://doi.org/10.1007/s11837-023-05894-4

[42] Nagabhushana P, Ramprasad S, Durga Prasad C, Vasudev H, Prakash C, "Numerical investigation on heat transfer of a nano-fluid saturated vertical composite porous channel packed between two fluid layers", *International Journal on Interactive Design and Manufacturing (IJIDeM)* (2023), https://doi.org/10.1007/s12008-023-01379-5

[43] Shanthala K, Hebbale A M, Durga Prasad C, Muralidhar Singh M, Harish H, Karigowda S, "Analysis of high velocity forming of metallic tubes", *Materials Today Proceedings* (2023), https://doi.org/10.1016/j.matpr.2023.04.647

[44] Praveen N, Mallik U S, Shivasiddaramaih A G, Suresh R, Shivaramu L, Durga Prasad C, Gupta M, "Design and analysis of shape memory alloys using optimization techniques", *Advances in Materials and Processing Technologies* (2023), https://doi.org/10.1080/2374068X.2023.2208021

[45] Madhusudana Reddy G, Durga Prasad C, Patil P, Kakur N, Ramesh M R, "High temperature erosion performance of NiCrAlY/Cr_2O_3/YSZ plasma spray coatings", *Transactions of the IMF* (2023), https://doi.org/10.1080/00202967.2023.2208899

[46] Muralidhar Singh M, Hebbale A M, Durga Prasad C, Harish H, Kumar M, Shanthala K, "Design and simulation of vertical axis windmill for streetlights", *Materials Today Proceedings* (2023), https://doi.org/10.1016/j.matpr.2023.03.729

[47] Poojari M, Hanumanthappa H, Durga Prasad C, Madhusoodan Jathanna H, Ksheerasagar A R, Shetty P, Shanmugam B K, Vasudev H, "Computational modelling for the manufacturing of solar-powered multifunctional agricultural robot", *International Journal on Interactive Design and Manufacturing (IJIDeM)* (2023), https://doi.org/10.1007/s12008-023-01291-y

[48] Manjunatha C J, Durga Prasad C, Hanumanthappa H, Rajesh Kannan A, Mohan D G, Shanmugam B K, Venkategowda C, "Influence of microstructural characteristics on wear and corrosion behaviour of Si_3N_4 reinforced Al2219 composites", *Advances in Materials Science and Engineering*, 2023 (2023) Article ID 1120569, https://doi.org/10.1155/2023/1120569

[49] Sharanabasva H, Durga Prasad C, Ramesh M R, "Characterization and wear behavior of NiCrMoSi microwave cladding", *Journal of Materials Engineering and Performance* (2023), https://doi.org/10.1007/s11665-023-07998-z

[50] Rajesh Kannan A, Rajkumar V, Durga Prasad C, Siva Shanmugam N, Jonghun Yoon, "Microstructure and hot corrosion performance of stainless steel 347 produced by wire arc additive manufacturing," *Vacuum* (2023), 111901, https://doi.org/10.1016/j.vacuum.2023.111901.

[51] Madhusudana Reddy G, Durga Prasad C, Patil P, Kakur N, Ramesh M R, "Investigation of plasma sprayed NiCrAlY/Cr_2O_3/YSZ coatings on erosion performance of MDN 420 steel substrate at elevated temperatures", *International Journal of Surface Science and Engineering* (2023), http://dx.doi.org/10.1504/IJSURFSE.2023.10054266

[52] Sharanabasva H, Durga Prasad C, Ramesh M R, "Effect of mo and SiC reinforced NiCr microwave cladding on microstructure, mechanical and wear properties", *Journal of The Institution of Engineers (India): Series D* (2023). https://doi.org/10.1007/s40033-022-00445-8

[53] Nithin H S, Nishchitha K M, Pradeep D G, Durga Prasad C, Mathapati M, "Comparative analysis of CoCrAlY coatings at high temperature oxidation behavior using different reinforcement composition profiles", *Welding in the World,* 67 (2023) 585–592. https://doi.org/10.1007/s40 194-022-01405-2

[54] Naveen D.C, Kakur N, Keerthi Gowda B.S, Madhu Sudana Reddy G, Durga Prasad C, Ragavanantham S, "Effects of polypropylene waste addition as coarse aggregate in concrete: experimental characterization and statistical analysis", *Advances in Materials Science and Engineering* (2022), Article ID 7886722, 11 pages, https://doi.org/10.1155/2022/7886722.

[55] Gowda V, Hanumanthappa H, Shanmugam B K, Durga Prasad C, Sreenivasa T N, Rajendra Kumar M S, "High-temperature tribological studies on hot forged Al6061-TiB$_2$ in-situ composites", *Journal of Bio and Tribo-Corrosion,* 8 (2022) 101. https://doi.org/10.1007/s40735-022-00699-5

Chapter 15

Advantages and applications of various surface engineering techniques

Mukhtiar Singh, Maninder Singh, Hitesh Vasudev and Amrinder Mehta

15.1 INTRODUCTION TO CLADDING TECHNOLOGY

Cladding technology refers to the use of surface coatings to improve the performance and appearance of materials (K. J. Khor 2000). Surface cladding involves the application of a layer of material onto the surface of a substrate to enhance its properties, such as corrosion resistance, wear resistance, and chemical resistance. This process can provide a cost-effective and durable alternative to using high-performance materials throughout the entire structure (Leyens et al., 2010). Surface cladding can be performed using various techniques such as thermal spraying, laser cladding, and electroplating, depending on the substrate material and the desired properties. The cladding material can be a metal, ceramic, polymer, or composite, and can be selected based on factors such as cost, performance, and environmental impact. Thermal spraying is a commonly used technique for surface cladding. This process involves heating the cladding material in a high-temperature flame or plasma stream, which melts and atomizes the material into tiny droplets. These droplets are then propelled onto the substrate surface, where they solidify and form a thin layer of cladding material. Laser cladding is another technique that uses a high-energy laser to melt and fuse the cladding material onto the substrate surface (Vaßen et al., 2016). Electroplating involves the deposition of a metal coating onto a substrate surface through an electrolytic process. Surface cladding technology can be applied to a wide range of materials, including metals, alloys, plastics, and composites. It can provide a durable and cost-effective alternative to using high-performance materials throughout the entire structure (Boccaccini et al., 2016). For example, a steel substrate can be coated with a layer of corrosion-resistant material to improve its longevity and performance in harsh environments. In addition to its functional benefits, surface cladding can also improve the aesthetic appearance of materials. The cladding material can be selected to match the color and texture of the substrate, or to create a contrasting effect. Surface cladding can also be used to create decorative patterns or designs, or to apply branding or signage. However, the selection and application of surface

DOI: 10.1201/9781032713830-15

cladding materials require careful consideration of the material properties, substrate properties, and the intended use and environment of the material. Improper selection or application of surface cladding materials can result in reduced performance, increased costs, and potential safety hazards. The history of surface cladding technology can be traced back to ancient times when craftsmen used various surface coatings to protect and enhance the appearance of metal objects. However, the development of modern surface cladding technology began in the early 20th century, with the discovery of new materials and the advancement of surface engineering techniques. The history of electroplating dates back to the early 19th century when Italian chemist Luigi V. Brugnatelli discovered the electroplating process by accident while conducting experiments with silver nitrate and electricity. However, it was not until the early 20th century that electroplating became widely used in industry. One of the early industrial applications of electroplating was the deposition of a thin layer of nickel onto brass or copper substrates, which created a surface that resembled solid nickel but was much less expensive (Guilemany et al., 2019). This process became known as "nickel plating" and was used in the production of various household items such as faucets, knobs, and handles. In the automotive industry, electroplating was used to create a shiny, corrosion-resistant finish on various metal components such as bumpers, grilles, and wheels. Chromium plating, in particular, became popular in the 1930s and 1940s and was used extensively in the design of automobiles.

15.2 SURFACE CLADDING TECHNIQUES

A sort of surface engineering approach used to improve the thermal performance of surfaces is called thermal cladding. These techniques entail covering an object's surface with a coating of material to enhance its thermal characteristics, such as heat transfer, insulation, or heat resistance. Some common thermal cladding systems are depicted in Figure 15.1. Surface cladding techniques refer to a group of processes used to apply a layer of material onto the surface of a substrate. The primary goal of surface cladding is to enhance the surface properties of the substrate, such as wear resistance, corrosion resistance, or aesthetic appearance.

Cladding techniques are widely used in various industries, including aerospace, automotive, biomedical, and construction. Surface cladding techniques can be classified into four main categories: physical vapor deposition (PVD), chemical vapor deposition (CVD), thermal spraying, and electroplating. Each technique has its unique advantages and disadvantages, making them suitable for specific applications. This chapter aims to provide a comprehensive overview of surface cladding techniques, including their principles, advantages, limitations, and current research trends.

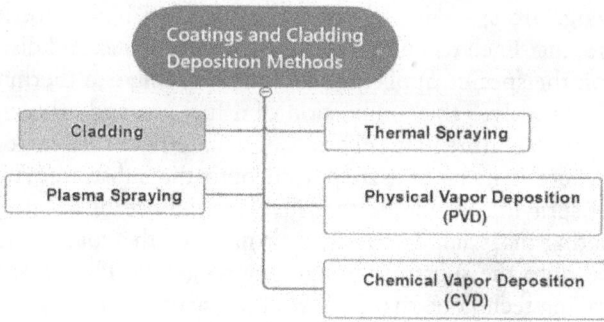

Figure 15.1 Some common thermal cladding technologies.

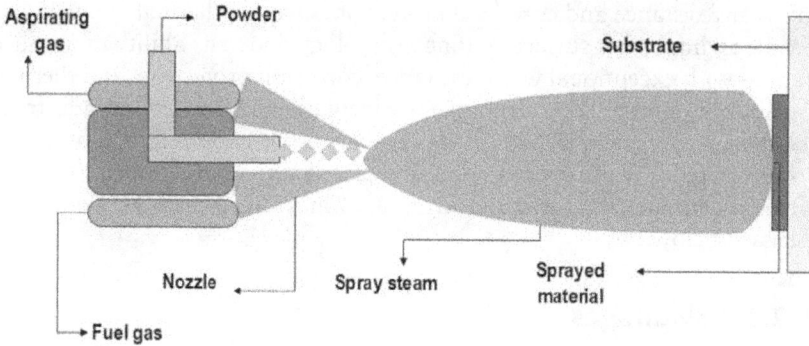

Figure 15.2 Schematic principle of thermal spraying process.

15.2.1 Thermal spraying process

Thermal spraying is a process of applying a coating material to a substrate using a high-temperature flame or plasma jet. The technique was first developed in the 1910s, and it became widely used in the 1940s for applications such as wear resistance, corrosion protection, and thermal insulation. As shown in Figure 15.2, the thermal spraying process begins with the preparation of a substrate, which is typically made of metal, ceramic, or plastic. The substrate is first cleaned and then prepared using a variety of techniques, such as grit blasting, to ensure proper adhesion of the coating material. The coating material is then fed into a high-temperature flame or plasma jet, where it is melted and accelerated toward the substrate. The molten droplets of the coating material then impact the substrate, where they solidify and form a coating.

Thermal spraying can be performed using a variety of techniques, including flame spraying, arc spraying, plasma spraying, and high-velocity oxy-fuel (HVOF) spraying. Each technique has its own advantages and disadvantages depending on the specific application. Flame spraying is a thermal spraying technique that involves the combustion of a fuel gas and an oxygen gas to create a high-temperature flame. The coating material is then introduced into the flame, where it melts and is sprayed onto the substrate. Arc spraying is a thermal spraying technique that involves the use of an electric arc to melt the coating material. The molten droplets of the coating material are then sprayed onto the substrate using compressed air. Plasma spraying is a thermal spraying technique that involves the use of a plasma jet to melt and accelerate the coating material. The high-velocity plasma stream provides excellent control over the coating process, allowing for precise coating thickness and composition. HVOF spraying is a thermal spraying technique that involves the use of a high-velocity combustion process to accelerate the coating material. The resulting coating is highly dense and provides excellent wear resistance and corrosion protection. Overall, thermal spraying has become an important surface coating technology due to its ability to produce coatings with exceptional wear resistance, corrosion protection, and thermal insulation properties. The technique is widely used in a variety of industries, including aerospace, automotive, and energy and is expected to continue to grow in popularity as new materials and techniques are developed.

This technique offers several advantages and disadvantages, which are discussed below.

15.2.2 Advantages

- *Wide range of materials*: Thermal spraying can be used with a wide range of materials, including metals, ceramics, polymers, and composites.
- *Thickness flexibility*: Thermal spraying can produce coatings with a wide range of thicknesses, making it suitable for a variety of applications.
- *Enhanced durability*: Thermal sprayed coatings can improve the wear resistance and hardness of a substrate, making it more durable.
- *Repair capabilities*: Thermal spraying can be used to repair damaged components, reducing the need for expensive replacements.
- *Cost-effective*: Thermal spraying is a cost-effective surface cladding technique, as it is relatively inexpensive compared to other methods such as chemical vapor deposition or physical vapor deposition.

15.2.3 Disadvantages

- *Porosity issues*: Thermal sprayed coatings can have porosity issues if the coating material is not properly prepared or applied.

- *Adhesion issues*: Thermal sprayed coatings may have adhesion issues if the substrate is not properly prepared before the spraying process.
- *Environmental concerns*: Thermal spraying can produce hazardous waste and emissions, which can pose environmental and health risks if not handled properly.
- *Limited coating uniformity*: Thermal spraying can produce non-uniform coatings, which may affect the performance of the substrate.

15.3 PHYSICAL VAPOR DEPOSITION (PVD)

The PVD is a process of depositing a thin film of material onto a substrate using a high-energy source such as a plasma or ion beam. The technique was first developed in the 1960s and became widely used in the 1980s for applications such as wear resistance, corrosion protection, and decorative coatings. The PVD process begins with the preparation of a substrate, which is typically made of metal, ceramic, or plastic. The substrate is first cleaned and then placed in a vacuum chamber along with the material to be deposited. The chamber is then evacuated to a high vacuum to remove any residual gases and other contaminants. Once the vacuum is achieved, a high-energy source such as a plasma or ion beam is used to bombard the material, causing it to vaporize and form a cloud of atoms or ions. The atoms or ions then travel through the vacuum and condense onto the substrate, forming a thin film of material (Staia et al., 2018). PVD can be performed using a variety of techniques, including sputtering, evaporation, and ion plating. Each technique has its own advantages and disadvantages, depending on the specific application. Sputtering is a PVD technique that involves bombarding a target material with high-energy ions, causing atoms to be ejected from the target and deposited onto the substrate. This technique is particularly useful for depositing metallic coatings and can be performed at relatively low temperatures. Evaporation is another PVD technique that involves heating a material until it vaporizes and condenses onto the substrate. This technique is particularly useful for depositing materials with high melting points and can be used to create thin films with precise thickness and composition. Ion plating is a PVD technique that involves bombarding the substrate with high-energy ions while simultaneously depositing the material. This technique can produce coatings with exceptional adhesion and wear resistance and is commonly used in the automotive and aerospace industries.

This technique offers several advantages and disadvantages, which are discussed below.

15.3.1 Advantages

- *High-quality coatings*: PVD can produce high-quality coatings with excellent adhesion, hardness, and wear resistance.

- *Thickness control*: PVD coatings can be deposited with high precision, allowing for precise thickness control and uniformity.
- *Improved surface properties*: PVD coatings can improve the surface properties of a substrate, such as reducing friction, improving corrosion resistance, and enhancing the appearance.
- *Environmentally friendly*: PVD is an environmentally friendly surface cladding technique, as it produces no hazardous waste or emissions.
- *Versatile*: PVD can be used with a wide range of materials, including metals, alloys, ceramics, and plastics.

15.3.2 Disadvantages

- *High equipment cost*: PVD equipment can be expensive to purchase and maintain, making it less accessible to smaller businesses.
- *Limited coating thickness*: PVD coatings are typically thin, with a maximum thickness of a few microns, making them unsuitable for some applications.
- *Limited substrate size*: PVD is limited by the size of the vacuum chamber used for the process, which can limit the size of the substrate that can be coated.
- *Limited material compatibility*: PVD coatings may not be compatible with all materials, and the substrate may need to be specially prepared to ensure adhesion.
- *Deposition rate limitations*: PVD can have a slow deposition rate, which may not be suitable for high-volume production.

15.4 CHEMICAL VAPOR DEPOSITION (CVD)

CVD is a process of depositing a thin film of material onto a substrate by chemical reaction. The technique was first developed in the 1940s and became widely used in the 1970s for applications such as semiconductor manufacturing, optical coatings, and protective coatings. The CVD process typically involves the use of a precursor gas, which is introduced into a reaction chamber containing the substrate. The precursor gas is then activated by a high-energy source, such as plasma or a heated filament, which causes it to dissociate and react with the substrate. The reaction between the precursor gas and the substrate results in the deposition of a thin film of material onto the substrate. The thickness and composition of the film can be controlled by adjusting the process parameters, such as the temperature, pressure, and precursor gas flow rate (Chen et al., 2019).CVD can be used to deposit a wide range of materials, including metals, ceramics, and semiconductors. The technique is widely used in the semiconductor industry for the deposition of thin films of silicon, silicon dioxide, and other materials used in the fabrication of microelectronic devices. CVD is also used in the production of optical coatings, such as anti-reflective coatings and mirror coatings, which are used in a variety of applications, including eyeglasses, camera lenses, and telescopes. Protective coatings produced by CVD are used in a variety of industrial applications to provide wear resistance, corrosion protection, and thermal insulation. For example, the coatings produced can be used to protect cutting tools and engine components from wear and corrosion.

The technique offers several advantages and disadvantages, which are discussed below:

15.4.1 Advantages

- *High-quality films*: CVD can produce high-quality and uniform films of various materials, including metals, semiconductors, and insulators. The films can have precise thickness, composition, and structure, making them suitable for various applications.
- *High growth rate*: CVD can produce films at a high growth rate, allowing for high throughput and cost-effectiveness in large-scale production.
- *Conformal coating*: CVD can coat complex shapes and structures with high conformality, enabling the deposition of uniform and precise films onto irregular and intricate surfaces.
- *Low-temperature processing*: CVD can deposit films at relatively low temperatures, making it suitable for temperature-sensitive substrates such as polymers and composites.

15.4.2 Disadvantages

- *Equipment and maintenance cost*: CVD equipment can be expensive and requires regular maintenance and calibration to ensure reproducibility and reliability.
- *Hazardous and toxic chemicals*: CVD involves the use of hazardous and toxic chemicals, such as metal–organic precursors, hydrogen, and nitrogen, which can pose safety and environmental risks.
- *Limited material range*: CVD can only deposit materials that can be synthesized in a gaseous form, limiting the range of available materials for deposition.
- *Precursor purity*: CVD requires high-purity precursors to ensure the quality and reproducibility of the deposited films, which can be costly and difficult to obtain

15.5 LASER CLADDING PROCESS

Laser cladding is a process of depositing a coating material onto a substrate using a high-power laser beam. The technique was first developed in the 1980s and became widely used in the 1990s for applications such as wear resistance, repair of damaged components, and aerospace coatings. One of the key advantages of laser cladding is its ability to deposit coatings with precise control over their microstructure and composition. This has made laser cladding an attractive option for the repair of damaged components and the production of high-performance coatings. Another advantage of laser cladding is its ability to produce coatings with high wear resistance and corrosion resistance. This has made the technique popular in the aerospace industry, where coatings with these properties are required for critical components. Laser cladding could be used to produce coatings with high wear resistance and low friction, making them suitable for use in aircraft engine components (Davis, 2003). Laser cladding has also been used to produce coatings with unique properties, such as super hydrophobicity and biocompatibility (Bhushan, 2017). Laser cladding could be used to produce coatings with superhydrophobic properties, which could be useful in applications such as self-cleaning surfaces and anti-icing coatings. Laser cladding could be used to produce biocompatible coatings for use in medical implants. In recent years, there has been growing interest in the use of additive manufacturing techniques, such as laser cladding, to produce complex components with customized geometries. Laser cladding could be used to produce complex components with high precision and accuracy, making it a promising option for the production of customized parts in a variety of industries.

15.5.1 Advantages

- *Precise and controlled deposition*: Laser cladding allows for precise and controlled deposition of the coating material, resulting in a high-quality and uniform coating.
- *Minimal heat input*: The heat input during laser cladding is minimal, which reduces the chances of thermal distortion or damage to the substrate.
- *Wide range of materials*: Laser cladding can be used with a wide range of materials, including metals, ceramics, and composites.
- *High deposition rate*: Laser cladding has a high deposition rate compared to other techniques, which makes it more efficient for large-scale applications.
- *Customizable properties*: The properties of the coating can be customized by controlling the process parameters, such as laser power, beam size, and scanning speed.

15.5.2 Disadvantages

- *High equipment cost*: The equipment required for laser cladding is expensive, which makes it less accessible for small-scale applications.

- *Limited thickness*: Laser cladding is limited in terms of the thickness of the coating that can be deposited. Thicker coatings require multiple passes, which can be time-consuming and costly.
- *Limited coating area*: The size of the coating area is limited by the size of the laser beam, which can be a disadvantage for large-scale applications.
- *Health and safety hazards*: Laser cladding can produce hazardous fumes and dust, which can be harmful to operators if proper safety measures are not taken.

15.6 MATERIALS FOR SURFACE CLADDING

Surface cladding involves the application of a protective coating to a material surface, and it is essential for the longevity and performance of various engineering applications. Different materials can be used for surface cladding, depending on the specific application requirements. The choice of cladding material is usually influenced by factors such as cost, durability, chemical and mechanical properties, and ease of application. In this review, we will explore some of the commonly used materials for surface cladding and their applications (see Figure 15.3).

Aluminium, titanium, nickel-based alloys, and stainless steel are widely used materials known for their strength, corrosion resistance, and durability. Aluminium, valued for its lightweight and strength-to-weight ratio,

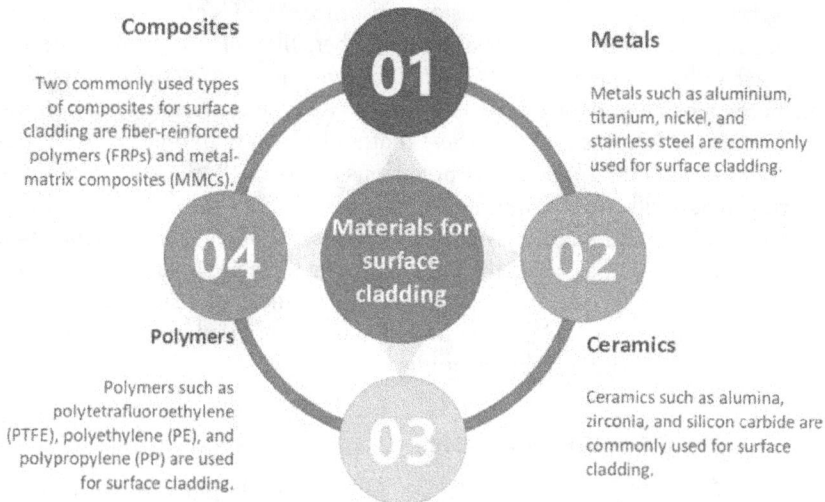

Composites

Two commonly used types of composites for surface cladding are fiber-reinforced polymers (FRPs) and metal-matrix composites (MMCs),

Metals

Metals such as aluminium, titanium, nickel, and stainless steel are commonly used for surface cladding.

01

04

Materials for surface cladding

02

Polymers

Polymers such as polytetrafluoroethylene (PTFE), polyethylene (PE), and polypropylene (PP) are used for surface cladding.

03

Ceramics

Ceramics such as alumina, zirconia, and silicon carbide are commonly used for surface cladding.

Figure 15.3 Materials for surface cladding.

finds applications in the aerospace and automotive industries (Callister & Rethwisch, 2018). It can be coated onto substrates using techniques like thermal spraying and electroplating. Titanium, with high strength, low density, and corrosion resistance, is used in aerospace, medical, and chemical sectors, and deposited onto substrates through techniques such as PVD and CVD. Nickel-based alloys are preferred in the oil and gas industry due to their wear resistance and high-temperature performance, applied using methods like thermal spraying and electroplating. Stainless steel's corrosion resistance makes it suitable for harsh environments in industries such as food, chemical, and oil and gas, coated via electroplating, PVD, or thermal spraying. Ceramics, like alumina, zirconia, and silicon carbide, are used for surface cladding due to their high-temperature stability, wear resistance, and chemical inertness. Alumina provides excellent wear resistance and stability and therefore used in various applications. Zirconia offers high melting point, hardness, and chemical stability, widely used as a thermal barrier and wear-resistant coating. Silicon carbide exhibits excellent hardness, wear resistance, and thermal shock resistance, commonly applied in abrasive applications. Polymers, including PTFE (polytetrafluoroethylene), PE (polyethylene), and PP (polypropylene), are increasingly used for surface cladding due to their unique properties. Polymers offer low friction coefficient, non-stick properties (PTFE), and chemical resistance (PE and PP). However, polymers have limitations in hardness, scratch resistance, and UV degradation, requiring careful consideration for specific applications. Fiber-reinforced polymers (FRPs) and metal matrix composites (MMCs) are commonly used in aerospace, marine, and high-temperature applications. FRPs comprise a polymer matrix reinforced with carbon, glass, or aramid fibers, providing a high strength-to-weight ratio and corrosion resistance. MMCs comprise a metallic matrix, such as aluminum, reinforced with fibers like silicon carbide, offering high strength, stiffness, and thermal conductivity. However, composites can be expensive and prone to brittleness (ASM International Handbook Committee, 1994). In summary, these materials, each with unique properties and limitations, find applications in various industries for surface cladding, catering to specific needs like strength, corrosion resistance, lightweight, wear resistance, and thermal stability. MMCs are increasingly being used in the aerospace and automotive industries due to their superior properties. They are also being explored for use in the medical industry, such as for implants and prostheses. Additionally, they are being used in the military for armor plating and other military applications.

15.5 SURFACE PREPARATION

Surface preparation is a critical step in surface cladding to ensure the quality and durability of the coating. It involves removing any impurities, contaminants, and oxidation from the substrate's surface to achieve a clean and smooth surface. The surface preparation method used depends on the type of substrate, the desired surface finish, and the type of coating material used. Cleaning is the first step in surface preparation, which involves removing any dirt, oil, grease, and other contaminants from the surface (Rohatgi, 1996). This can be achieved using solvents, detergents, or mechanical cleaning methods such as brushing, wiping, or blasting. Cleaning ensures that the surface is free from any organic or inorganic impurities that can affect the bonding of the coating material to the substrate. Several studies have investigated the effect of cleaning on the quality of surface cladding. To investigate the effect of cleaning on the adhesion strength of a zinc coating deposited by thermal spraying on steel substrates. The authors found that cleaning the surface with a combination of solvent cleaning and grit blasting resulted in a higher adhesion strength compared to other cleaning methods. Similarly, a study investigated the effect of cleaning on the quality of a nickel coating deposited by electroplating on aluminum substrates. The authors found that cleaning the surface with a combination of alkaline degreasing, acid pickling, and rinsing resulted in a smoother and cleaner surface, which led to a higher quality coating with improved corrosion resistance. Blasting is another common method used in surface preparation, which involves the use of abrasive media to remove any rust, scale, or oxidation from the surface. The abrasive media can be in the form of sand, grit, or shot, depending on the type of substrate and the desired surface finish. Blasting creates a rough surface profile that enhances the adhesion of the coating material to the substrate. On investigating the effect of blasting on the adhesion strength of a copper coating deposited by electroplating on steel substrates, the authors found that blasting the surface with aluminium oxide particles improved the surface roughness and led to higher adhesion strength of the coating (Qian et al., 2018). Coating is the final step in surface preparation, which involves applying a thin layer of primer or adhesive to the substrate's surface. The coating material enhances the adhesion of the cladding material to the substrate and protects the surface from further oxidation or contamination (Kalpakjian & Schmid,2014). The type of coating material used depends on the type of substrate and the type of cladding material used. The effectiveness of surface preparation methods depends on several factors, including the type of substrate, the type of coating material, and the desired surface finish. Inadequate surface preparation can result in poor adhesion, blistering, or delamination of the cladding material, leading to premature failure of the coating (Kelly & Groves, 1993). Therefore, it is essential to select the appropriate surface preparation method based on the substrate's condition and the desired surface finish to achieve a high-quality surface cladding. On investigating the effect of coating on the corrosion resistance of a magnesium alloy, the authors found that coating the surface with a nickel coating using electroplating resulted in a higher corrosion resistance compared to uncoated surface.

15.7 DESIGN CONSIDERATIONS

Design considerations are essential in surface cladding to ensure that the coating is effective and durable. Some key design considerations are shown in Figure 15.4. The thickness of the coating is critical in determining the protective properties of the cladding. A thicker coating may offer better protection against wear, corrosion, and erosion but may also result in higher production costs. The coating thickness is influenced by factors such as the intended application, the properties of the substrate, and the properties of the coating material (Matthews & Rawlings, 1994). Proper selection of the coating thickness is important to ensure that it can withstand the intended environmental conditions.

The adhesion between the substrate and the coating is also a crucial factor in the design of surface cladding. Proper adhesion ensures that the coating remains firmly attached to the substrate, preventing delamination, cracking, and peeling. The strength of the adhesion is influenced by factors such as the surface preparation, the type of coating material, and the environmental conditions. Techniques such as surface roughening, chemical treatment, and interlayer bonding are used to improve the adhesion between the substrate and the coating (Hull & Clyne, 1996). Surface cladding can affect the mechanical properties of the substrate. This is particularly important in load-bearing applications such as structural components. The mechanical properties of the substrate must be considered when selecting the coating material, as the coating may affect the substrate's strength, stiffness, and toughness. Careful selection of the coating material can help maintain the desired mechanical properties of the substrate.

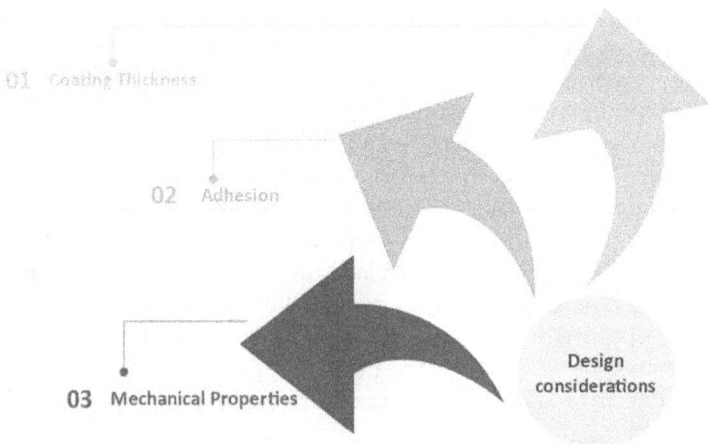

01 Coating Thickness

02 Adhesion

03 Mechanical Properties

Design
considerations

Figure 15.4 Design considerations in surface cladding.

15.8 APPLICATIONS OF SURFACE CLADDING

Surface cladding has a wide range of applications across various industries. Some important applications of the surface cladding are shown in Figure 15.5.

- *Aerospace*: Surface cladding is commonly used in the aerospace industry for repairing and enhancing the wear resistance of components such as engine blades, landing gear, and turbine blades. Materials such as nickel-based alloys and titanium are often used for their high-temperature performance and corrosion resistance. Techniques such as laser cladding and thermal spraying are commonly used for their precision and efficiency.
- *Automotive*: Surface cladding is used in the automotive industry for improving the wear resistance and corrosion resistance of engine components, such as camshafts, crankshafts, and cylinder liners. Materials such as aluminum, magnesium, and zinc are used for their lightweight and high strength-to-weight ratio. Techniques such as electroplating and physical vapor deposition are commonly used for their high-quality finishes and precision.
- *Biomedical applications*: Surface cladding is used in the biomedical industry for enhancing the biocompatibility and wear resistance of implants such as hip and knee replacements. Materials such as titanium, cobalt–chrome, and stainless steel are used for their biocompatibility and corrosion resistance. Techniques such as plasma spraying and chemical vapor deposition are commonly used for their ability to create bioactive coatings that promote bone growth and integration.
- *Oil and gas*: Surface cladding is used in the oil and gas industry for repairing and enhancing the wear resistance of components such as drill bits, valves, and pumps. Materials such as tungsten carbide and nickel-based alloys are used for their wear resistance and

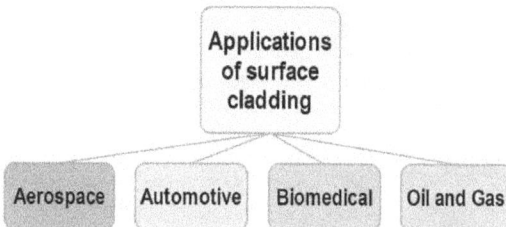

Figure 15.5 Applications of surface cladding.

high-temperature performance. Techniques such as thermal spraying and laser cladding are commonly used for their efficiency and precision.

15.9 CHALLENGES AND FUTURE TRENDS

Surface cladding technology has evolved significantly over the years, offering a wide range of benefits in terms of improving the mechanical, chemical, and thermal properties of materials. However, there are still some challenges that need to be addressed in order to improve the effectiveness and efficiency of surface cladding technology. Figure 15.6 illustrates some of the more prevalent difficulties encountered with surface cladding.

One of the major challenges in surface cladding technology is the development of new materials that can withstand high-temperature, high-pressure, and corrosive environments. This requires the development of new alloys and composites that can provide better mechanical and chemical properties, as well as better adhesion to the substrate. Another challenge is the improvement of coating adhesion, as this can affect the performance and durability of the coating. Improving the adhesion between the coating and the substrate requires the development of new surface preparation techniques and the optimization of coating parameters such as the thickness and composition of the coating. The reduction of environmental impact is also a major

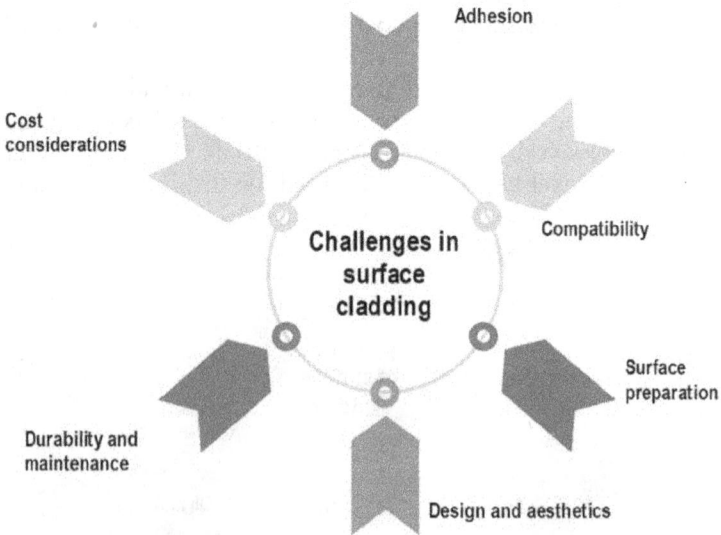

Figure 15.6 Some common challenges in surface cladding.

challenge in surface cladding technology. Many surface cladding techniques involve the use of hazardous chemicals and materials, which can have negative environmental impacts. Therefore, it is important to develop more sustainable and environmentally friendly surface cladding techniques and materials. In terms of future trends, the use of additive manufacturing is expected to revolutionize surface cladding technology. Additive manufacturing allows for the creation of complex geometries and the deposition of multiple materials, which can be used to create customized coatings for specific applications. Another future trend is the development of self-healing coatings, which can repair themselves in response to mechanical or chemical damage. Self-healing coatings can significantly increase the durability and lifespan of surface coatings, reducing the need for frequent maintenance and replacement. In conclusion, while surface cladding technology has come a long way, there are still challenges that need to be addressed to improve its effectiveness and efficiency. By developing new materials, improving coating adhesion, and reducing environmental impact, as well as exploring new trends such as additive manufacturing and self-healing coatings, surface cladding technology has the potential to transform a wide range of industries and applications.

REFERENCES

ASM International Handbook Committee, *ASM Handbook: Volume 5, Surface Engineering*, ASM International, 1994.

Bhushan, B. (Ed.). (2017). *Handbook of Surface and Nanometrology* (4th ed.), Springer International Publishing.

Boccaccini, A. R., M. Braic, M. Brito Correia, & S. Virtanen, "Surface coatings in biomedical applications," *Surface Coatings and Technology*, vol. 309, pp. 110–138, 2016.

Callister, W. D., & D. G. Rethwisch (2018). *Materials Science and Engineering: An Introduction* (10th ed.), Wiley.

Callister, W. D., and D. G. Rethwisch. *Materials Science and Engineering: An Introduction* (9th ed.), Wiley, 2011.

Chen, Y., et al. (2019). "Recent advances in chemical vapor deposition for coatings." *Journal of Materials Chemistry A*, 7, 9294–9316.

Davis, J.R. (2003). *Surface Engineering for Wear Resistance*, ASM International.

Guilemany, J.M., et al. (2019). "Recent advances in thermal spraying for surface engineering." *Surface Engineering*, 35(6), 459–464.

Hull, D., & T. W. Clyne, *An Introduction to Composite Materials*, Cambridge University Press, 1996.

Kalpakjian, S., & Schmid, S.R. (2014). *Manufacturing Engineering and Technology* (7th ed.), Pearson Education.

Kelly, A., & G. W. Groves, *The Properties of Materials Used in Engineering*, Longman, 1993.

Khor, K. J. "Surface coatings and functional finishes: a survey," *Journal of Materials Processing Technology*, vol. 107, pp. 1–15, 2000.

Leyens, C., & M. Peters, Eds., *Surface Engineering and Technology for Biomedical Implants*, Woodhead Publishing, 2010.

Matthews, F. L., & J. Rawlings, *Composite Materials: Engineering and Science*, Chapman and Hall, 1994.

Qian, L., et al. (2018). "Recent advances in laser cladding technology: A review." *Journal of Laser Applications*, 30(2), 022303.

Rohatgi, P. K. *Composite Materials: Science and Engineering*, Wiley, 1996.

Staia, M.H., et al. (2018). "Advancements in physical vapor deposition (PVD) coatings." *Materials Science Forum*, 915, 24–33.

Vaßen, R., K. Bobzin, N. Bagcivan, & A. Öte, "Thermal spraying for surface engineering: A review," *Journal of Thermal Spray Technology*, vol. 25, pp. 249–283, 2016.

Index

For Product Safety Concerns and Information please contact our EU
representative GPSR@taylorandfrancis.com
Taylor & Francis Verlag GmbH, Kaufingerstraße 24, 80331 München, Germany

www.ingramcontent.com/pod-product-compliance
Lightning Source LLC
Chambersburg PA
CBHW060805220326
41598CB00022B/2542

9 781032 744681